Computational Chemistry Using the PC

Third Edition

Computational Chemistry Using the PC

Third Edition

Donald W. Rogers

WILEY-INTERSCIENCE

A John Wiley & Sons, Inc., Publication

Library of Congress Cataloging-in-Publication Data:

Rogers, Donald, 1932–
    Computational chemistry using the PC / Donald W. Rogers. – 3rd ed.
        p. cm.
Includes Index.
    ISBN 0-471-42800-0 (pbk.)
1. Chemistry–Data processing. 2. Chemistry–Mathematics.   I. Title.
    QD39.3.E46R64   1994
541.2′2′02855365—dc21   2003011758

Printed in the United States of America.

10 9 8 7 6 5 4 3 2 1

Live joyfully with the wife whom thou lovest all the days of the life of thy vanity, which He hath given thee under the sun, all the days of thy vanity: for that is thy portion in this life, and in thy labor which thou takest under the sun.

Ecclesiastes 9:9

THIS BOOK IS DEDICATED TO KAY

Contents

Preface to the Third Edition   xv

Preface to the Second Edition   xvii

Preface to the First Edition   xix

Chapter 1. Iterative Methods   1

Iterative Methods   1
An Iterative Algorithm   2
Blackbody Radiation   2
Radiation Density   3
Wien's Law   4
The Planck Radiation Law   4

  COMPUTER PROJECT 1-1 | *Wien's Law*   5

  COMPUTER PROJECT 1-2 | *Roots of the Secular Determinant*   6

The Newton–Raphson Method   7

    Problems   9

Numerical Integration   9
Simpson's Rule   10

Efficiency and Machine Considerations   13
Elements of Single-Variable Statistics   14
The Gaussian Distribution   15

    COMPUTER PROJECT 1-3 | *Medical Statistics*   17

Molecular Speeds   19

    COMPUTER PROJECT 1-4 | *Maxwell–Boltzmann Distribution Laws*   20

    COMPUTER PROJECT 1-5 | *Elementary Quantum Mechanics*   23

    COMPUTER PROJECT 1-6 | *Numerical Integration of Experimental Data Sets*   24

    Problems   29

Chapter 2. Applications of Matrix Algebra   31

Matrix Addition   31
Matrix Multiplication   33
Division of Matrices   34
Powers and Roots of Matrices   35
Matrix Polynomials   36
The Least Equation   37
Importance of Rank   38
Importance of the Least Equation   38
Special Matrices   39
The Transformation Matrix   41
Complex Matrices   42
What's Going On Here?   42

    Problems   44

Linear Nonhomogeneous Simultaneous Equations   45
Algorithms   47
Matrix Inversion and Diagonalization   51

    COMPUTER PROJECT 2-1 | *Simultaneous Spectrophotometric Analysis*   52

    COMPUTER PROJECT 2-2 | *Gauss–Seidel Iteration: Mass Spectroscopy*   54

    COMPUTER PROJECT 2-3 | *Bond Enthalpies of Hydrocarbons*   56

    Problems   57

Chapter 3. Curve Fitting   59

Information Loss   60
The Method of Least Squares   60

Least Squares Minimization   61
Linear Functions Passing Through the Origin   62
Linear Functions Not Passing Through the Origin   63
Quadratic Functions   65
Polynomials of Higher Degree   68
Statistical Criteria for Curve Fitting   69
Reliability of Fitted Parameters   70

COMPUTER PROJECT 3-1 | *Linear Curve Fitting: KF Solvation*   73

COMPUTER PROJECT 3-2 | *The Boltzmann Constant*   74

COMPUTER PROJECT 3-3 | *The Ionization Energy of Hydrogen*   76

Reliability of Fitted Polynomial Parameters   76

COMPUTER PROJECT 3-4 | *The Partial Molal Volume of $ZnCl_2$*   77

Problems   79

Multivariate Least Squares Analysis   80
Error Analysis   86

COMPUTER PROJECT 3-5 | *Calibration Surfaces Not Passing
Through the Origin*   88

COMPUTER PROJECT 3-6 | *Bond Energies of Hydrocarbons*   89

COMPUTER PROJECT 3-7 | *Expanding the Basis Set*   90

Problems   90

Chapter 4. Molecular Mechanics: Basic Theory   93

The Harmonic Oscillator   93
The Two-Mass Problem   95
Polyatomic Molecules   97
Molecular Mechanics   98
Ethylene: A Trial Run   100
The Geo File   102
The Output File   103
TINKER   108

COMPUTER PROJECT 4-1 | *The Geometry of Small Molecules*   110

The GUI Interface   112
Parameterization   113
The Energy Equation   114
Sums in the Energy Equation: Modes of Motion   115

COMPUTER PROJECT 4-2 | *The MM3 Parameter Set*   117

COMPUTER PROJECT 4-3 | *The Butane Conformational Mix*   125
Cross Terms   128

    Problems   129

Chapter 5. Molecular Mechanics II: Applications   **131**

Coupling   131
Normal Coordinates   136
Normal Modes of Motion   136
An Introduction to Matrix Formalism for Two Masses   138
The Hessian Matrix   140
Why So Much Fuss About Coupling?   143
The Enthalpy of Formation   144
Enthalpy of Reaction   147

    COMPUTER PROJECT 5-1 | *The Enthalpy of Isomerization of*
                                  cis- and trans-2-Butene   148
Enthalpy of Reaction at Temperatures $\neq$ 298 K   150
Population Energy Increments   151
Torsional Modes of Motion   153

    COMPUTER PROJECT 5-2 | *The Heat of Hydrogenation*
                                    of Ethylene   154

Pi Electron Calculations   155

    COMPUTER PROJECT 5-3 | *The Resonance Energy of Benzene*   157

Strain Energy   158
False Minima   158
Dihedral Driver   160
Full Statistical Method   161
Entropy and Heat Capacity   162
Free Energy and Equilibrium   163

    COMPUTER PROJECT 5-4 | *More Complicated Systems*   164

    Problems   166

Chapter 6. Huckel Molecular Orbital Theory I: Eigenvalues   **169**

Exact Solutions of the Schroedinger Equation   170
Approximate Solutions   172
The Huckel Method   176
The Expectation Value of the Energy: The Variational Method   178

    COMPUTER PROJECT 6-1 | *Another Variational Treatment of the*
                                    Hydrogen Atom   181

Huckel Theory and the LCAO Approximation   183
Homogeneous Simultaneous Equations   185
The Secular Matrix   186
Finding Eigenvalues by Diagonalization   187
Rotation Matrices   188
Generalization   189
The Jacobi Method   191
Programs QMOBAS and TMOBAS   194

    COMPUTER PROJECT 6-2 | *Energy Levels (Eigenvalues)*   195

    COMPUTER PROJECT 6-3 | *Huckel MO Calculations of*
                              Spectroscopic Transitions   197

    Problems   198

Chapter 7. Huckel Molecular Orbital Theory II: Eigenvectors   201

Recapitulation and Generalization   201
The Matrix as Operator   207
The Huckel Coefficient Matrix   207
Chemical Application: Charge Density   211
Chemical Application: Dipole Moments   213
Chemical Application: Bond Orders   214
Chemical Application: Delocalization Energy   215
Chemical Application: The Free Valency Index   217
Chemical Application: Resonance (Stabilization) Energies   217

    LIBRARY PROJECT 7-1 | *The History of Resonance and*
                         Aromaticity   219

Extended Huckel Theory—Wheland's Method   219
Extended Huckel Theory—Hoffman's EHT Method   221
The Programs   223

    COMPUTER PROJECT 7-1 | *Larger Molecules: Calculations*
                         using SHMO   225

    COMPUTER PROJECT 7-2 | *Dipole Moments*   226

    COMPUTER PROJECT 7-3 | *Conservation of Orbital Symmetry*   227

    COMPUTER PROJECT 7-4 | *Pyridine*   228

    Problems   229

Chapter 8. Self-Consistent Fields   231

Beyond Huckel Theory   231
Elements of the Secular Matrix   232

The Helium Atom   235
A Self-Consistent Field Variational Calculation of
    IP for the Helium Atom   236

    COMPUTER PROJECT 8-1 | *The SCF Energies of First Row*
                                             Atoms and Ions   240

    COMPUTER PROJECT 8-2 | *A High-Level ab initio Calculation of SCF*
                                             First IPs of the First Row Atoms   241

The STO-xG Basis Set   242
The Hydrogen Atom: An STO-1G "Basis Set"   243
Semiempirical Methods   248
PPP Self-Consistent Field Calculations   248
The PPP-SCF Method   249
Ethylene   252
Spinorbitals, Slater Determinants, and Configuration Interaction   255
The Programs   256

    COMPUTER PROJECT 8-3 | *SCF Calculations of Ultraviolet*
                                             Spectral Peaks   256

    COMPUTER PROJECT 8-4 | *SCF Dipole Moments*   258

    Problems   259

Chapter 9. Semiempirical Calculations on Larger Molecules   263

The Hartree Equation   263
Exchange Symmetry   266
Electron Spin   267
Slater Determinants   269
The Hartree–Fock Equation   273
The Fock Equation   276
The Roothaan–Hall Equations   278
The Semiempirical Model and Its Approximations:
    MNDO, AM1, and PM3   279
The Programs   283

    COMPUTER PROJECT 9-1 | *Semiempirical Calculations on Small*
                                             Molecules: HF to HI   284

    COMPUTER PROJECT 9-2 | *Vibration of the Nitrogen Molecule*   284

Normal Coordinates   285
Dipole Moments   289

    COMPUTER PROJECT 9-3 | *Dipole Moments (Again)*   289

Energies of Larger Molecules   289

COMPUTER PROJECT 9-4 | *Large Molecules: Carcinogenesis* 291

Problems 293

Chapter 10. *Ab Initio* Molecular Orbital Calculations **299**

The GAUSSIAN Implementation 299
How Do We Determine Molecular Energies? 301
Why Is the Calculated Energy Wrong? 306
Can the Basis Set Be Further Improved? 306
Hydrogen 308
Gaussian Basis Sets 309

COMPUTER PROJECT 10-1 | *Gaussian Basis Sets: The HF Limit* 311

Electron Correlation 312
G2 and G3 313
Energies of Atomization and Ionization 315

COMPUTER PROJECT 10-2 | *Larger Molecules: G2, G2(MP2), G3,*
and G3(MP2) 316

The GAMESS Implementation 317

COMPUTER PROJECT 10-3 | *The Bonding Energy Curve of H_2:*
GAMESS 318

The Thermodynamic Functions 319
Koopmans's Theorem and Photoelectron Spectra 323
Larger Molecules I: Isodesmic Reactions 324

COMPUTER PROJECT 10-4 | *Dewar Benzene* 326

Larger Molecules II: Density Functional Theory 327

COMPUTER PROJECT 10-5 | *Cubane* 330

Problems 330

Bibliography **333**

Appendix A. Software Sources **339**

Index **343**

Preface to the Third Edition

It is a truism (cliche?) that microcomputers have become more powerful on an almost exponential curve since their advent more than 30 years ago. Molecular orbital calculations that I ran on a supercomputer a decade ago now run on a fast desktop microcomputer available at a modest price in any popular electronics store or by mail order catalog. With this has come a comparable increase in software sophistication.

There is a splendid democratization implied by mass-market computers. One does not have to work at one of the world's select universities or research institutes to do world class research. Your research equipment now consists of an off-the-shelf microcomputer and your imagination.

At the first edition of this book, in 1990, I made the extravigant claim that "a quite respectable academic program in chemical microcomputing can be started for about $1000 per student". The degree of difficulty of the problems we solve has increased immeasurably since then but the price of starting a good teaching lab is probably about half of what it was. To equip a workstation for two students, one needs a microcomputer connected to the internet, a BASIC interpreter and a beginner's bundle of freeware which should include the utility programs suggested with this book, a Huckel Molecular Orbital program, TINKER, MOPAC, and GAMESS.

There are 42 Computer Projects included in this text. Several of the Computer Projects connect with the research literature and lead to extensions suitable for undergraduate or MS thesis projects. All of the computer projects in this book have been successfully run by the author. Unfortunately, we still live in an era of system incompatibility. The instructor using these projects in a teaching laboratory is urged to run them first to sort out any system specific difficulties. In this, the projects here are no different from any undergraduate experiment; it is a foolish instructor indeed who tries to teach from untested material.

The author wishes to acknowledge the unfailing help and constructive criticism of Frank Mc Lafferty, the computer tips of Nikita Matsunaga and Xeru Li. Some of the research which gave rise to Computer Projects in the latter half of the book were carried out under a grant of computer time from the National Science Foundation through the National Center for Supercomputing Applications both of which are gratefully acknowledged.

Donald W. Rogers
Greenwich Village, NY
July 2003

Preface to the Second Edition

A second edition always needs an excuse, particularly if it follows hard upon the first. I take the obvious one: a lot has happened in microcomputational chemistry in the last five years. Faster machines and better software have brought more than convenience; there are projects in this book that we simply could not do at the time of the first edition.

Along with the obligatory correction of errors in the first edition, this one has five new computer projects (two in high-level *ab initio* calculations), and 49 new problems, mostly advanced. Large parts of Chapters 9 and 10 have been rewritten, more detailed instructions are given in many of the computer projects, and several new illustrations have been added, or old ones have been redrawn for clarity. The BASIC programs on the diskette included here have been translated into ASCII code to improve portability, and each is written out at the end of the chapter in which it is introduced. Several illustrative input and output files for Huckel, self-consistent field, molecular mechanics, *ab initio*, and semiempirical procedures are also on the disk, along with an answer section for problems and computer projects.

One thing has not changed. By shopping among the software sources at the end of this book, and clipping popular computer magazine advertisements, the prudent instructor can still equip his or her lab at a starting investment of about $2000 per workstation of two students each.

Preface to the First Edition

This book is an introduction to computational chemistry, molecular mechanics, and molecular orbital calculations, using a personal microcomputer. No special computational skills are assumed of the reader aside from the ability to read and write a simple program in BASIC. No mathematical training beyond calculus is assumed. A few elements of matrix algebra are introduced in Chapter 3 and used throughout.

The treatment is at the upperclass undergraduate or beginning graduate level. Considerable introductory material and material on computational methods are given so as to make the book suitable for self-study by professionals outside the classroom. An effort has been made to avoid logical gaps so that the presentation can be understood without the aid of an instructor. Forty-six self-contained computer projects are included.

The book divides itself quite naturally into two parts: The first six chapters are on general scientific computing applications and the last seven chapters are devoted to molecular orbital calculations, molecular mechanics, and molecular graphics. The reader who wishes only a tool box of computational methods will find it in the first part. Those skilled in numerical methods might read only the second. The book is intended, however, as an entity, with many connections between the two parts, showing how chapters on molecular orbital theory depend on computational techniques developed earlier.

Use of special or expensive microcomputers has been avoided. All programs presented have been run on a 8086-based machine with 640 K memory and a math coprocessor. A quite respectable academic program in chemical microcomputing can be started for about $1000 per student. The individual or school with more expensive hardware will find that the programs described here run faster and that more visually pleasing graphics can be produced, but that the results and principles involved are the same. Gains in computing speed and convenience will be made as the technology advances. Even now, run times on an 80386-based machine approach those of a heavily used, time-shared mainframe.

Sources for all program packages used in the book are given in an appendix. All of the early programs (Chapters 1 through 7) were written by the author and are available on a single diskette included with the book. Programs HMO and SCF were adapted and modified by the author from programs in FORTRAN II by Greenwood (*Computational Methods for Quantum Organic Chemistry*, Wiley Interscience, New York, 1972). The more elaborate programs in Chapters 10 through 13 are available at moderate price from Quantum Chemistry Program Exchange, Serena Software, Cambridge Analytical Laboratories and other software sources [see Appendix].

I wish to thank Dr. A. Greenberg of Rutgers University, Dr. S. Topiol of Burlex Industries, and Dr. A. Zavitsas of Long Island University for reading the entire manuscript and offering many helpful comments and criticisms. I wish to acknowledge Long Island University for support of this work through a grant of released time and the National Science Foundation for microcomputers bought under grant #CSI 870827.

Several chapters in this book are based on articles that appeared in *American Laboratory* from 1981 to 1988. I wish to acknowledge my coauthors of these papers, F. J. McLafferty, W. Gratzer, and B. P. Angelis. I wish to thank the editors of *American Laboratory*, especially Brian Howard, for permission to quote extensively from those articles.

CHAPTER

1

Iterative Methods

Some things are simple but hard to do.

—A. Einstein

Most of the problems in this book are simple. Many of the methods used have been known for decades or for centuries. At the machine level, individual steps in the procedures are at the grade school level of sophistication, like adding two numbers or comparing two numbers to see which is larger. What makes them hard is that there are very many steps, perhaps many millions. The computer, even the once "lowly" microcomputer, provides an entry into a new scientific world because of its incredible speed. We are now in the enviable position of being able to arrive at practical solutions to problems that we could once only imagine.

Iterative Methods

One of the most important methods of modern computation is solution by iteration. The method has been known for a very long time but has come into widespread use only with the modern computer. Normally, one uses iterative methods when ordinary analytical mathematical methods fail or are too time-consuming to be

Computational Chemistry Using the PC, Third Edition, by Donald W. Rogers
ISBN 0-471-42800-0   Copyright © 2003 John Wiley & Sons, Inc.

practical. Even relatively simple mathematical procedures may be time-consuming because of extensive algebraic manipulation.

A common iterative procedure is to solve the problem of interest by repeated calculations that do not initially give the correct answer but get closer to it as the calculation is repeated, perhaps many times. The approximate solution is said to *converge* on the correct solution. Although no human would be willing to repeat an iterative calculation thousands of times to converge on the right answer, the computer does, and, because of its speed, it often arrives at the answer in a reasonable amount of time.

An Iterative Algorithm

The first illustrative problem comes from quantum mechanics. An equation in radiation density can be set up but not solved by conventional means. We shall guess a solution, substitute it into the equation, and apply a test to see whether the guess was right. Of course it isn't on the first try, but a second guess can be made and tested to see whether it is closer to the solution than the first. An iterative routine can be set up to carry out very many guesses in a methodical way until the test indicates that the solution has been approximated within some narrow limit.

Several questions present themselves immediately: How good does the initial guess have to be? How do we know that the procedure leads to better guesses, not worse? How many steps (how long) will the procedure take? How do we know when to stop? These questions and others like them will play an important role in this book. You will not be surprised to learn that answers to questions like these vary from one problem to another and cannot be set down once and for all. Let us start with a famous problem in quantum mechanics: blackbody radiation.

Blackbody Radiation

We can sample the energy density of radiation $\rho(v, T)$ within a chamber at a fixed temperature T (essentially an oven or furnace) by opening a tiny transparent window in the chamber wall so as to let a little radiation out. The amount of radiation sampled must be very small so as not to disturb the equilibrium condition inside the chamber. When this is done at many different frequencies v, the *blackbody spectrum* is obtained. When the temperature is changed, the area under the spectral curve is greater or smaller and the curve is displaced on the frequency axis but its shape remains essentially the same. The chamber is called a blackbody because, from the point of view of an observer within the chamber, radiation lost through the aperture to the universe is perfectly absorbed; the probability of a photon finding its way from the universe back through the aperture into the chamber is zero.

Radiation Density

If we think in terms of the particulate nature of light (wave-particle duality), the number of particles of light or other electromagnetic radiation (photons) in a unit of frequency space constitutes a *number density*. The blackbody radiation curve in Fig. 1-1, a plot of radiation energy density ρ on the vertical axis as a function of frequency ν on the horizontal axis, is essentially a plot of the number densities of light particles in small intervals of frequency space.

We are using the term *space* as defined by one or more coordinates that are not necessarily the x, y, z Cartesian coordinates of space as it is ordinarily defined. We shall refer to 1-space, 2-space, etc. where the number of dimensions of the space is the number of coordinates, possibly an n-space for a many dimensional space. The ρ and ν axes are the coordinates of the *density–frequency space*, which is a 2-space.

Radiation energy density is a function of both frequency and temperature $\rho(\nu,T)$ so that the single curve in Fig. 1-1 implies one and only one temperature. Because frequency ν times *wavelength* λ is the velocity of light $c = \nu\lambda = 2.998 \times 10^8$ m s^{-1} (a constant), an equivalent functional relationship exists between energy density and wavelength. The energy density function can be graphed in a different but equivalent form $\rho(\lambda,T)$. The intensity I of electromagnetic radiation within any narrow frequency (or wavelength) interval is directly proportional to the number density of photons. It is also directly proportional to the power output of a light sensor or photomultiplier; hence both I and ρ are measurable quantities. Whenever one plots some function of radiation intensity I vs. ν or λ, the resulting curve is called a *spectrum*.

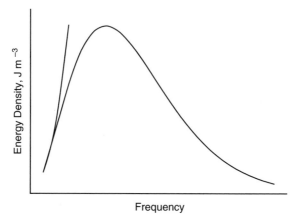

Figure 1-1 The Blackbody Radiation Spectrum. The short curve on the left is a Rayleigh function of frequency.

Wien's Law

In the late nineteenth century, Wien analyzed experimental data on blackbody radiation and found that the maximum of the blackbody radiation spectrum λ_{max} shifts with the temperature according to the equation

$$\lambda_{max}T = 2.90 \times 10^{-3} \text{ m K} \tag{1-1}$$

where λ is in meters and T is the temperature in kelvins.

The Planck Radiation Law

As Lord Rayleigh pointed out, the classical expression for radiation

$$\rho(v, T)dv = \left\{8\pi k_B T/c^3\right\}v^2 dv \tag{1-2}$$

where k_B is Boltzmann's constant and c is the speed of light, must fail to express the blackbody radiation spectrum because $\rho = \text{const.} \times v^2$ is a segment of a parabola open upward (the short curve to the left in Fig. 1-1) and does not have a relative maximum as required by the experimental data. In late 1900, Max Planck presented the equation

$$\rho(v, T)dv = \left\{8\pi hv/c^3\right\}v^2 \frac{dv}{e^{hv/k_B T} - 1} \tag{1-3}$$

where the units of $\rho(v, T)$ are joules per cubic meter, as appropriate to an energy in joules per unit volume and $h = 6.626 \times 10^{-34}$ J s (joule seconds) is a new constant, now called Planck's constant. This equation expressed in terms of wavelength λ is

$$\rho(\lambda, T)\, d\lambda = \left\{8\pi hc/\lambda^5\right\} \frac{d\lambda}{e^{hc/\lambda k_B T} - 1} \tag{1-4}$$

By setting $d\rho/d\lambda = 0$, one can differentiate Eq. (1-4) and show that the equation

$$e^{-x} + \frac{x}{5} = 1 \tag{1-5}$$

holds at the maximum of Fig. 1-1 where

$$x = \frac{hc}{\lambda k_B T} \tag{1-6}$$

Exercise 1-1

Given that $c = v\lambda$, show that Eqs. (1-3) and 1-4) are equivalent.

Exercise 1-2

Obtain Eq. (1-5) from Eq. (1-4).

COMPUTER PROJECT 1-1 | Wien's Law

The first computer project is devoted to solving Eq. (1-5) for x iteratively. When x has been determined, the remaining constants can be substituted into

$$\lambda T = \frac{hc}{k_B x} \tag{1-7}$$

where h is Planck's constant, c is the velocity of light in a vacuum, $2.998 \times 10^8 \, \mathrm{m \, s^{-1}}$, and $k_B = 1.381 \times 10^{-23} \, \mathrm{J \, K^{-1}}$ is the Boltzmann constant. The result is a test of agreement between Planck's theoretical quantum law and Wien's displacement law [Eq. 1-1], which comes from experimental data.

Procedure. One approach to the problem is to select a value for x that is obviously too small and to increment it iteratively until the equation is satisfied. This is the method of program WIEN, where the initial value of x is taken as 1 (clearly, $e^{-1} + \frac{1}{5} < 1$ as you can show with a hand calculator).

Program

```
PRINT "Program QWIEN"
x = 1
10 x = x + .1
a = EXP(-x) + (x / 5)
IF (a -1) < 0 THEN 10
PRINT a, x
END
```

In Program QWIEN (written in **QBASIC**, Appendix A), x is *initialized* at 1 and *incremented* by 0.1 in line 3, which is given the *statement number* 10 for future reference. Be careful to differentiate between a statement number like $10 \, x = x + .1$ and the product 10 times x which is 10^*x. A number a is calculated for $x = 1.1$ that is obviously too small so $(a - 1)$ is less than 0 and the IF statement in line 5 sends control back to the statement numbered 10, which increments x by 0.1 again. This continues until $(a - 1) \geq 0$, whereupon control exits from the loop and prints the result for a and x.

There are, of course, many variations that can be written in place of Program QWIEN. You are urged to try as many as you can. Some suggestions are as follows:

a. Vary the size of the increment in x in program statement 10. Tabulate the increment size, the computed result for x, and the calculated Wien constant. Comment on the relationship among the quantities tabulated.

b. Change Program QWIEN so that the second term on the right of the line below statement 10 is x instead of $x/5$. Solve for this new equation. Change the line below statement 10 so that the second term on the left is $x/2$. Repeat with $x/3$, $x/4$, etc. Tabulate the values of x and the values of the denominator. Is x a sensitive function of the denominator in the second term of Program WIEN?

c. Devise and discuss a scheme for more efficient convergence. For example, some scheme that uses large increments for x when x is far away from convergence and small values for the increment in x when x is near its true value would be more efficient than the preceding schemes. How, in more detail, could this be done? Try coding and running your scheme.

d. Another coding scheme can be used in *True BASIC* (Appendix A)

Program

```
PRINT "Program TWIEN"
let X = 1
do
let X = X + .1
let A = exp(-X) + X / 5
loop until (A - 1) > 0
PRINT A, X
END
```

The program contains a "do loop" that iterates the statements within the loop until the condition $(A - 1) < 0$ is true. Try moving the "do" statement around in the program to see what changes in the output. Explain. If you encounter an "infinite loop," *True BASIC* has a STOP statement to get you out.

It is good practice to translate programs in one BASIC (*QBASIC* or *True BASIC*) to programs in the other if you have both interpreters. Note that the statement $X = X + .1$ in both programs makes no sense algebraically, but in BASIC it means, "take the number in memory register X, add 0.1 to it and store the result back in register X." If you are not familiar with coding in BASIC, an hour or so with an instruction manual should suffice for the simple programs used in the first half of this book. By all means, look at the programs on the Wiley website.

COMPUTER PROJECT 1-2 | *Roots of the Secular Determinant*

Later in this book, we shall need to find the roots of the secular matrix

$$\begin{bmatrix} 210 - 42x & 42 - 9x \\ 42 - 9x & 12 - 2x \end{bmatrix} \tag{1-8}$$

One way of obtaining the roots is to expand the determinantal equation

$$\begin{vmatrix} 210 - 42x & 42 - 9x \\ 42 - 9x & 12 - 2x \end{vmatrix} = 0 \tag{1-9}$$

To do this, multiply the binomials at the top left and bottom right (the principal diagonal) and then, from this product, subtract the product of the remaining two elements, the off-diagonal elements $(42 - 9x)$. The difference is set equal to zero:

$$(210 - 42x)(12 - 2x) - (42 - 9x)^2 = 0 \tag{1-10}$$

This equation is a quadratic and has two roots. For quantum mechanical reasons, we are interested only in the lower root. By inspection, $x = 0$ leads to a large number on the left of Eq. (1-10). Letting $x = 1$ leads to a smaller number on the left of Eq. (1-10), but it is still greater than zero. Evidently, increasing x approaches a solution of Eq. (1-10), that is, a value of x for which both sides are equal. By systematically increasing x beyond 1, we will approach one of the roots of the secular matrix. Negative values of x cause the left side of Eq. (1-10) to increase without limit; hence the root we are approaching must be the lower root.

Program

```
PRINT "Program QROOT"
x = 0
20 x = x + 1
a = (210 - 42 * x) * (12 - 2 * x) - (42 - 9 * x)^2
IF a > 0 GOTO 20
PRINT x: END
```

Program QROOT increments x by 1 on each iteration. It prints out 5 when the polynomial on the right of line 4 is greater than 0. We have gone past the root because x is too large. The program did not exit from the loop on $x = 4$, but it did on $x = 5$, so x is between 4 and 5. By letting $x = 4$ in the second line and changing the third line to increment x by 0.1, we get 5 again so x is between 4.9 and 5.0. Letting $x = 4.9$ with an increment of 0.01 yields 4.94 and so on, until the increment 0.00001 yields the lower root $x = 4.93488$.

Although we will not need it for our later quantum mechanical calculation, we may be curious to evaluate the second root and we shall certainly want to check to be sure that the root we have found is the smaller of the two. Write a program to evaluate the left side of Eq. (1-10) at integral values between 1 and 100 to make an approximate location of the second root. Write a second program to locate the second root of matrix Eq. (1-10) to a precision of six digits. Combine the programs to obtain both roots from one program run.

The Newton–Raphson Method

The root-finding method used up to this point was chosen to illustrate iterative solution, not as an efficient method of solving the problem at hand. Actually, a more efficient method of root finding has been known for centuries and can be traced back to Isaac Newton (1642–1727) (Fig. 1-2).

Suppose a function of x, $f(x)$, has a first derivative $f'(x)$ at some arbitrary value of x, x_0. The slope of $f(x)$ is

$$f'(x) = \frac{f(x_0)}{(x_0 - x_1)} \tag{1-11}$$

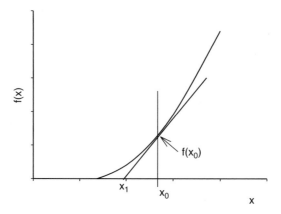

Figure 1-2   The First Step in the Newton–Raphson Method.

whence

$$x_1 = x_0 - \frac{f(x_0)}{f'(x)}$$

$(1\text{-}12)$

The intersection of the slope and the x axis at x_1 is closer to the root $f(x) = 0$ than x_0 was. By repeating this process, one can arrive at a point x_n arbitrarily close to the root.

Exercise 1-3

Carry out the first two iterations of the Newton–Raphson solution of the polynomial Eq. (1-10).

Solution 1-3

The polynomial (1-10) can be written

$$x^2 - 56x + 252 = 0$$

$(1\text{-}13)$

The first derivative is

$$2x - 56 = 0$$

Starting at $x_0 = 0$

$$x_1 = x_0 - \left(-\frac{252}{56}\right) = 4.5$$

and the second step yields

$$x_2 = 4.5 - \left(-\frac{20.25}{47}\right) = 4.93085$$

$(1\text{-}14)$

This approximates the root $x = 4.93488$ from Program QROOT in only two steps. Solution by the quadratic equation yields $x = 4.93487$.

PROBLEMS

1. Show that Eq. (1-12) is the same as Eq. (1-11).
2. The energy of radiation at a given temperature is the integral of radiation density over all frequencies

$$E = \int_0^v \rho(v, T)dv$$

Find E from the known integral

$$\int_0^\infty \frac{x^3}{e^x - 1} dx = \frac{\pi^4}{15}$$

and compare the result with the Stefan–Boltzmann law

$$E = \left(\frac{4\sigma}{c}\right)T^4$$

where c is the velocity of light and σ is an empirical constant equal to 5.67×10^{-8} $J\,m^{-2}\,s^{-1}$. Just in case the value of the "known integral" is not obvious to you (it isn't to me, either), we shall determine it numerically in another problem.

3. Analysis of the electromagnetic radiation spectrum emanating from the star Sirius shows that $\lambda_{max} = 260$ nm. Estimate the surface temperature of Sirius.

Numerical Integration

The term "quadrature" was used by early mathematicians to mean finding a square with an area equal to the area of some geometric figure other than a square. It is used in numerical integration to indicate the process of summing the areas of some number of simple geometric figures to approximate the area under some curve, that is, to approximate the integral of a function. We include numerical integration among the iterative methods because the integration program we shall use, following Simpson's rule (Kreyszig, 1988), iteratively calculates small subareas under a curve $f(x)$ and then sums the subareas to obtain the total area under the curve.

This discussion will be limited to functions of one variable that can be plotted in 2-space over the interval considered and that constitute the upper boundary of a well-defined area. The functions selected for illustration are simple and *well-behaved*; they are smooth, single valued, and have no discontinuities. When discontinuities or singularities do occur (for example the cusp point of the 1s hydrogen orbital at the nucleus), we shall integrate up to the singularity but not include it.

Contrary to the impression that one might have from a traditional course in introductory calculus, well-behaved functions that cannot be integrated in closed form are not rare mathematical curiosities. Examples are the Gaussian or standard error function and the related function that gives the distribution of molecular or atomic speeds in spherical polar coordinates. The famous blackbody radiation curve, which inspired Planck's quantum hypothesis, is not integrable in closed form over an arbitrary interval.

Heretofore, the integral of a function of this kind was usually approximated by expressing it as an infinite series and evaluating some arbitrarily limited number of terms of the series. This always leads to a truncation error that depends on the number of terms retained in the sum before it is cut off (truncated). Numerical integration may be used instead of series solution when the analytical form of the function is known but not integrable or when the analytical form of the function is not known because the functional relationship exists as an instrument plot or a collection of paired measurements. This is the common case for data that have been obtained in an experimental setting. An example is the function describing a chromatographic peak, which may or may not approximate a Gaussian function.

We shall use the term *analytical form* to indicate a closed algebraic expression such as

$$y = x^2 \tag{1-15}$$

as contrasted to functions that are expressed as an infinite series, for example,

$$C_P = a + bT + cT^2 + dT^3 + \cdots \tag{1-16}$$

Equation (1-15) is an analytical form that has a closed integral. The Gaussian function

$$f(x) = (2\pi)^{-1/2} \, e^{z^2/2} \tag{1-17}$$

is a closed analytical form but it has no closed integral. (Try to integrate it!)

Several related "rules" or algorithms for numerical integration (rectangular rule, trapezoidal rule, etc.) are described in applied mathematics books, but we shall rely on Simpson's rule. This method can be shown to be superior to the simpler rules for well-behaved functions that occur commonly in chemistry, both functions for which the analytical form is not known and those that exist in analytical form but are not integrable.

Simpson's Rule

In applying Simpson's rule, over the interval [a, b] of the independent variable, the interval is partitioned into an *even* number of subintervals and three consecutive points are used to determine the unique parabola that "covers" the area of the first

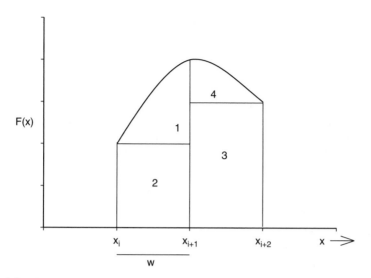

Figure 1-3 Areas Under a Parabolic Arc Covering Two Subintervals of a Simpson's Rule Integration.

subinterval pair (see Fig. 1-3). The area under this parabolic arc is $\frac{1}{3}w(f(x_i) + 4f(x_{i+1}) + f(x_{i+2}))$. Summing for successive subinterval pairs over the entire interval constitutes the method known as Simpson's rule. Looking at the formula below, one anticipates that an iterative loop will implement it on a microcomputer

$$\int_a^b f(x)dx = \frac{1}{3}w(f(x_0) + 4f(x_1) + 2f(x_2) + 4f(x_3) + \cdots$$

$$+ 2f(x_{n-2}) + 4f(x_{n-1}) + f(x_n)) \tag{1-18}$$

Exercise 1-14

Show that the area under a parabolic arc that is convex upward is $\frac{1}{3}w(f(x_i) + 4f(x_{i+1}) + f(x_{i+2}))$, where w is the width of the subinterval $x_{i+1} - x_i$.

Solution 1-4

The area under a parabolic arc concave upward is $\frac{1}{3}bh$, where b is the base of the figure and h is its height. The area of a parabolic arc concave downward is $\frac{2}{3}bh$. The areas of parts of the figure diagrammed for Simpson's rule integration are shown in Fig. 1-3.

The area A under the parabolic arc in Fig. 1-3 is given by the sum of four terms:

$$A = \frac{2}{3}w(f(x_{i+1}) - f(x_i)) + wf(x_i) + w(f(x_{i+2}) + \frac{2}{3}w(f(x_{i+1}) - f(x_{i+2}))$$
$$= w(\frac{2}{3}f(x_{i+1}) + \frac{1}{3}f(x_i) + \frac{2}{3}f(x_{i+1}) + \frac{1}{3}f(x_{i+2}))$$
$$= \frac{1}{3}w(f(x_i) + 4f(x_{i+1}) + f(x_{i+2}))$$

which was to be proven.

Our Simpson's rule program is written in **QBASIC** (Appendix A). Today's computer world is full of complicated and expensive software, some of which we shall use in later chapters. Unfortunately, it is not hard to find software that is overpriced and overwritten (which we shall not use). Although it is not appropriate to recommend software in a book of this kind, the simple software used here has been used for several years in both a teaching and a research setting. It works.

More complicated and expensive programs are not necessarily better programs. One author recently described BASIC as a "primitive" language. Be that as it may, BASIC is ideal for solving simple problems. A hammer is a primitive tool. I wonder what our author friend would use to drive a nail.

Program QSIM is more general than any of the programs we have used to this point. By changing the *define function* statement DEF fna in line 8 of Program QSIM, one can obtain the integral of any well-behaved function between the limits *a* and *b*, which are specified in the *interactive* input to line 5. The term "interactive" is used here to denote interaction between the system and the operator (you). Line 6 is part of an INPUT statement requiring a response from you. The program will not run until you have specified the limits of integration, *a* and *b* along with *n*, the number of subintervals you wish to break the interval into. (The input numbers are separated by commas.) Note that statement 7 takes the subintervals in pairs so *n* must be an even number for the *system* to produce the correct integral. We are using the term "system" to denote both the hardware and software (hardware + software = system).

As an interesting beginning integration, let us determine the integral

$$\int_a^b f(x)dx = \int_0^{10} 100 - x^2 \, dx$$

over the interval [0, 10] We can solve this integral by conventional means as a check on the result of numerical integration.

$$\int_0^{10} 100 - x^2 \, dx = 100x - \frac{x^3}{3}\Big|_0^{10} = 1000 - \frac{1000}{3} = 666.667$$

Program

```
CLS
PRINT "Program QSIM"
PRINT "Simpson's Rule integration of the area under y = f(x)"
DEF fna (x) = 100 - x ^ 2 '***DEF fna lets you put any function you like here.
PRINT "input limits a, and b, and the number of iterations
    desired n"
INPUT a, b, n
d = (b - a) / n
FOR x = a + d TO b STEP 2 * d
```

```
sum = sum + 4 * fna(x) + 2 * fna(x - d): NEXT x
PRINT: PRINT: PRINT "RESULTS": PRINT
PRINT: PRINT "The interval is" ; a; "to"; b; " "
PRINT: PRINT "The number of iterations is =" ; n; " "
a = d / 3 * (fna(a) + sum - fna(b))
PRINT: PRINT "Numerical integration yields", a: END
```

Names of programs written in **QBASIC** begin with Q. Programs written in **True BASIC** begin with T. Program QSIM differs from Programs QWIEN and QROOT in having more documentation. Documentation is used to make the program and the output easier for the operator to read. It is useful when a program is passed along to a colleague who was not in on the writing and may have difficulty understanding the logic of it. The CLS statement clears the screen, followed by a number of PRINT statements that should be obvious from context. Note that a full colon : is equivalent to a new line. Nothing enclosed in full quotes " influences the functioning of the numerical part of the program. The prime or apostrophe ' in line 4 instructs the system to ignore anything following it on the same line.

Efficiency and Machine Considerations

We selected a simple test function for integration. The function $f(x) = 100 - x^2$ is a smooth, monotonically decreasing parabolic curve over the interval [0, 10]. It has a closed definite integral over this interval of 666.667 units. The function is well-behaved, and integration is easy over the first half of the interval but not so easy over the second half of the interval owing to its increasing steepness. (Note that steep functions can be integrated by an algorithm that sums horizontal slices of the area under the curve rather than vertical ones.)

The approximation to the closed integral improves as the number of iterations increases up to a point. The actual values in Table 1-1 may be *system specific*, that is, different hardware and software combinations may give slightly different results because of different ways of storing numbers. One is tempted to think of approximations as getting better without limit, the sum approaching the integral

Table 1-1 Approach of the Area Sum of Program QSIM to 666.667

Iterations (Subintervals)*					
10	100	1000	10000	100000	1000000
Area sum*					
733.73	673.33	667.33	666.75	666.84	665.82

* Large numbers may be input as exponentials, for example, $1e6 = 1 \times 10^6$.
** May be system specific.

as Achilles approached the tortoise. This does not occur, however, because of machine rounding error. (Only so many digits can be stored on a chip.) The last few entries in Table 1-1 show that for very many iterations, the area sum begins to diverge from, rather than approach, the integral it is supposed to represent (see also Norris, 1981). Keep rounding error in mind when writing programs with many iterations.

Elements of Single-Variable Statistics

When we report the result of a measurement x, there are two things a person reading the report wants to know: the *magnitude* (size) of the measurement and the *reliability* of the measurement (its "scatter"). If measuring errors are random, as they very frequently are, the magnitude is best expressed as the arithmetic mean μ of N repeated trials x_i

$$\mu = \frac{\sum x_i}{N} \tag{1-19}$$

and the reliability is best expressed as the *standard deviation*

$$\sigma = \sqrt{\frac{\sum (x_i - \mu)^2}{N}} \tag{1-20}$$

These equations apply when an entire *population* is available for measurement. The most common situation in practical problems is one in which the number of measurements is smaller than the entire population. A group of selected measurements smaller than the population is called a *sample*. Sample statistics are slightly different from population statistics but, for large samples, the equations of sample statistics approach those of population statistics.

If very many measurements are made of the same variable x, they will not all give the same result; indeed, if the measuring device is sufficiently sensitive, the surprising fact emerges that *no* two measurements are exactly the same. Many measurements of the same variable give a distribution of results x_i clustered about their arithmetic mean μ. In practical work, the assumption is almost always made that the distribution is random and that the distribution is Gaussian (see below).

Decision Making

A simple decision-making problem is: I measure variable x of a population A and the same variable x of a population B. I get (slightly) different results. Is there a real difference between populations A and B based on the difference in measurements, or am I only seeing different parts of the distributions of identical populations?

A similar decision-making problem consists of very many measurements of variable x on a large sample from population A, followed by a single measurement of the same property x of an individual. The single measurement will not be precisely at the arithmetic mean of the large population. The question is whether the difference between μ for the large population and measurement x indicates that the individual is not from the test population (is abnormal) or whether the deviation can be ascribed to a normal statistical fluctuation.

The second decision-making situation is very close to the problem presented in medical diagnosis in which we wish to know whether a patient is a member of the healthy general population or not. We shall apply Gaussian statistics to a diagnostic problem involving risk to a patient of atherosclerosis, given the blood cholesterol analysis of very many normal patients to which we compare the blood cholesterol analysis of the individual patient. In Computer Project 1-3, the patient is known to have a high blood cholesterol level but the problem is whether the measured level is sufficiently far from the mean of the normal population to be dangerous or whether it is only the random fluctuation we expect to see in some normal patients.

The Gaussian Distribution

The Gaussian distribution for the probability of random events is

$$p(x) = \frac{1}{\sqrt{2\pi}\sigma} \exp\left(-\frac{(x_i - \mu)^2}{2\sigma^2}\right) \tag{1-21}$$

It is widely used in experimental chemistry, most commonly in statistical treatment of experimental uncertainty (Young, 1962). For convenience, it is common to make the substitution

$$z = \frac{x_i - \mu}{\sigma} \tag{1-22}$$

With this substitution, distributions having different μ and σ can be compared by using the same curve, frequently called the *normal curve* (Fig. 1-4).

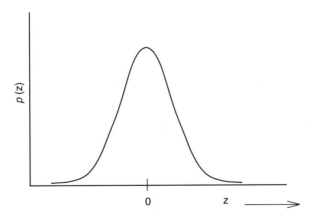

Figure 1-4 The Gaussian Normal Distribution.

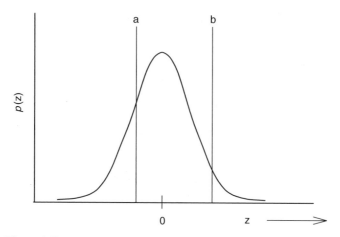

Figure 1-5   An Interval [a, b] on the Gaussian Normal Distribution.

The integral of the Gaussian function over the interval [a, b] in a one-dimensional probability space z is

$$p(z) = \frac{1}{\sqrt{2\pi}} \int_a^b e^{-z^2/2} dz \qquad (1\text{-}23)$$

Equation (1-23) gives the probability of an event occurring within an arbitrary interval [a, b] (Fig. 1-5). Equation (1-23) has been "normalized" by choosing the right premultiplying constant $\frac{1}{\sqrt{2\pi}}$ to make the integral over all space $[-\infty, \infty]$ come out to 1.00 . . . . (see Problems) so the probability over any smaller interval [a, b] has a value not less than zero and not more than one.

The integral of the Gaussian distribution function does not exist in closed form over an arbitrary interval, but it is a simple matter to calculate the value of $p(z)$ for any value of z, hence numerical integration is appropriate. Like the test function, $f(x) = 100 - x^2$, the accepted value (Young, 1962) of the definite integral (1-23) is approached rapidly by Simpson's rule. We have obtained four-place accuracy or better at millisecond run time. For many applications in applied probability and statistics, four significant figures are more than can be supported by the data.

The iterative loop for approximating an area can be nested in an outer loop that prints the area under the Gaussian distribution curve for each of many increments in z. If the output is arranged in appropriate rows and columns, a table of areas under one half of the Gaussian curve can be generated, for example, from 0.0 to 3.0 z, resulting in printed values of the area at intervals of 0.01 z. This is suggested to the interested reader as an exercise. We generated a 400-entry table in a negligible run time. The Gaussian function is symmetrical, so knowing one half of the curve means that we know the other half as well. The practical value of generating a table of Gaussian areas is small because many such tables are available in statistics books. The method, however, can be applied to derivative functions of the Gaussian function with only minor modifications, resulting in generation of tables of considerable practical importance (see below).

COMPUTER PROJECT 1-3 | *Medical Statistics*

The first application of the Gaussian distribution is in medical decision making or diagnosis. We wish to determine whether a patient is at risk because of the high cholesterol content of his blood. We need several pieces of input information: an expected or normal blood cholesterol, the standard deviation associated with the normal blood cholesterol count, and the blood cholesterol count of the patient. When we apply our analysis, we shall arrive at a diagnosis, either yes or no, the patient is at risk or is not at risk.

But decision making in the real world isn't that simple. Statistical decisions are not absolute. No matter which choice we make, there is a probability of being wrong. The converse probability, that we are right, is called the *confidence level*. If the probability for error is expressed as a percentage, $100 - (\%$ probability for error) $= \%$ confidence level.

The Problem.   Suppose that the total serum cholesterol level in normal adults has been established as 200 mg/100 mL (mg%) with a standard deviation of 25 mg%, that is, $\mu = 200$ and $\sigma = 25$. (Please distinguish between mg% and % probability.) A patient's serum is analyzed for cholesterol and found to contain 265 mg% total cholesterol.

a.   May we say at the 0.95 (95%) confidence level that the patient's cholesterol is abnormally elevated, or is this just a chance fluctuation in a normal patient? To do this, we must first calculate z and then show that the patient's cholesterol level is greater than or less than that of 95% of normal patients. For the reading to be abnormally elevated with 95% confidence, the z-value must be in an area above the 95% limit of the z-curve. The 95% limit of the z-curve is that point on the z-axis with 95% of normal cholesterol measurements below it and 5% of the measurements above it (Fig. 1-6).

b.   May we reach the same conclusion at the 0.99% confidence level?

c.   If the patient's cholesterol level is just at the 95% level, there is a 5% probability that his cholesterol is randomly high and not indicative of pathology. What is the probability that the cholesterol reading obtained for this patient (265 mg%) resulted from chance factors and does not indicate a genuine atherosclerosis risk factor?

d.   The relative consequences of predictive errors cannot be ignored. In alerting the patient to risk, recommending reduction in eggs, meat, and fats, the diagnostician may be wrong, and this will certainly annoy the patient. Conversely, an erroneous failure to issue a warning carries the risk of the

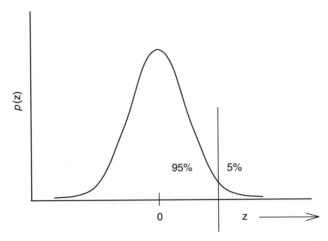

Figure 1-6   The Gaussian Distribution with the 95% Limit Indicated.

patient's death. Relative severity of outcome error should be a factor in evaluating the statistical results once they are known.

e.   What is the 95% limiting cholesterol level (in mg%) in normal patients?

Procedure.   Calculate z for the patient, his "z-score," numerically from the integral in Eq. (1-23). Compare this with the % probability of finding the same z-score in a normal patient. Once knowing the probability of the patient's z-score, one knows the probability that his cholesterol reading is due to chance factors and not indicative of risk. Note that the integral over the interval $[-\infty, 0]$ on the z-axis is 0.5000, so we know everything we need to know by calculating our integrals from 0 to some upper limit. We are not worried about whether the patient's cholesterol level is low; we already know that it is well above the arithmetic mean. The probability that x_i will fall in the normal interval is the same as the probability of a random z in the normal interval. We can then arrive at decisions **a** through **e** with their relative confidence levels (and risk levels).

Determine the probability of a random z using Program QSIM by substituting the two lines

```
m = 1 / (SQR(2 * 3.14159))
DEF FNA (X) = EXP(-X * X / 2)      '**** define function
```

in place of the single DEF fna line of Program QSIM. Notice the convenience substitution of X for z. Multiply a by m in the final line

```
PRINT: PRINT "Numerical integration yields," m * a: END
```

Use the results of your integrations to answer questions **a–e**. Turn in the results of this experiment with a short discussion.

Molecular Speeds

The Maxwell–Boltzmann *distribution function* (Levine, 1983; Kauzmann, 1966) for atoms or molecules (particles) of a gaseous sample is

$$F(\mathbf{v}_x) = \left(\frac{m}{2\pi k_B T}\right)^{1/2} e^{(-m\mathbf{v}_x^2/2k_B T)} \tag{1-24}$$

for molecular *velocity* vectors $\mathbf{v}_x$ about their arithmetic mean $\mathbf{v}_x = 0$ along an arbitrarily selected x-axis. The temperature is T, the mass of the particles (assumed identical to one another) is m, and k_B is the Boltzmann constant, 1.381×10^{-23} J K^{-1}.

The Maxwell–Boltzmann velocity distribution function resembles the Gaussian distribution function because molecular and atomic velocities are randomly distributed about their mean. For a hypothetical particle constrained to move on the x-axis, or for the x-component of velocities of a real collection of particles moving freely in 3-space, the peak in the velocity distribution is at the mean, $\mathbf{v}_x = 0$. This leads to an apparent contradiction. As we know from the kinetic theory of gases, at $T > 0$ all molecules are in motion. How can all particles be moving when the most probable velocity is $\mathbf{v}_x = 0$?

The answer lies in the meaning of the probability curve. The maximum at $\mathbf{v}_x = 0$ arises not because we have maximized our probability of guessing the right velocity but because we have minimized the square of our probable error. (Using the square of the error makes its sign irrelevant.) If we guess a velocity at some value of $\mathbf{v}_x$ other than zero, say a positive value, we will be right some of the time but the square of our error will be large for all negative velocities (half of them). If we guess $\mathbf{v}_x = 0$, we will be wrong all of the time but the sum of squares of our errors (positive and negative) will be least. In essence, the maximum of the velocity probability curve is at zero because we are completely ignorant of the *direction* of motion, and we had best make the guess that specifies no direction at all, namely, zero. This is an application of the *principle of least squares*.

The distribution function for molecular *speeds* v is

$$G(v) = \left(\frac{m}{2\pi k_B T}\right)^{3/2} e^{(-mv^2/2k_B T)} 4\pi v^2 \tag{1-25}$$

where $v = \sqrt{v_x^2 + v_y^2 + v_z^2}$. These lead to the familiar speed distribution curves like those in Fig. 1-7. Unlike the velocity vector, which can be negative, speed v is a scalar and is always positive. The probability of finding $\mathbf{v}_x$ between the limits $[a, b]$ is

$$p(\mathbf{v}_x) = \int_a^b F(\mathbf{v}_x) d\mathbf{v}_x \tag{1-26}$$

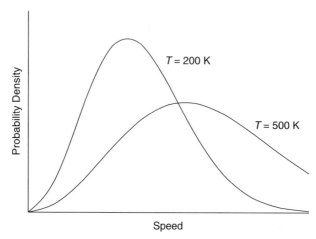

Figure 1-7 A Molecular Speed Distribution. The probability density is the expected number of speeds within an infinitesimal speed interval dv.

and the probability of finding v in the interval $[a, b]$ is

$$p(v) = \int_a^b G(v)dv \tag{1-27}$$

The most probable value of the speed v_{mp} can be obtained by differentiation of the distribution function and setting $dG(v)/dv = 0$ (Kauzmann, 1966; Atkins 1990) to obtain

$$v_{mp} = \left(\frac{2k_B T}{m}\right)^{1/2} \tag{1-28}$$

which is the particle speed at the peak of the curve in Fig. 1-7.

COMPUTER PROJECT 1-4 | *Maxwell–Boltzmann Distribution Laws*
In chemical kinetics, it is often important to know the proportion of particles with a velocity that exceeds a selected velocity v'. According to collision theories of chemical kinetics, particles with a speed in excess of v' are energetic enough to react and those with a speed less than v' are not. The probability of finding a particle with a speed from 0 to v' is the integral of the distribution function over that interval

$$\int_0^{v'} G(v)dv = \left(\frac{m}{2\pi k_B T}\right)^{3/2} \int_0^{v'} e^{(-mv^2/2k_B T)} 4\pi v^2 dv \tag{1-29}$$

The probability of finding a particle with a molecular speed somewhere between 0 and ∞ is 1.0 because negative molecular speeds are impossible; hence, the relative frequency of speeds in excess of v' is $1.0 - \int_0^{v'} G(v)dv$.

It is convenient to reason in terms of the fraction of particles having a velocity in excess of v_{mp}. The most probable velocity works as a normalizing factor, permitting us to generate one curve that pertains to all gases rather than having a different curve for each molecular weight and temperature. The integral of $G(v)dv$ over an arbitrary interval, however, cannot be obtained in closed form. It is usually integrated by parts (Levine, 1989) with the use of a scaling factor, to yield a three-term equation that is evaluated to give the fraction $f(v)$ of particles with speeds in excess of v'/v_{mp} as a function of v'/v_{mp}. This technique does not really escape the problem of nonintegrable functions because the second term in the evaluation for the frequency factor is a nonintegrable Gaussian.

It is also possible to integrate Eq. (1-29) directly by numerical means and to subtract the result from 1.0 to obtain the proportion of particles with speeds in excess of v'/v_{mp}. In this project we shall use numerical integration of $G(v)dv$ over various intervals to obtain $f(v)$ as a function of v'/v_{mp}. Because $v_{mp} = (2k_B T/m)^{1/2}$ [Eq. (1-28)], $\int_0^{v'} G(v)dv$ can be written

$$\int_0^{v'} G(v)dv = \frac{4}{\sqrt{\pi}} X^2 e^{-X^2} dX = 2.25626 X^2 e^{-X^2} dX \tag{1-30}$$

where $X = v'/v_{mp}$. This is the function we shall integrate in this project.

Procedure. Modify Program QSIM by substituting

```
DEF fna (X) = X * X * EXP(-X * X) * 2.25626
```
in place of the DEF fna line of Program QSIM and put

```
(1-a)
```
in place of

```
a
```
in the last line.

a. Using Program QSIM, generate the fraction of particles $f(v)$ with a speed in excess of v'/v_{mp} as a function of v'/v_{mp} by numerical evaluation of the integral for intervals from 0 to 0.2, 0.4, etc. up to 2.0. Compare your plot of $f(v)$ vs. v'/v_{mp} with the literature (Kauzmann, 1966, Rogers and Gratzer, 1984).

b. Find the speed below which 75% of N_2 molecules move at 500 K. On average, one in four N_2 molecules is moving faster than the calculated value of v' at 500 K. Why is $f(v)$ near but not equal to 0.5000 when $v' = v_{mp}$? The *median speed* is that speed at which half the particles in a collection are gong faster than v_{med} and half are going slower. Use Program QSIM to determine the ratio of v_{med} to v_{mp}.

Because the computer cannot store an infinite number of bits, computations leading to very small and very large numbers are often inaccurate unless special precautions are taken. Results of the present calculation are poor at high velocities because of limitations imposed on handling very small exponential numbers. Fortunately, an approximation formula for $\frac{v'}{v_{mp}} \gg 1.0$ is known (Kauzmann, 1966)

$$f(v) = \left(\frac{1}{\sqrt{\pi}}\right) e^{-v^2} \left(2v + \frac{1}{v}\right) \tag{1-31}$$

for the fraction of molecular velocities that are substantially in excess of v_{mp}. Particles moving with these extreme velocities are rare but important because, in many reactions, only very fast-moving molecules react. The proportion of very energetic molecules relative to ordinary molecules, say those with speeds in excess of $4v_{mp}$, increases rapidly with temperature. This is the cause of an exponential rise of reaction rate with temperature observed in many reactions (Arrhenius' rate law).

Atomic Orbitals

Once a numerical integration scheme that permits easy insertion of defined functions and convenient setting of the limits of integration has been set up and debugged, we may wish to use numerical integration for convenience rather than necessity. For example, establishing that hydrogenic wave functions have been correctly normalized and distinguishing between normalized and nonnormalized wave functions are common exercises in introductory quantum mechanic courses and can be mathematically difficult for all but the lowest atomic orbitals. Because the square of the wave function ψ^2 at r is proportional to the probability of finding an electron within an infinitesimal interval $r + dr$, the integral over the entire range $0 < r < \infty$ must be a certainty, $p(r) = 1.0$.

Normalization is the process of finding a multiplicative constant for the wave function such that the integral of ψ^2 over all space is 1.0. "All space" in this calculation is nonnegative because r cannot be less than 0.

The 1s orbital $\psi_{1s} = e^{-r}$ is correct but not normalized. The normalized function governing the probability of finding an electron at some distance r along a fixed axis measured from the nucleus in units of the Bohr radius $a_0 = 5.292 \times 10^{-11}$ m is

$$\psi_{1s} = \frac{1}{\sqrt{\pi}} \left(\frac{1}{a_0}\right) e^{-r/a_0} \tag{1-32}$$

The probability function (1-33 below) governs the probability of finding the electron at some distance r from the nucleus in *any* direction. Owing to the factor r^2, this function gives us the probability of finding the electron anywhere within the interval $r + dr$ on the surface of a sphere of radius r. The radial function (1-32) is monotonically decreasing, but the function in spherical polar coordinates [Eq. (1-33)] goes through a maximum similar to that of the Maxwell–Boltzmann function of the last computer project.

Spherically symmetric (radial) wave functions depend only on the radial distance r between the nucleus and the electron. They are the $1s, 2s, 3s \ldots$ orbitals of atomic hydrogen. For spherically symmetric wave functions, simply typing FNA(X) as the wave function in question and integrating its square over the interval $[0, \infty]$ approximates 1.0 for normalized wave functions and something else for nonnormalized functions. We cannot really integrate to an upper limit of infinity, so we select an upper limit that is large relative to electronic excursions. If the upper limit is not self-evident, it can be systematically incremented until a self-consistent integral is found. When the integral no longer increases for a small increase in the upper limit of integration, the limit is, for all practical purposes, "infinite."

Evaluation of the integral $\int_{r_1}^{r_2} \psi^2(r)dr$, where $\psi(r)$ is a normalized radial wave function, yields the probability density for finding an electron within a finite interval $r_1 < r < r_2$ from the nucleus. A common assigned problem in elementary quantum chemistry (McQuarrie, 1983; Hanna, 1981) is to determine the probability of finding an electron in the $1s$ orbital of a hydrogen atom at a radial distance of one Bohr radius or less from the nucleus. This problem is usually solved by integration in closed form (ans. $p(r) = 0.323$), but the wave function can easily be introduced into an iterative procedure, such as a Simpson's rule integration program, that calculates the probability between any stipulated limits on r

$$p(r) = \frac{4}{a_0^3} \int_0^r r^2 e^{\frac{-2r}{a_0}} dr = 4 \int_0^x x^2 e^{-2x} dx \qquad (1\text{-}33)$$

where $x = r/a_0$ and a_0 is the Bohr radius. The value π in the normalization constant of Eq. (1-33) cancels with the π in $4\pi r^2$ as the surface of a sphere. Check the algebra to see that this is true.

COMPUTER PROJECT 1-5 | *Elementary Quantum Mechanics*

Procedure. Modify Program QSIM to perform the integration in Eq. (1-33) so as to generate the probability of finding the electron within radial distances of 0.1, 0.2, 0.3, ... 5.0 Bohr radii from the nucleus of the hydrogen atom. Check the function (1-33) to verify that the program is working and that the function is normalized. (It is.) Note that a correctly modified program run between limits of 0 and 1 at, say, $n = 1000$ subintervals, gives the probability of finding the electron anywhere within a sphere having a radius of 1.0 bohr with the nucleus at its center. This is a double check on the program. You should get 0.323 in agreement with the analytical integration mentioned above. In what radius interval of 0.10 bohr is the probability of finding the electron greatest? What is the probability within that interval?

Draw a cumulative probability curve $p(x)$ vs. x for finding an electron within any given radius. The curve resembles an *ogive* or S-shaped curve common in chemical applications, but it is flattened at the top owing to the non-Gaussian nature of the square of the $1s$ wave function. An extension of this project is to set up probability limits so that critical radii can be generated that contain the electron with a probability of 0.1, 0.2, ... 0.9. When these radii are known, probability contour maps can be drawn (Gerhold, 1972). Draw the appropriate contour map for the hydrogen atom. What is the probability of finding an electron between a and $2a$, where a is the Bohr radius? As a further extension of this project, repeat the procedure for the $2s$ and $3s$ orbitals of the hydrogen atom available in most physical chemistry and quantum chemistry textbooks (e.g., House, 1998).

Entropy

The subject of entropy is introduced here to illustrate treatment of experimental data sets as distinct from continuous theoretical functions like Eq. (1-33). Thermodynamics and physical chemistry texts develop the equation

$$S_2 = S_1 + \int_{T_1}^{T_2} \frac{C_P}{T} \, dT \tag{1-34}$$

where C_P is the heat capacity at constant pressure, as the fundamental equation for determining the enthalpy change $S_2 - S_1$ of a substance that is heated from T_1 to T_2 but does not suffer a phase change over that temperature interval. The alternative form

$$S_2 = S_1 + \int_{\ln T_1}^{\ln T_2} C_P \, d(\ln T) \tag{1-35}$$

is also used. Armed with the third law of thermodynamics, heat capacities, and thermal data that permit calculation of accurate entropies of intervening phase changes, these integrations permit one to determine *absolute entropies*.

Several examples have been given (Norris, 1981) in which the entropy change of a diatomic gas at 500 K is determined from a knowledge of its entropy at 298.15 K by numerical integration of accurate heat capacity data from 298 to 500 K. Several other chemical applications of numerical integration are given, including determination of the equilibrium constant at an arbitrary temperature T_2 from the integrated van't Hoff equation (Cox and Pilcher, 1970) and a knowledge of K_1 at T_1. Supporting algorithms, data tables, references, and commentaries on the calculations are given.

In the first part of this project, the analytical form of the functional relationship is not used because it is not known. Integration is carried out directly on the experimental data themselves, necessitating a rather different approach to the programming of Simpson's method. In the second part of the project, a curve fitting program (**TableCurve**, Appendix A) is introduced. **TableCurve** presents the area under the fitted curve along with the curve itself.

COMPUTER PROJECT 1-6 | Numerical Integration of Experimental Data Sets

For the first part of this project, we suppose that we are presented with the following experimental data on the heat capacity at constant pressure C_P of solid lead at various temperatures up to and including 298 K (Table 1-2).

We shall assume that $C_P = 0$ at $T = 0$ K. We wish to obtain the absolute entropy of solid lead at 298 K. Each entry in Table 1-2 leads to a value of C_P/T. The

Table 1-2 Experimental Heat Capacities at Constant Pressure for Lead

T, K	0	5	10	15	20	25	30	50	70
C_P, J K^{-1} mol^{-1}	0	0.305	2.80	7.00	10.8	14.1	16.5	21.4	23.3
	100	150	200	250	298				
	24.5	25.4	25.8	26.2	26.5				

experimental data set can be entered into a Simpson's rule integration program in the form of a DATA statement consisting of 14 number pairs, T first and C_P/T second, in each pair. Note that spaces are not used in the data statement. There must be an even number of data pairs for Simpson's rule integration because the subintervals are chosen in pairs.

A Shortcut. The spreadsheet *Excel* (Appendix A) is available on many microcomputer systems. It is designed for business applications, not science, but it can be useful for handling large data sets. In this problem, we have a set of 14 C_P values at corresponding T values and we would like to enter C_P/T and T values into the program. Carrying out the repeated divisions by hand calculator is not very time-consuming or error-prone for this small problem, but it would be in a research project generating hundreds of data points.

Once entered into a spreadsheet, data can be manipulated column at a time. For example, let us take the "top cells" in Table 1-3 as cells A3 and B3 (columns A and B, line 3 in Table 1-3) containing 5 and 0.305 to avoid dividing 0 by 0. Using the *easycalc* option of the *tools* menu in *Excel*, divide the contents of B3 by A3 and place the results in cell C3. Now select C3 and the remaining 12 unfilled cells in the column, C3 to C15, and *fill down* using the mouse. The results of the calculation of C_P/T appear for all remaining cells in the C column.

Table 1-3 Excel Output for Entropy Calculations

T	C_P	C_P/T
A1	B1	
0	0	
5	0.305	0.061
10	2.8	0.28
15	7	0.46667
20	10.8	0.54
25	14.1	0.564
30	16.5	0.55
50	21.4	0.428
70	23.3	0.33286
100	24.5	0.245
150	25.4	0.16933
200	25.8	0.129
250	26.2	0.1048
298	26.5	0.08893

Note that different spreadsheets and different versions of the same spreadsheet vary in the details of the calculation but that the basic idea for all is to carry out the calculation for the top cell and "fill in" the remaining cells in the same column with the mouse—a very convenient technique for simple calculations on large data sets. Consult the **Help** section of your spreadsheet for specific details.

Program

```
Program QENTROPY
DIM X(100), Y(100)
DATA 0,0,5,.061,10,.28,15,.4666,20,.54,25,.564,30,.55,50,
    .428, 70,.333,100,.245
DATA 150,.169,200,.129,250,.105,298,.089
N = 14
FOR I = 1 TO N: READ X(I), Y(I)          'this module reads the data set
PRINT X(I), Y(I)
NEXT I
FOR I = 0 TO N − 2 STEP 2                 'this module calculates the
S = S + (X(I + 1) − X(I)) * (Y(I))        'area of the rectangular
S = S + (X(I + 2) − X(I + 1)) * Y(I + 2)  'blocks
NEXT I
FOR I = 0 TO N − 2 STEP 2                 'this module calculates the area under
S = S + (X(I + 1) − X(I)) *              'the parabolas 1, 4 in Fig. 1-3.
   (Y(I + 1) − Y(I)) *.6667
S = S + (X(I + 2) − X(I + 1)) * (Y(I + 1) − Y(I + 2)) *.6667
NEXT I: PRINT
PRINT "THE ENTROPY (CHANGE) IS: ": PRINT S: END
```

The DIM statement in Program QENTROPY sets aside 100 memory locations for the experimental data points. It is necessary for any data set having more than 12 data pairs. What is the entropy of Pb at 100 and 200 K? Make a rough sketch of the curve of C_p vs. T for lead. Sketch the curve of C_p/T vs. T for lead.

Sigmaplot and Tablecurve

Jandel Scientific produces two programs that have many features useful in data processing. Both are rather complicated, intended for professional rather than student use, consequently some learning time must be invested to become proficient. This time is amply repaid later and the learning curve is not steep, so one can put these programs to practical use on relatively simple problems while learning how to handle more difficult ones. We shall give two examples here: curve plotting using **SigmaPlot** and curve fitting with numerical integration using **TableCurve**.

On entering **SigmaPlot** (we use version 5.0), one is presented with a data table that is essentially a spreadsheet. Enter T as the independent or x-variable into the first column of the **SigmaPlot** data table and C_P/T as the dependent or y-variable into the second column. The **SigmaPlot** data table should resemble columns 1 and 3 of Table 1-3. Rounding to three significant figures is permissible.

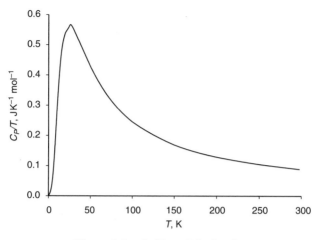

Figure 1-8 C_P/T vs. T for Lead.

After the data set has been entered and saved, one has several plotting options represented by icons in square boxes at the left of the data table. Click on the icon with a single zig-zag line to select a single plot on rectangular coordinates. After selection of the single plot option, one is presented with several suboptions. Select the option represented by the wavy line for a *spline fit* (similar to Simpson's rule) to give a single continuous curve through the points. After selection of the spline fit, one is presented with a "plotting wizard" that asks if you want an *x-y* plot. Click yes. Now specify the *x* variable as column 1 and the *y* variable as column 2. The wizard will present you with the option **Finish**. Click on **Finish** to obtain a plot that is in all essential respects Fig. 1-8 except for some cosmetic changes that you can make according to the instructions in the ***SigmaPlot*** manual or the **Help** file.

Different systems may require different protocols to obtain one of many possible graphs, and several protocols in one system often achieve the same result. At entry level, all this may seem a bit bewildering, but to anyone who has struggled with mechanical drawing tools to make a simple line drawing like Fig. 1-8, ***SigmaPlot*** seems a miracle.

To anyone who has carried out curve-fitting calculations with a mechanical calculator (yes, they once existed) ***TableCurve*** (Appendix A) is equally miraculous. ***TableCurve*** fits dozens, hundreds, or thousands of equations to a set of experimental data points and ranks them according to how well they fit the points, enabling the researcher to select from among them. Many will fit poorly, but usually several fit well.

We shall find the equation that best fits the points in columns 1 and 3 of Table 1-3 with ***TableCurve***. On opening ***TableCurve***, one is presented with a blank desktop with several commands at the top. The command to enter data is not **Enter** but **Edit**. Two formats are available, the ***TableCurve*** editor and the ASCII editor. The ***TableCurve*** format is probably a little simpler than the ASCII format, but they are both fairly self-evident and either should yield a data file resembling the data file for **SigmaPlot**. Each x-variable should be entered as an entry in the first column, followed by the y-variable as an entry in the second column. The statistical weight of each data point is 1, which is automatically entered in the **TableCurve** format. Click on **Process** ⇒ **Fit all equations** and wait a moment while the curve fitting takes place. The formula of the best fit will appear with a graph showing the curve of the equation and the data points for comparison. In this case, the fit of the first ranked equation is very good. The first ranked equation turns out to be a quotient of polynomials

$$y^{0.5} = (a + cx + ex^2)/(1 + bx + dx^2 + fx^3) \qquad (1\text{-}36a)$$

that is,

$$\frac{C_p}{T} = \frac{(a + cT + eT^2)^2}{(1 + bT + dT^2 + fT^3)^2} \qquad (1\text{-}36b)$$

where constants a through f are empirical fitting constants given by the program. Equations fitting the curve with a lower ranking according to closeness of fit are also given.

Along with the curve fitting process, **TableCurve** also calculates the area under the curve. According to the previous discussion, this is the entropy of the test substance, lead. To find the integral, click on the **numeric** at the left of the desktop and find 65.06 as the area under the curve over the range of x. The literature value depends slightly on the source; one value (CRC Handbook of Chemistry and Physics) is 64.8 J K^{-1} mol^{-1}.

Mathcad

Before posing the problem for this computer project, we shall introduce another very useful piece of microcomputer software by repeating the integration of Eq. (1-36a) with **Mathcad** (Appendix A). Like other software of this kind, there is a short learning process before **mathcad** can be used with ease. Once one has entered the equation of interest, **mathcad** solves it with a click on the = sign. In the present example, the constants of (Eq. 1-36a) are entered followed by the desired integral

a := 0.013003   b := 0.017052   c := 0.068394   d := 0.002334   e := 0.000745
f := 0.000002912

$$\int_0^{298} \frac{(a + c \cdot x + e \cdot x^2)^2}{(1 + b \cdot x + d \cdot x^2 + f \cdot x^3)^2} \, dx = 65.061$$

Note that the constants must be defined equal to their numerical values (defined = is ; on the keyboard). These definitions must be above the integral you wish to solve. **Mathcad** operates top down. **Mathcad** produces the same value for the integral that we obtained from **TableCurve**. This calculation is redundant with the calculations already performed in this section to introduce new software by solving a problem for which we already know the answer.

Table 1-4   Experimental Heat Capacities at Constant Pressure for an
Unknown Metal

0,0,5,0.24,10,0.64,15,1.36,20,2.31,25,3.14,30,4.48,50,9.64,70,15.7,100,20.1,150,22.0,
200,23.4,250,24.3,298,25.5

The Problem (at last).
A lustrous metal has the heat capacities as a function of temperature shown in
Table 1-4 where the integers are temperatures and the floating point numbers
(numbers with decimal points) are heat capacities. Print the curve of C_P vs. T and
C_P/T vs. T and determine the entropy of the metal at 298 K assuming no phase
changes over the interval [0, 298]. Use as many of the methods described above as
feasible. If you do not have a plotting program, draw the curves by hand. Scan a
table of standard entropy values and decide what the metal might be.

PROBLEMS

1. Show that the area under a parabolic arc similar to Fig. 1-3 but that is concave
   upward is $\frac{1}{3}w(f(x_i) + 4f(x_{i+1}) + f(x_{i+2}))$.
2. Compute the probability of finding a randomly selected experimental measurement between the limits of ± 0.5 standard deviations from the mean.
3. Given experimental measurements with $\mu = 123.4$ and $\sigma = 12.9$, draw the
   entire probability distribution curve for the population of all experimental
   measurements in the class studied.
4. Write a program in BASIC to generate the area under the normal curve over the
   interval [0, 4] at intervals of $0.01z$.
5. The program in Problem 4 gives final values for the integral under the normal
   curve that are obviously too large. The last entry is 0.5002, whereas, from the
   nature of the problem, we know that the integral cannot exceed 0.5000. Suggest
   a reason for this.
6. If Eq. (1-22) is normalized to 1.0, then

$$f(z) = \int_a^b e^{-z^2/2} dz$$

   should be $\sqrt{2\pi}$ for $[-\infty, \infty]$. Find out if this is true by numerical integration
   using limits on the integral that are wide enough that the area under the curve
   doesn't change by more than a part per thousand or so for a small change in the
   limits of integration.
7. (a) Is the atomic wave function

$$\Psi = \frac{1}{\sqrt{\pi}} e^{-r}$$

   normalized to 1?

(b) The probability that the electron in the H atom will be found at a radial distance r from the nucleus is

$$p(r) = \int_0^r 4r^2 e^{-2r} dr$$

where r is measured in units of bohr (1 bohr $= 52.92$ pm). What is the probability that the electron will be found within 2 bohr radii?

(c) At approximately what radial distance is the probability of finding the H atom electron less than 1%?

8. What is the probability of finding an electron between 0.6 and 1.2 Bohr radii of the nucleus. Assume the electron to be in the $1s$ orbital of hydrogen.

9. The $2s$ orbital of hydrogen can be written

$$\Psi = (2 - r)e^{-r}$$

Plot this orbital with appropriate scale factors to determine the behavior of Ψ in rectangular coordinates. Describe its behavior in spherical polar coordinates.

10. Plot the probability density obtained from Ψ in Problem 9 as a function of r, that is, simply square the function above with an appropriate scale factor as determined by trial and error. Comment on the relationship between your plot and the shell structure of the atom.

11. Sketch the probability of finding an electron in the $2s$ orbital of hydrogen at distance r from a hydrogen nucleus as a function of r as a contour map with heavy lines at high probability and light lines at low probability. How does this distribution differ from the $1s$ orbital?

12. Draw the curve of C_P vs. T and C_P/T vs. T from the following heat capacity data for solid chlorine and determine the absolute entropy of solid chlorine at 70.0 K

T	5	10	15	20	25	30	35	40	50	60	70
C_P	0.14	1.10	3.72	7.74	12.09	16.69	20.79	23.97	29.25	33.47	36.32

13. Which of the following two integrals is wrong?

1. $\int_0^\infty x^3 e^{-ax^2} dx = \dfrac{1}{2a^2}$

2. $\int_{-\infty}^\infty x^3 e^{-ax^2} dx = \dfrac{1}{2a}$

14. A function for which $f(x) = -f(-x)$ over a specific intereval is called an odd function over that interval. If $f(x) = f(-x)$, the function is even. For example, $y = x$, is an odd function over $[-2, 2]$. The interval $[-2, 2]$ is symmetrical about $x = 0$. Write some odd functions. Write some even functions. Find a general rule for the integrals of odd functions over a symmetrical interval. Find a general rule for the integral of the product of an odd function and an even function over an interval that is symmetrical for both.

CHAPTER

2

Applications of Matrix Algebra

A matrix is a rectangular array of elements, for example,

$$\mathbf{A} = \begin{bmatrix} a_{11} & a_{12} \\ a_{21} & a_{22} \end{bmatrix}$$

Each element is designated with a double subscript; in general, an element is called a_{ij} where j is its horizontal position in the ith row of the matrix. A matrix with m rows and n elements in each row is an $m \times n$ matrix. A square matrix with n elements in each row is an $n \times n$ matrix.

Matrix Addition

Matrices obey an algebra of their own that resembles the algebra of ordinary numbers in some respects and not in others. The elements of a matrix may be numbers, operators, or functions. We shall deal primarily with matrices of numbers in this chapter, but matrices of operators and functions will be important later.

Addition and subtraction of matrices is carried out by adding or subtracting corresponding elements. With matrices denoted by boldface capital letters and matrix elements by lower case letters, if

$$\mathbf{C} = \mathbf{A} + \mathbf{B} \tag{2-1}$$

Computational Chemistry Using the PC, Third Edition, by Donald W. Rogers
ISBN 0-471-42800-0   Copyright © 2003 John Wiley & Sons, Inc.

then each element in **C** is the sum of the corresponding elements in **A** and **B**

$$c_{ij} = a_{ij} + b_{ij} \qquad (2\text{-}2)$$

It should be evident that there must be the same number of elements in two matrices to be added and that the elements must be arranged in the same way, so that there is a match of one element in matrix **A** with its corresponding element in matrix **B**. Such matrices are said to be *conformable* to addition.

Exercise 2-1

Give an example of matrices that are conformable to addition and an example of matrices that are not.

Solution 2-1

The matrices

$$\begin{pmatrix} 2 & 7 \\ 1 & 1 \end{pmatrix} \quad \text{and} \quad \begin{pmatrix} -4 & -1 \\ 0 & -3 \end{pmatrix}$$

are conformable to addition and have the sum

$$\begin{pmatrix} -2 & 6 \\ 1 & -2 \end{pmatrix}$$

The matrices

$$\begin{pmatrix} 2 & 7 & 3 \\ 1 & 1 & 5 \end{pmatrix} \quad \text{and} \quad \begin{pmatrix} 4 & -1 \\ 0 & -3 \end{pmatrix}$$

are not conformable to addition.

Subtraction of matrices is the inverse of addition. If

$$\mathbf{D} = \mathbf{A} - \mathbf{B} \qquad (2\text{-}3)$$

then

$$d_{ij} = a_{ij} - b_{ij} \qquad (2\text{-}4)$$

where matrices **A** and **B** must be conformable to subtraction.

The normal rules of association and commutation apply to addition and subtraction of matrices just as they apply to the algebra of numbers. The zero matrix has zero as all its elements; hence addition to or subtraction from **A** leaves **A** unchanged

$$\mathbf{A} + \mathbf{0} = \mathbf{A} \qquad (2\text{-}5)$$

We shall denote the zero matrix as **0**, not 0 or O. The zero matrix is sometimes called the *null matrix*.

Matrix Multiplication

Multiplication of a matrix $\mathbf{A}$ by a scalar x follows the rules one would expect from the algebra of numbers: Each element of $\mathbf{A}$ is multiplied by the scalar. If

$$\mathbf{E} = x\mathbf{A} \tag{2-6}$$

then

$$e_{ij} = xa_{ij} \tag{2-7}$$

Multiplication of two matrices, however, is quite different from multiplication of two numbers. The first row of the premultiplying matrix is multiplied element by element into the first column of the postmultiplying matrix, and the resulting sum is the first element in the product matrix. This process is repeated with the first row of the premultiplying matrix and the second column of the postmultiplying matrix to obtain the second element in the product matrix and so on, until all of the elements of the product matrix have been filled in. If

$$\mathbf{F} = \mathbf{AB} \tag{2-8}$$

where $\mathbf{A}$ is the *premultiplying* matrix and $\mathbf{B}$ is the *postmultiplying* matrix, then

$$f_{ij} = \sum_{k=1}^{n} a_{ik}b_{kj} \tag{2-9}$$

To be conformable to multiplication, the horizontal dimension of $\mathbf{A}$ must be the same as the vertical dimension of $\mathbf{B}$, that is, $n_A = m_B$. Square matrices of the same size are always conformable to multiplication. This unusual definition of multiplication, with its rules for dimensions, will become clear with repeated use. The matrices we shall be interested in will usually be square; you should assume that the matrices discussed below are square unless otherwise stipulated. The rules for rectangular matrices and column and row matrices will be developed as needed.

Except in special cases, matrix multiplication is not commutative,

$$\mathbf{AB} \neq \mathbf{BA}_{\text{general case}} \tag{2-10}$$

which is why we are careful to distinguish between the premultiplying and postmultiplying matrices.

Exercise 2-2

Find the product $\mathbf{AB}$ and the product $\mathbf{BA}$ where

$$\mathbf{A} = \begin{pmatrix} 1 & 2 \\ 3 & 4 \end{pmatrix} \quad \text{and} \quad \mathbf{B} = \begin{pmatrix} 5 & 6 \\ 7 & 8 \end{pmatrix}$$

Solution 2-2

$$\mathbf{AB} = \begin{pmatrix} 19 & 22 \\ 43 & 50 \end{pmatrix} \quad \text{and} \quad \mathbf{BA} = \begin{pmatrix} 23 & 34 \\ 31 & 42 \end{pmatrix}$$

Division of Matrices

Division of matrices is not defined, but the equivalent operation of multiplication by an inverse matrix (if it exists) is defined. If a matrix $\mathbf{A}$ is multiplied by its own inverse matrix, $\mathbf{A}^{-1}$, the *unit matrix* $\mathbf{I}$ is obtained. The unit matrix has 1s on its principal diagonal (the longest diagonal from upper left to lower right) and 0s elsewhere; for example, a 3×3 unit matrix is

$$\mathbf{I} = \begin{pmatrix} 1 & 0 & 0 \\ 0 & 1 & 0 \\ 0 & 0 & 1 \end{pmatrix}$$

The unit matrix plays the same role in matrix algebra that 1 plays in ordinary algebra. Multiplication of a matrix by the unit matrix leaves it unchanged:

$$\mathbf{AI} = \mathbf{A} \tag{2-11}$$

Inverse matrices are among the special matrices that commute

$$\mathbf{AA}^{-1} = \mathbf{A}^{-1}\mathbf{A} = \mathbf{I} \tag{2-12}$$

Among the ordinary numbers, only 0 has no inverse. Many matrices have no inverse. The question of whether a matrix $\mathbf{A}$ has or does not have a defined inverse is closely related to the question of whether a set of simultaneous equations has or does not have a unique set of solutions. We shall consider this question more fully later, but for now recall that if one equation in a pair of simultaneous equations is a multiple of the other,

$$\begin{aligned} x + 2y &= 4 \\ 2x + 4y &= 8 \end{aligned} \tag{2-13}$$

no unique solution exists. Similarly for matrices, if one row (or column) of elements is a multiple of any other in the matrix, for example,

$$\mathbf{A} = \begin{pmatrix} 1 & 2 \\ 2 & 4 \end{pmatrix} \tag{2-14}$$

no inverse exists.

Exercise 2-3

Obtain the product matrix **AB** where

$$\mathbf{A} = \begin{pmatrix} 1 & 2 & 3 \\ 4 & 5 & 6 \\ 7 & 8 & 9 \end{pmatrix} \quad \text{and} \quad \mathbf{B} = \begin{pmatrix} 1 & 0 & 1 \\ 2 & 2 & 2 \\ 4 & 1 & 3 \end{pmatrix}$$

Solve the problem by hand. The operation requires 27 individual multiplications and 9 additions.

Exercise 2-4

Write a short BASIC program to solve for **AB** above. Solve for **BA**. Do **AB** and **BA** commute? Solve the same problem using *Mathcad*.

Solutions 2-3 and 2-4

Both problems can be solved by hand, by writing a short BASIC program, or by *Mathcad* as follows:

$$\mathbf{A} := \begin{pmatrix} 1 & 2 & 3 \\ 4 & 5 & 6 \\ 7 & 8 & 9 \end{pmatrix} \qquad \mathbf{B} := \begin{pmatrix} 1 & 0 & 1 \\ 2 & 1 & 2 \\ 4 & 1 & 3 \end{pmatrix}$$

$$\mathbf{A} \cdot \mathbf{B} = \begin{pmatrix} 17 & 5 & 14 \\ 38 & 11 & 32 \\ 59 & 17 & 50 \end{pmatrix}$$

$$\mathbf{B} \cdot \mathbf{A} = \begin{pmatrix} 8 & 10 & 12 \\ 20 & 25 & 30 \\ 29 & 37 & 45 \end{pmatrix}$$

Note that in *Mathcad*, both matrices must be defined *above* the problem to be worked. In *Mathcad*, the symbol $\mathbf{A} :=$ means "matrix **A** is set equal to."

Powers and Roots of Matrices

If two square matrices of the same size can be multiplied, then a square matrix can be multiplied into itself to obtain $\mathbf{A}^2, \mathbf{A}^3$, or $\mathbf{A}^n$. **A** is the square root of $\mathbf{A}^2$ and the nth root of $\mathbf{A}^n$. A number has only two square roots, but a matrix has infinitely many square roots. This will be demonstrated in the problems at the end of this chapter.

Matrix Polynomials

Polynomial means "many terms." Now that we are able to multiply a matrix by a scalar and find powers of matrices, we can form matrix polynomial equations, for example,

$$\mathbf{A}^2 + 4\mathbf{A} + 5\mathbf{I} = \mathbf{0} \qquad (2\text{-}15)$$

There are infinitely many matrices that satisfy this polynomial equation; hence, the polynomial has infinitely many roots.

Exercise 2-5

Show that the matrix

$$\mathbf{A} = \begin{pmatrix} 2 & 3 \\ 3 & 2 \end{pmatrix}$$

satisfies the polynomial

$$\mathbf{A}^2 - 4\mathbf{A} - 5\mathbf{I} = \mathbf{0}$$

Solution 2-5

Using *Mathcad* we get

$$\mathbf{A} := \begin{pmatrix} 2 & 3 \\ 3 & 2 \end{pmatrix} \qquad \mathbf{I} := \begin{pmatrix} 1 & 0 \\ 0 & 1 \end{pmatrix}$$

$$\mathbf{A}^2 - 4 \cdot \mathbf{A} - 5 \cdot \mathbf{I} = \begin{pmatrix} 0 & 0 \\ 0 & 0 \end{pmatrix}$$

Notice that the matrix **A** does not have to be squared before entering it into the *Mathcad* equation. *Mathcad* does the work of squaring **A** as part of the solution of the matrix equation. (keystroke : translates as := in *Mathcad*.)

Exercise 2-6

Find the roots of the ordinary polynomial

$$a^2 - 4a - 5 = 0 \qquad (2\text{-}16)$$

Exercise 2-7

Note that the matrix polynomial in **A** can be factored to give

$$(\mathbf{A} - 5\mathbf{I}) \quad \text{and} \quad (\mathbf{A} + \mathbf{I})$$

Perform the subtraction and addition above and multiply the resultant matrices to show that the null matrix is obtained.

Solution 2-7

Mathcad yields

$$A := \begin{pmatrix} 2 & 3 \\ 3 & 2 \end{pmatrix} \qquad I := \begin{pmatrix} 1 & 0 \\ 0 & 1 \end{pmatrix}$$

$$(A - 5I) \cdot (A + I) = \begin{pmatrix} 0 & 0 \\ 0 & 0 \end{pmatrix} \tag{2-17}$$

The Least Equation

The general form for a matrix polynomial equation satisfied by $\mathbf{A}$ is

$$c_m\mathbf{A}^m + c_{m-1}\mathbf{A}^{m-1} + \cdots + c_0\mathbf{I} = \mathbf{0} \tag{2-18}$$

The *least equation* is the polynomial equation satisfied by $\mathbf{A}$ that has the smallest possible degree. There is only one least equation

$$\mathbf{A}^k + c_{k-1}\mathbf{A}^{k-1} + \cdots + c_0\mathbf{I} = \mathbf{0} \tag{2-19}$$

The *degree* of the least equation, k, is called the *rank* of the matrix $\mathbf{A}$. The degree k is never greater than n for the least equation (although there are other equations satisfied by $\mathbf{A}$ for which $k > n$). If $k = n$, the size of a square matrix, the inverse $\mathbf{A}^{-1}$ exists. If the matrix is not square or $k < n$, then $\mathbf{A}$ has no inverse.

One method of finding the least equation for the simple second degree case is illustrated. Find a number r such that

$$\mathbf{A}^2 - r\mathbf{I}$$

is a matrix that has 0 as the lead element (the element in the 1,1 position). Now, find a number s such that

$$\mathbf{A} - s\mathbf{I}$$

has 0 as the lead element. Find a number t such that

$$(\mathbf{A}^2 - r\mathbf{I}) - t(\mathbf{A} - s\mathbf{I}) = \mathbf{0}$$

This leads to the least equation

$$\mathbf{A}^2 - t\mathbf{A} + (ts - r)\mathbf{I} = \mathbf{0} \tag{2-20}$$

where the coefficients $c_0 = ts - r$ and $c_1 = -t$ in Eq. (2-18). If the coefficient $c_0 = 0$, the matrix $\mathbf{A}$ is *singular* and has no inverse. The method can be extended to higher degrees, but it soon becomes tedious.

Exercise 2-8

Use the method given above to find the least equation of the matrix

$$\mathbf{A} = \begin{pmatrix} 2 & 1 \\ 1 & 3 \end{pmatrix}$$

Does $\mathbf{A}$ have an inverse?

Solution 2-8

$$\mathbf{A}^2 - 5\mathbf{A} + 5\mathbf{I} = \mathbf{0}$$
$$c_0 \neq 0$$

$\mathbf{A}^{-1}$ exists and is $\begin{pmatrix} 0.6 & -0.2 \\ -0.2 & 0.4 \end{pmatrix}$

Verify this solution by calculating and substituting $\mathbf{A}^2$ and $5\mathbf{A}$ to prove the equality. We can see that $\mathbf{A}^{-1}$ exists because neither row nor column can be obtained from the other by simple multiplication. They are *linearly independent*.

Importance of Rank

The degree of the least polynomial of a square matrix $\mathbf{A}$, and hence its rank, is the number of linearly independent rows in $\mathbf{A}$. A linearly independent row of $\mathbf{A}$ is a row that cannot be obtained from any other row in $\mathbf{A}$ by multiplication by a number. If matrix $\mathbf{A}$ has, as its elements, the coefficients of a set of simultaneous nonhomogeneous equations, the rank k is the number of independent equations. If $k = n$, there are the same number of independent equations as unknowns; $\mathbf{A}$ has an inverse and a unique solution set exists. If $k < n$, the number of independent equations is less than the number of unknowns; $\mathbf{A}$ does not have an inverse and no unique solution set exists. The matrix $\mathbf{A}$ is square, hence $k > n$ is not possible.

Importance of the Least Equation

A number s for which

$$\mathbf{A} - s\mathbf{I} \tag{2-21}$$

has no reciprocal is called an *eigenvalue* of $\mathbf{A}$. The equation

$$\mathbf{A}\mathbf{V} = s\mathbf{V} \tag{2-22}$$

where $\mathbf{V}$ is a vector (or vector function), is called the *eigenvalue equation*. If

$$\mathbf{A}^k + c_{k-1}\mathbf{A}^{k-1} + \cdots + c_0\mathbf{I} = \mathbf{0} \tag{2-23}$$

is the least equation satisfied by $\mathbf{A}$, then s is an eigenvalue only if

$$s^k + c_{k-1}s^{k-1} + \cdots + c_0 = 0 \tag{2-24}$$

This is one way of finding eigenvalues. All atomic and molecular energy levels are eigenvalues of a special eigenvalue equation called the *Schroedinger equation*.

Exercise 2-9

Perform the matrix subtraction

$$\mathbf{A} - E\mathbf{I}$$

where

$$\mathbf{A} = \begin{pmatrix} \alpha & \beta \\ \beta & \alpha \end{pmatrix}$$

What is the condition on the resulting matrix that must be met if E is to be an eigenvalue of $\mathbf{A}$?

Solution 2-9

$$\mathbf{A} - E\mathbf{I} = \begin{pmatrix} \alpha - E & \beta \\ \beta & \alpha - E \end{pmatrix}$$

The matrix $\begin{pmatrix} \alpha-E & \beta \\ \beta & \alpha-E \end{pmatrix}$ must have no inverse.

Historical Note.   It is interesting to note (Pauling and Wilson, 1935) that the very first systematic approach to what we now call *quantum mechanics* was made by Heisenberg, who began to develop his own algebra to describe the frequencies and intensities of spectral transitions. It was soon seen by Born and Jordan that Heisenberg's "new" algebra is really matrix algebra. Heisenberg's *eigenfunctions* were later called *wave functions* by Schroedinger in an independent but equivalent method. Schroedinger's method is now called *wave mechanics* and is the method most familiar to chemists. Heisenberg's method is called *matrix mechanics*.

Special Matrices

The transpose $\mathbf{A}^{\mathbf{T}}$ of a matrix is obtained by reflecting the matrix through its principal diagonal:

$$a_{ij}^{\mathbf{T}} = a_{ji} \tag{2-25}$$

Properties of the transpose include

$$(\mathbf{A} + \mathbf{B})^{\mathbf{T}} = \mathbf{A}^{\mathbf{T}} + \mathbf{B}^{\mathbf{T}} \tag{2-26}$$

and

$$(\mathbf{AB})^{\mathbf{T}} = \mathbf{B}^{\mathbf{T}}\mathbf{A}^{\mathbf{T}} \tag{2-27}$$

(note the order of $\mathbf{A}$ and $\mathbf{B}$).

Exercise 2-10

Demonstrate that properties (2-26) and (2-27) hold for arbitrarily selected matrices $\mathbf{A}$ and $\mathbf{B}$.

A *symmetric* matrix equals its own transpose.

$$\mathbf{A} = \mathbf{A}^{\mathbf{T}} \tag{2-28}$$

Exercise 2-11

Give three examples of symmetric matrices.

The transpose of an *orthogonal matrix* is equal to its inverse

$$\mathbf{A}^{\mathbf{T}} = \mathbf{A}^{-1} \tag{2-29}$$

The *trace* of a matrix is the sum of the elements on its principal diagonal

$$tr(\mathbf{A}) = \sum a_{ii} \tag{2-30}$$

Exercise 2-12

What is the trace of a unit matrix of size n?

A *diagonal* matrix has nonzero elements only on the principal diagonal and zeros elsewhere. The unit matrix is a diagonal matrix. Large matrices with small matrices symmetrically lined up along the principal diagonal are sometimes encountered in computational chemistry.

A *tridiagonal* matrix has nonzero elements only on the principal diagonal and on the diagonals on either side of the principal diagonal. If the diagonals on either side of the principal diagonal are the same, the matrix is a symmetric tridiagonal matrix.

Triangular matrices have nonzero elements only on and above the principal diagonal (upper triangular) or on and below the principal diagonal (lower triangular). Some of the more important numerical methods are devoted to transforming a general matrix into its equivalent diagonal or triangular form.

A *column matrix* is an ordered set of numbers; therefore, it satisfies the definition of a *vector*. The 2×1 array

$$\mathbf{x} = \begin{pmatrix} 1 \\ 2 \end{pmatrix}$$

is both a matrix and a vector in 2-space. An $m \times 1$ matrix has one element in each of m rows; therefore, it is one way of representing a vector in an m-dimensional space. An $m \times n$ matrix may be thought of as representing n vectors in m-space where each vector is a column in the matrix. The transpose of a column matrix is a *row matrix*, which can also represent a vector.

The Transformation Matrix

If a vector $\mathbf{x}$ is transformed into a new vector $\mathbf{x}'$ by a matrix multiplication

$$\mathbf{x}' = \mathbf{A}\mathbf{x} \tag{2-31}$$

then $\mathbf{A}$ is a *transformation matrix*. If several vectors are transformed in the same operation, where $\mathbf{X}$ is the matrix consisting of the column vectors $\mathbf{x_i}$, we write

$$\mathbf{X}' = \mathbf{A}\mathbf{X}$$

If the transformation matrix is orthogonal, then the transformation is *orthogonal*. If the elements of $\mathbf{A}$ are numbers (as distinct from functions), the transformation is *linear*. One important characteristic of an orthogonal matrix is that none of its columns is linearly dependent on any other column. If the transformation matrix is orthogonal, $\mathbf{A}^{-1}$ exists and is equal to the transpose of $\mathbf{A}$. Because $\mathbf{A}^{-1} = \mathbf{A}^T$

$$\mathbf{A}\mathbf{A}^T = \mathbf{A}\mathbf{A}^{-1} = \mathbf{A}^{-1}\mathbf{A} = \mathbf{A}^T\mathbf{A} = \mathbf{I} \tag{2-32}$$

Orthogonal transformations preserve the lengths of vectors. If the same orthogonal transformation is applied to two vectors, the angle between them is preserved as well. Because of these restrictions, we can think of orthogonal transformations as rotations in a plane (although the formal definition is a little more complicated).

If two matrices are related as

$$\mathbf{B} = \mathbf{C}^{-1}\mathbf{A}\mathbf{C} \tag{2-33}$$

then $\mathbf{B}$ and $\mathbf{A}$ are *similar matrices*. If the squares of the coefficients of each of two or more orthogonal vectors add up to 1, the vectors are *orthonormal*. If $\mathbf{A}$ is symmetric, the vectors of $\mathbf{A}$ are or can be chosen to be orthonormal and $\mathbf{X}$ in the equation

$$\mathbf{A}\mathbf{X} = \mathbf{X}\mathbf{D}$$

$$\mathbf{D} = \mathbf{X}^{-1}\mathbf{A}\mathbf{X} = \mathbf{X}^T\mathbf{A}\mathbf{X} \tag{2-34}$$

holds, where the vectors comprising the matrix $\mathbf{X}$ are called *eigenvectors*. $\mathbf{D}$ has been chosen to be a diagonal matrix with the eigenvalues of $\mathbf{A}$ on the principal diagonal. The question is whether we can find $\mathbf{X}$. If we can, we have successfully converted $\mathbf{A}$ into a *similar matrix* $\mathbf{D}$ that has only one element in each row or column. If $\mathbf{A}$ was the matrix of coefficients of (possibly many) simultaneous equations, $\mathbf{D}$ is the matrix of coefficients of a mathematically similar set of equations, each equation containing only one term. Thus the entire set of equations has been solved if we can find $\mathbf{X}$ in Eqs. 2-34. We shall go into the details of this problem later. The point here is that matrix $\mathbf{A}$ can be reduced to a very simple form $\mathbf{D}$ if we can find or approximate the matrix of eigenvectors $\mathbf{X}$.

Complex Matrices

Numbers may be real, a, imaginary, ic, or complex, $a \pm ic$, where $i = \sqrt{-1}$. The elements in a matrix may be complex numbers. If so, the matrix is complex

$$\mathbf{A} = \mathbf{B} + i\mathbf{C} \qquad (2\text{-}35)$$

(For a real matrix, $\mathbf{C} = \mathbf{0}$.) The *complex conjugate* of a complex matrix $\mathbf{A}$ is $\mathbf{A}^*$. In $\mathbf{A}^*$, each element in $\mathbf{A}$ replaced by its complex conjugate; $a \pm ic$ becomes $a \mp ic$. The complex conjugate $\mathbf{A}^*$ of $\mathbf{A}$ is

$$\mathbf{A}^* = \mathbf{B} - i\mathbf{C} \qquad (2\text{-}36)$$

The *Hermetian conjugate* of $\mathbf{A}$ is the transpose of $\mathbf{A}^*$

$$\mathbf{A}^{\mathbf{H}} = (\mathbf{A}^*)^{\mathbf{T}} \qquad (2\text{-}37)$$

The Hermetian conjugate plays the same role for complex matrices that the symmetric matrix plays for real matrices.

If the Hermetian conjugate of a square complex matrix is equal to its inverse,

$$\mathbf{U}^{\mathbf{H}} = \mathbf{U}^{-1} \qquad (2\text{-}38)$$

the matrix $\mathbf{U}$ is called a *unitary matrix*. A Hermetian matrix is reduced to diagonal form by a *unitary transformation*

$$\mathbf{D} = \mathbf{U}^{\mathbf{H}}\mathbf{A}\mathbf{U} = \mathbf{U}^{-1}\mathbf{A}\mathbf{U} \qquad (2\text{-}39)$$

where $\mathbf{D}$ is real with elements equal to the eigenvalues of $\mathbf{A}$. $\mathbf{U}$ has columns that are eigenvectors of $\mathbf{A}$.

What's Going On Here?

The best way to avoid losing the physics of these procedures is to think of a particle describing an elliptical path about an origin. If we choose our coordinate system in an arbitrary way, the result might look like Fig. 2-1 (left).

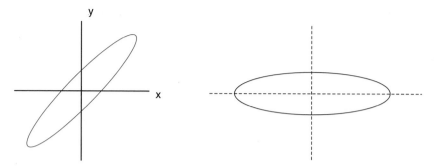

Figure 2-1 A Particle on an Elliptical Orbit.

In general, the equation describing an elliptical path

$$ax^2 + 2bxy + cy^2 = Q$$

contains mixed terms, $2bxy$. For example, the equation of the ellipse on the left in Fig. 2-1 might look something like

$$5x^2 + 8xy + 5y^2 = 9 \qquad (2\text{-}40)$$

If we can find the appropriate **X** matrix to carry out a similarity transformation on the coefficient matrix for the quadratic equation (2-40)

$$\mathbf{A} = \begin{pmatrix} a & b \\ b & c \end{pmatrix} = \begin{pmatrix} 5 & 4 \\ 4 & 5 \end{pmatrix} \qquad (2\text{-}41)$$

we get

$$\mathbf{A}' = \begin{pmatrix} 1 & 0 \\ 0 & 9 \end{pmatrix} \qquad (2\text{-}42)$$

which leads to the equation of the ellipse as represented on the right of Fig. 2-1

$$x'^2 + 9y'^2 = 9 \qquad (2\text{-}43)$$

It turns out that the "appropriate **X** matrix" of the eigenvectors of **A** rotates the axes $\pi/4$ so that they coincide with the *principle axes* of the ellipse. The ellipse itself is unchanged, but in the new coordinate system the equation no longer has a mixed term. The matrix **A** has been diagonalized. Choice of the coordinate system has no influence on the physics of the situation, so we choose the simple coordinate system in preference to the complicated one.

The physical meaning of the sections on transformation matrices and unitary matrices is that we can try to rotate our coordinate system so that each component of the motion is independent of all the rest. We may be successful or nearly successful. Note well that there is no restriction on the number of dimensions of the n-dimensional space the coordinate system *spans*. When, in future work, we seek to diagonalize an n-dimensional matrix, we are seeking to rotate a set of orthonormal axes, one in each dimension of an n-space, such that each axis is a principal axis of the matrix.

The unitary transform does the same thing as a similarity transform, except that it operates in a complex space rather than a real space. Thinking in terms of an added imaginary dimension for each real dimension, the space of the unitary matrix is a $2m$-dimensional space. The unitary transform is introduced here because atomic or molecular wave functions may be complex.

PROBLEMS

1. Carry out hand calculations to find the products **AB** and **BA**

$$\mathbf{A} = \begin{pmatrix} 3 & 0 & 3 \\ 4 & -1 & -1 \\ 1 & 2 & 5 \end{pmatrix} \qquad \mathbf{B} = \begin{pmatrix} 1 & 1 & 1 \\ -2 & 1 & 6 \\ 3 & 4 & 5 \end{pmatrix}$$

   Do **A** and **B** commute?
2. Write a program in BASIC to carry out the multiplications in Problem 1. Cross-check your program results with your hand calculations from Problem 1.
3. Invert **A** in Problem 1. Systematic methods exist for inverting matrices and will be discussed in the next chapter. For now, use *Mathcad* if it is available to you.
4. Find $\mathbf{AA}^{-1}$. Is it true that $\mathbf{AA}^{-1} = \mathbf{I}$? Does **A** commute with $\mathbf{A}^{-1}$?
5. Transpose **A** and **B** in Problem 1.
6. Transpose the product **AB** to find $(\mathbf{AB})^{\mathrm{T}}$.
7. Find the product $\mathbf{A}^{\mathrm{T}}\mathbf{B}^{\mathrm{T}}$; compare it with the transpose of the product from Problem 6. Deduce a rule for $\mathbf{A}^{\mathrm{T}}\mathbf{B}^{\mathrm{T}}$.
8. Is the matrix **Q** conformable to multiplication into its own transpose? What about $\mathbf{Q}^{\mathrm{T}}\mathbf{Q}$? What is the dimension of $\mathbf{QQ}^{\mathrm{T}}$? What is the dimension of $\mathbf{Q}^{\mathrm{T}}\mathbf{Q}$?

$$\mathbf{Q} = \begin{pmatrix} p & q \\ r & s \\ t & u \end{pmatrix}$$

9. The problem of a mass suspended by a spring from another mass suspended by another spring, attached to a stationary point (Kreyszig, 1989, p. 159ff) yields the matrix equation

$$\mathbf{Ax} = \lambda \mathbf{x}$$

   where **x** is a vector. For a certain combination of masses and springs,

$$\mathbf{A} = \begin{pmatrix} 5 & 3 \\ 3 & 5 \end{pmatrix}$$

Is $\lambda = 7$ an eigenvalue of the system? Is $\lambda = 8$ an eigenvalue of the system? There is another eigenvalue. What is it?

10. The transpose of $\begin{pmatrix} x \\ y \end{pmatrix}$ is $(x \quad y)$. Carry out the multiplication

$$(x \quad y)\begin{pmatrix} 5 & 4 \\ 4 & 5 \end{pmatrix}\begin{pmatrix} x \\ y \end{pmatrix}$$

11. The eigenvector matrix for a $\pi/4$ rotation is

$$\mathbf{X} = \frac{1}{\sqrt{2}}\begin{pmatrix} 1 & 1 \\ -1 & 1 \end{pmatrix}$$

Carry out the rotation

$$\mathbf{X}^{\mathrm{T}}\begin{pmatrix} 5 & 4 \\ 4 & 5 \end{pmatrix}\mathbf{X}$$

Linear Nonhomogeneous Simultaneous Equations

The problem of n linear independent *nonhomogeneous* equations in n real unknowns

$$
\begin{aligned}
a_{11}x_1 + a_{12}x_2 &= b_1 \\
a_{21}x_1 + a_{22}x_2 &= b_2
\end{aligned}
\tag{2-44}
$$

is often encountered in an experimental context. We have taken two equations in two unknowns for notational simplicity, but the equation set may be extended to the n-variable case. Many coded programs have been published in FORTRAN (e.g., Carnahan and Luther, 1969; Isenhour and Jurs, 1979) for each of the algorithms discussed here. Most are short and easily translated into other computer languages. The problem of more than n equations in n real unknowns is called the *multivariate* problem. Linearly dependent *homogeneous* equations frequently occur in some branches of quantum mechanics. These problems will be treated later.

Linear independence implies that no equation in the set can be obtained by multiplying any other equation in the set by a constant. The $n \times n$ matrix populated by n^2 elements a_{ij}

$$\mathbf{A} = \begin{pmatrix} a_{11} & a_{12} \\ a_{21} & a_{22} \end{pmatrix} \tag{2-45}$$

is called the *coefficient matrix*. For the set to be linearly independent, the rank of $\mathbf{A}$ must be n. An ordered set of numbers is a vector; hence the ordered number pair $\mathbf{x} = \begin{pmatrix} x_1 \\ x_2 \end{pmatrix}$ is called the *solution vector* or *solution set* and the ordered set of constants

$\mathbf{b} = \begin{pmatrix} b_1 \\ b_2 \end{pmatrix}$ is called the *constant vector*. I like the term *nonhomogeneous vector*, because existence of any nonzero element b_i causes the equation set to be nonhomogeneous. A convenient term is the b *vector*. To be conformable for multiplication by a matrix, the dimension of a row vector must be the same as the column dimension of the matrix. A column vector must have the same dimension as the row dimension of the matrix.

Designating the two vectors and one matrix just defined by boldface letters, the set of equations (2-44) is

$$\mathbf{Ax} = \mathbf{b} \tag{2-46}$$

Equation (2-46) is a matrix equation because vectors $\mathbf{x}$ and $\mathbf{b}$ are properly regarded as one-column matrices. Vectors are often differentiated from matrices by writing them as lower case letters.

Multiplication by the rules of matrix algebra produces equation set (2-44) from Eq. (2-46), demonstrating their equivalence. Equation (2-46) is an economical way of expressing Eqs. (2-44), especially where n is large, but it is more than that: Systematic methods of solving Eqs. (2-44) really depend on the properties of the coefficient matrix and on what we can do with it. For example, if the set of Eqs. (2-44) is linearly dependent, $\mathbf{A}$ is *singular*, which means, among other things, that its determinant is zero and it has no inverse $\mathbf{A}^{-1}$. In practical terms, this means that no unique solution set exists for Eqs. (2-44). We already knew that, but less obvious operations on Eqs. (2-44) such as triangularization and diagonalization can be more easily visualized and programmed in terms of operations on the coefficient matrix $\mathbf{A}$ than in terms of the entire set.

Exercise 2-13

Show that a vector in a plane can be unambiguously represented by an ordered number pair and hence that any ordered number pair can be regarded as a vector.

Solution 2-13

Consider a vector as an arrow in two-dimensional space. Now superimpose $x-y$ coordinates on the 2-space, arbitrarily placing the origin on the "tail" of the arrow.

The vector in Fig. 2-2 happens to fall in the fourth quadrant as drawn. The number pair giving the point that coincides with the tip of the arrow gives its magnitude and direction relative to the coordinate system chosen. Magnitude and direction are all that you can know about a vector; hence it is completely defined by the number pair $(5, -1)$.

In general, a vector in an n-space can be represented by an n-tuple of numbers; for example, a vector in 3-space can be represented as a number triplet.

The *determinant* having the same form as matrix $\mathbf{A}$,

$$\det \mathbf{A} = \begin{vmatrix} a_{11} & a_{12} \\ a_{21} & a_{22} \end{vmatrix}$$

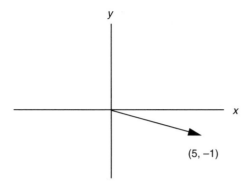

Figure 2-2 A Vector in
Two-Dimensional x-y Space.

is not the same as **A**, because a matrix is an operator and a determinant is a scalar; a matrix is irreducible but a determinant can, if it satisfies some restrictions, be written as a single number.

$$\det \mathbf{A} = \begin{vmatrix} a_{11} & a_{12} \\ a_{21} & a_{22} \end{vmatrix} = a_{11}a_{22} - a_{12}a_{21}$$

Algorithms

An algorithm is a recipe for solving a computational problem. It gives the general approach but does not go into specific detail. Although there are many algorithms for simultaneous equation solving in the literature, they can be separated into two classes: elimination and iterative substitution. Elimination methods are closed methods; in principle, they are capable of infinite accuracy. Iterative methods converge on the solution set, and so, strictly speaking, they are never more than approximations. In practice, the distinction is not so great as it might seem, because iterative approximations can be made highly self-consistent, that is, nearly identical from one iteration to the next, and closed elimination methods suffer the same machine word-size limitations that prevent infinite accuracy in any fairly involved computer procedure.

Gaussian Elimination. In the most elementary use of *Gaussian elimination*, the first of a pair of simultaneous equations is multiplied by a constant so as to make one of its coefficients equal to the corresponding coefficient in the second equation. Subtraction eliminates one term in the second equation, permitting solution of the equation pair.

Solving several equations by the method of Gaussian elimination, one might divide the first equation by a_{11}, obtaining 1 in the a_{11} position. Multiplying a_{21} into the first equation makes $a_{11} = a_{21}$. Now subtracting the first equation from the second, a zero is produced in the a_{21} position. The same thing can be done to produce a zero in the a_{31} position and so on, until the first column of the coefficient matrix is filled with zeros except for the a_{11} position.

Attacking the a_{22} position in the same way, but leaving the first horizontal row of the coefficient matrix alone, yields a matrix with zeros in the first two columns except for the triangle

$$
\begin{pmatrix}
a_{11} & a_{12} & a_{13} & \cdots \\
0 & a_{22} & a_{23} & \cdots \\
0 & 0 & a_{33} & \cdots \\
\vdots & \vdots & \vdots & \ddots
\end{pmatrix}
$$

This is continued $n - 1$ times until the entire coefficient matrix has been converted to an *upper triangular matrix*, that is, a matrix with only zeros below the principal diagonal. The b vector is operated on with exactly the same sequence of operations as the coefficient matrix. The last equation at the very bottom of the triangle, $a_{nn}x_n = b_n$, is one equation in one unknown. It can be solved for x_n, which is back-substituted into the equation above it to obtain x_{n-1} and so on, until the entire solution set has been generated.

Exercise 2-14

In Exercise 2-8, we obtained the least equation of the matrix

$$
\mathbf{A} = \begin{pmatrix} 2 & 1 \\ 1 & 3 \end{pmatrix}
$$

Solve the simultaneous equation set by Gaussian elimination

$$2x + y = 4$$
$$x + 3y = 7$$

Note that the matrix from Exercise 2-8 is the matrix of coefficients in this simultaneous equation set. Note also the similarity in method between finding the least equation and Gaussian elimination.

Solution 2-14

The triangular matrix $\mathbf{A}^{\mathbf{G}}$ resulting from Gaussian elimination is

$$
\mathbf{A}^{\mathbf{G}} = \begin{pmatrix} 1 & 0.5 \\ & 2.5 \end{pmatrix}
$$

In the process of obtaining the upper triangular matrix, the nonhomogeneous vector has been transformed to $\begin{pmatrix} 2 \\ 5 \end{pmatrix}$. The bottom equation of $\mathbf{Ax} = \mathbf{b}$

$$x + 0.5y = 2$$
$$2.5y = 5$$

$$(2\text{-}47)$$

yields $y = 2$. Back-substitution into the top equation yields $x = 1$. The solution set, as one could have seen by inspection, is $\left({1 \atop 2} \right)$. **Mathcad**, after the matrix **A** and the vector **b** have been defined by using the full colon keystroke, solves the problem using the **lsolve** command

$$\mathbf{A} := \begin{pmatrix} 2 & 1 \\ 1 & 3 \end{pmatrix} \qquad \mathbf{b} := \begin{pmatrix} 4 \\ 7 \end{pmatrix}$$

$$\text{lsolve}(\mathbf{A}, \mathbf{b}) = \begin{pmatrix} 1 \\ 2 \end{pmatrix}$$

Exercise 2-15

Write a program in BASIC for solving linear nonhomogeneous simultaneous equations by Gaussian elimination and test it by solving the equation set in Exercise 2-13.

In the computer algorithm, division by the diagonal element, multiplication, and subtraction are usually carried out at the same time on each target element in the coefficient matrix, leading to some term like $a_{jk} - a_{ik}\left(\frac{a_{ji}}{a_{ii}}\right)$. Next, the same three combined operations are carried out on the elements of the b vector. The arithmetic statements are simple, as is the procedure for back-substitution. The trick in writing a successful Gaussian elimination program is in constructing a looping structure and keeping the variable indices straight so that the right operations are being carried out on the right elements in the right sequence.

Gauss–Jordan Elimination. It is possible to continue the elimination process to remove nonzero elements above the principal diagonal, leaving only $a_{ii} \neq 0$ in the coefficient matrix. This extension of the Gaussian elimination method is called the *Gauss–Jordan* method. By exchanging columns (which does not change the solution set) one can switch the largest element in each row of the coefficient matrix into the pivotal position a_{ii}, and most Gauss–Jordan programs do this. Once the coefficient matrix has been completely *diagonalized* so that the a_{ii} are the only nonzero elements, and the same operations have been carried out on the b vector, the original system of n equations in n unknowns has been reduced to n equations, each in *one* unknown. The solution set follows routinely.

Exercise 2-16

Extend the matrix triangularization procedure in Exercise 2-14 by the Gauss–Jordan procedure to obtain the fully diagonalized matrix $\left({1 \atop 0} \; {0 \atop 0.5} \right)$ and the b vector $\left({1 \atop 1} \right)$. The solution set follows routinely.

By *Cramer's rule*, each solution of Eqs. (2-44) is given as the ratio of determinants

$$x_i = \frac{D_i}{D} \tag{2-48}$$

where D is the determinant of the coefficients a_{ij} and D_i is a similar determinant in which the ith column has been replaced by the b vector. The method is open-ended, that is, it can be applied to any number of equations containing the same number of unknowns, resulting in $n + 1$ determinants of dimension $n \times n$.

Exercise 2-17

Solve the equation set of Eqs. (2-44) using Cramer's rule.

Although apparently quite different from the Gauss and Gauss–Jordan methods, it turns out that the most efficient method of reducing large determinants is mathematically equivalent to Gaussian elimination. As far as the computer programming is concerned, the method of Cramer's rule is only a variant on the Gaussian elimination method. It is slower because it requires evaluation of several determinants rather than triangularization or diagonalization of one matrix; hence it is not favored, except where the determinants are needed for something else. Determinants have one property that is very important in what will follow. If a row is exchanged with another row or a column is exchanged with another column, the determinant changes sign.

Exercise 2-18

Verify the preceding statement for the determinant

$$\det \mathbf{M} = \begin{vmatrix} 1 & 2 \\ 3 & 4 \end{vmatrix}$$

Solution 2-18

$$\begin{vmatrix} 1 & 2 \\ 3 & 4 \end{vmatrix} = 4 - 6 = -2 \qquad \begin{vmatrix} 2 & 1 \\ 4 & 3 \end{vmatrix} = 6 - 4 = 2$$

The Gauss–Seidel Iterative Method.   The *Gauss-Seidel iterative method* uses substitution in a way that is well suited to machine computation and is quite easy to code. One guesses a solution for x_1 in Eqs. (2-44)

$$
\begin{aligned}
a_{11}x_1 + a_{12}x_2 &= b_1 \\
a_{21}x_1 + a_{22}x_2 &= b_2
\end{aligned}
\tag{2-49}
$$

and substitutes this guess into the first equation, which leads to a solution for x_2. The solution is wrong, of course, because x_1 was only a guess, but, when substituted into the second equation, it gives a solution for x_1. That solution is also wrong, but, under some circumstances, it is less wrong than the original guess. The new approximation to x_1 is substituted to obtain a new x_2 and so on in an iterative loop, until self-consistency is obtained within some small predetermined limit.

The drawback of the Gauss–Seidel method is that the iterative series does not always converge. Nonconvergence can be spotted by printing the approximate solution on each iteration. A favorable condition for convergence is dominance of the principal diagonal. Some Gauss–Seidel programs arrange the rows and columns of the coefficient matrix so that this condition is, insofar as possible, satisfied. A more detailed discussion of convergence is given in advanced texts (Rice, 1983; Norris, 1981).

Matrix Inversion and Diagonalization

Looking at the matrix equation $\mathbf{Ax} = \mathbf{b}$, one would be tempted to divide both sides by matrix $\mathbf{A}$ to obtain the solution set $\mathbf{x} = \mathbf{b}/\mathbf{A}$. Unfortunately, division by a matrix is not defined, but for some matrices, including nonsingular coefficient matrices, the inverse of $\mathbf{A}$ is defined.

The unit matrix, $\mathbf{I}$, with $a_{ii} = 1$ and $a_{ij} = 0$ for $i \neq j$, plays the same role in matrix algebra that the number 1 plays in ordinary algebra. In ordinary algebra, we can perform an operation on any number, say 5, to reduce it to 1 (divide by 5). If we do the same operation on 1, we obtain the inverse of 5, namely, 1/5. Analogously, in matrix algebra, if we carry out a *series* of operations on $\mathbf{A}$ to reduce it to the unit matrix and carry out the same series of operations on the unit matrix itself, we obtain the inverse of the original matrix $\mathbf{A}^{-1}$.

One series of mathematical operations that may be carried out on the coefficient matrix to diagonalize it is the Gauss–Jordan procedure. If each row is then divided by a_{ij}, the unit matrix is obtained. Generally, $\mathbf{A}$ and the unit matrix are subjected to identical *row operations* such that as $\mathbf{A}$ is reduced to $\mathbf{I}$, $\mathbf{I}$ is simultaneously converted to $\mathbf{A}^{-1}$. The computer program written to do this is essentially a Gauss–Jordan program as far as coding and machine considerations are concerned (Isenhour and Jurs, 1971). Alternatively, both reduction of $\mathbf{A}$ to $\mathbf{I}$ and conversion of $\mathbf{I}$ to $\mathbf{A}^{-1}$ may be done by the Gauss–Seidel iterative method (Noggle, 1985).

The attractive feature in matrix inversion is seen by premultiplying both sides of $\mathbf{Ax} = \mathbf{b}$ by $\mathbf{A}^{-1}$,

$$\mathbf{A}^{-1}\mathbf{Ax} = \mathbf{A}^{-1}\mathbf{b} = \mathbf{Ix} = \mathbf{x} = \mathbf{A}^{-1}\mathbf{b} \tag{2-50}$$

This means that once $\mathbf{A}^{-1}$ is known, it can be multiplied into *several* b vectors to generate a solution set $\mathbf{x} = \mathbf{A}^{-1}\mathbf{b}$ for each b vector. It is easier and faster to multiply a matrix into a vector than it is to solve a set of simultaneous equations over and over for the same coefficient matrix but different b vectors.

Exercise 2-19

Invert the matrix in Exercise 2-8 by row operations.

Solution 2-19

Place the original coefficient matrix next to the unit matrix. Divide row 1 by 2 and subtract the result from row 2. This is a *linear* operation, and it does not change the solution set.

$$\begin{pmatrix} 2 & 1 & 1 & 0 \\ 1 & 3 & 0 & 1 \end{pmatrix} \Rightarrow \begin{pmatrix} 1 & \frac{1}{2} & \frac{1}{2} & 0 \\ 0 & \frac{5}{2} & -\frac{1}{2} & 1 \end{pmatrix}$$

This gives you a zero in the 2,1 position of the original matrix, which is one step along the way of diagonalization. Continue with the necessary operations to diagonalize the coefficient matrix and at the same time perform the same operations on the adjacent unit matrix. On completion of this stepwise procedure, you have the unit matrix on the left and some matrix other than the unit matrix on the right.

$$\begin{pmatrix} 1 & 0 & z_{11} & z_{12} \\ 0 & 1 & z_{21} & z_{22} \end{pmatrix}$$

Show that the matrix on the right, $\mathbf{Z}$, is the inverse of the original matrix $\mathbf{A}$ by multiplying it into $\mathbf{A}$ to find $\mathbf{ZA} = \mathbf{I}$. Now generate the solution set to the equations for which $\mathbf{A}$ is the coefficient matrix by multiplying $\mathbf{Z}$ (which is $\mathbf{A}^{-1}$) into the nonhomogeneous vector to obtain the solution set $\{1, 2\}$.

COMPUTER PROJECT 2-1 | *Simultaneous Spectrophotometric Analysis*
A spectrophotometric problem in simultaneous analysis (Ewing, 1985) is taken from the original research of Weissler (1945), who reacted hydrogen peroxide with Mo, Ti, and V ions in the same solution to produce compounds that absorb light strongly in overlapping peaks with absorbances at 330, 410, and 460 nm, respectively as shown in (Fig. 2-3).

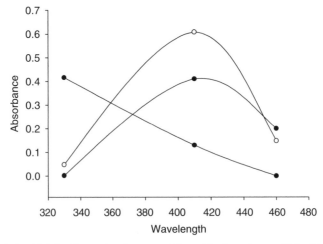

Figure 2-3   Visible Absorption Spectra of Peroxide Complexes of Mo, Ti, and V.

The absorbance A of dissolved complex is given by Beer's law

$$A = abc \qquad (2\text{-}51)$$

where a is the absorptivity, a function of wavelength, which is characteristic of the complex; b is the length of the light path through the absorbing solution in centimeters; and c is the concentration of the absorbing species in grams per liter. If more than one complex is present, the absorbance at any selected wavelength is the sum of contributions of each constituent.

Individual solutions of Mo, Ti, and V ions were complexed by hydrogen peroxide, and each spectrum in the visible region was taken with a 1.00-cm cell, with the results shown in Fig. 2-3. The absorbance of solutions containing a single complex was recorded at one of the wavelengths shown. The remaining two complexes were measured at the same wavelength, yielding three measurements. This was repeated with the other two complexes, each at its selected wavelength, yielding a total of nine measurements. The concentrations of the metal complex solutions were all the same: 40.0 mg L^{-1}.

The absorbance table at λ for each of the metal complexes constitutes a matrix with rows of absorbances, at one wavelength, of Mo, Ti, and V complexes, in that order. Each column comprises absorbances for one metal complex at 330, 410, and 460 nm, in that order:

$$\mathbf{C} = \begin{pmatrix} 0.416 & 0.130 & 0.000 \\ 0.048 & 0.608 & 0.148 \\ 0.002 & 0.410 & 0.200 \end{pmatrix} \qquad (2\text{-}52)$$

Dividing each entry in the table by 0.040 (to convert $\mathbf{C}$ to units of L g^{-1} cm^{-1}) yields the absorptivity matrix

$$\mathbf{A} = \begin{pmatrix} 10.4 & 3.25 & 0.00 \\ 1.20 & 15.2 & 3.70 \\ 0.050 & 10.25 & 5.00 \end{pmatrix} \qquad (2\text{-}53)$$

Notice that the matrix has been arranged so that it is as nearly diagonal dominant as the data permit.

The Problem. An unknown solution containing Mo, Ti, and V ions was treated with hydrogen peroxide, and its absorbance was determined with a 1.00-cm cell at the three wavelengths, in the same order (lowest to highest), that were used to generate the absorbance matrix for the single complexes. The absorbance of the unknown solution at the three wavelengths was 0.284, 0.857, and 0.718. The ordered set of absorbances of any mixture of the complexes constitutes a b vector, in this case, $\mathbf{b} = \{0.284, 0.857, 0.718\}$ where the brackets {} indicate a column vector written horizontally to save space. A set of simultaneous equations results. Taking M, T, and V to be the concentrations of the three metals involved, the unknown concentration vector $\mathbf{c} = \{M, T, V\}$ can be obtained by solving the simultaneous equation set

$$a_{1M}M + a_{1T}T = 0.284$$
$$a_{2M}M + a_{2T}T + a_{2V}V = 0.857 \qquad\qquad (2\text{-}54)$$
$$a_{3M}M + a_{3T}T + a_{3V}V = 0.718$$

where the third term in the first equation is missing because $a_{1V} = 0$. Equation set (2-54) is the same as the matrix equation

$$\mathbf{ac} = \mathbf{b} \qquad\qquad (2\text{-}55)$$

where $\mathbf{a}$ is the absorptivity matrix, $\mathbf{b}$ is the nonhomogeneous vector, and $\mathbf{c}$ is the solution vector of this equation set (a vector of concentrations). The light path was 1.00 cm, so it drops out of the calculation. The notation [M], etc. is not used. Rather, this notation is reserved for concentrations in units of moles per liter.

Procedure. Write a program for solving simultaneous equations by the Gaussian elimination method and enter the absorptivity matrix above to solve Eqs. (2-51). Set up and solve the problem resulting from a new set of experimental observations on a new unknown solution leading to the nonhomogeneous vector $\mathbf{b} = \{0.327, 0.810, 0.673\}$.

Write an iterative program for Gauss–Seidel solution of these two problems to three digits of self–consistency, that is, answers that agree with each other to three significant digits or better. Do the same thing for five–digit self-consistency. Write a counter (for example, $I = I + 1$ on each iteration) into each program to determine how many iterations each program takes. Note that, although you can obtain five digits in your answer, only three of them are significant digits because the experimental data going into the calculation have only three significant digits.

Even this number of significant digits is open to debate. Do we really know λ to an accuracy of ± 1 nm? Does it matter for the relatively flat peaks in Fig. 2-3? What about $a_{2M} = 0.048$? Is it better to report the concentration vector to two significant digits? Discuss these questions in your report.

COMPUTER PROJECT 2-2 | *Gauss–Seidel Iteration:*
Mass Spectroscopy

The purpose of this project is to gain familiarity with the strengths and limitations of the Gauss–Seidel iterative method (program QGSEID) of solving simultaneous equations.

The Problem. (a) Solve the prototypical equations

$$x + y = 3$$
$$x + 2y = 5 \qquad\qquad (2\text{-}56)$$

by the Gauss–Seidel iterative method. How many iterations are necessary to reach a self-consistent solution set?

(b) Try to solve the set

$$x + 2y = 5$$
$$3x + y = 5 \tag{2-57}$$

Reverse the order of the equations

$$3x + y = 5$$
$$x + 2y = 5 \tag{2-58}$$

and try again. Comment on the results of this experiment in relation to the diagonal dominance of the coefficient matrix. Is the first set linearly dependent? Convergence of the Gauss–Seidel method is guaranteed if the sum of the off-diagonal elements in each column is less than the diagonal element in that column. Convergence varies from one system to another.

Solve the same two problems with **Mathcad**. Is there a noticeable difference between the two sets? **Mathcad** uses a variant on the Gaussian substitution method called *LU Factorization* (Kreyzig, 1988).

(c) Solve the set

$$2x + y = 1$$
$$4x + 2.01y = 2 \tag{2-59}$$

This set is said to be *ill-conditioned* because the second equation is almost an exact multiple of the first. The matrix of coefficients is almost singular.

(d) An example similar to Computer Project 2-1 involves quantitative analysis by mass spectrometry of four cyclic hydrocarbons (Isenhour and Jurs, 1985). The 4×4 matrix of sensitivity coefficients (analogous to absorptivities) has as its columns ethylcyclopentane (Etcy), cyclohexane (Cy6), cycloheptane (Cy7), and methylcyclohexane (Mecy), at mass-to-charge ratios (*m/e*) of 69, 83, 84, and 98, respectively. Each *m/e* value constitutes a row; each entry in a given row represents the sensitivity coefficient at a specific *m/e* for the designated hydrocarbon.

	Etcy	Cy6	Cy7	Mecy
69	121.0	9.35	1.38	20.2
83	22.4	4.61	74.9	0.0
84	27.1	20.7	1.30	32.8
98	23.0	100.0	6.57	43.8

The peak at $m/e = 98$ is taken as the arbitrary standard. The height of the other peaks is measured relative to it. Once this matrix has been established, ordered sets of mass spectral peak heights at $m/e = 69$, 83, 84, and 98 constitute the experimental **b** vector for an unknown mixture that contains or may contain the four hydrocarbons in the standardization set. The solution vectors are ordered sets of relative molar concentrations of the components.

(e) Solve for the mol fraction $X_i = n_i / \sum n_i$ for each of the four components, given the experimental vector $\mathbf{b} = \{84.4, 58.8, 47.2, 100.0\}$. Solve this problem with **Mathcad** and as many of the BASIC programs as you have written for simultaneous equations (Gaussian elimination, etc.). Comment on the relative merits of the methods for the problem. Try rearranging the rows of the coefficient matrix and b vector to achieve diagonal dominance.

COMPUTER PROJECT 2-3 | *Bond Enthalpies of Hydrocarbons*
Derivation of bond enthalpies from thermochemical data involves a system of simultaneous equations in which the sum of unknown bond enthalpies, each multiplied by the number of times the bond appears in a given molecule, is set equal to the enthalpy of atomization of that molecule (Atkins, 1998). Taking a number of molecules equal to the number of bond enthalpies to be determined, one can generate an $n \times n$ set of equations in which the matrix of coefficients is populated by the (integral) number of bonds in the molecule and the set of n atomization enthalpies in the b vector. (Obviously, each bond must appear at least once in the set.)

Carrying out this procedure for propane and butane, $CH_3-CH_2-CH_3$ and $CH_3-CH_2-CH_2-CH_3$, yields the bond matrix and enthalpies of atomization:

$$\mathbf{A} = \begin{pmatrix} 2 & 8 \\ 3 & 10 \end{pmatrix} \qquad \mathbf{b} = \begin{pmatrix} 3994 \\ 5166 \end{pmatrix}$$

The bond matrix expresses 2 C—C bonds plus 8 C—H bonds for propane and 3 C—C bonds plus 10 C—H bonds for *n*-butane. Each enthalpy of atomization is obtained by subtracting the enthalpy of formation of the alkane from the sum of atomic atomization enthalpies (C: 716; H: 218 kJ mol^{-1}) for that molecule. For example, the molecular atomization enthalpy of propane is $3(716) + 8(218) - (-104) = 3996$ kJ mol^{-1}. Enthalpies of formation are available from Pedley et al. (1986) or on-line at **www.webbook.nist.gov**.

Procedure.   Run one or more simultaneous equation programs to determine the C—C and C—H bond energies and interpret the results. The error vector is the vector of calculated values minus the vector of bond enthalpies taken as "true" from an accepted source. Calculate the error vector using a standard source of bond enthalpies (e.g., Laidler and Meiser, 1999 or Atkins, 1994). Expand the method for 2-butene $[\Delta_f H^{298}$ (2-butene) $= -11$ kJ mol$^{-1}]$ and so obtain the C—H, C—C, and C=C bond enthalpies.

Solve the same problem for propane and isobutane (2-methylpropane). The bond matrix is the same as it is for *n*-butane, but the enthalpy of formation is somewhat different $[\Delta_f H^{298}$ (*n*-butane) $= -127.1$ kJ mol^{-1} vs. $\Delta_f H^{298}$ (isobutane) $= -134.2$ kJ mol$^{-1}]$, which must necessarily lead to different bond enthalpies. Benson has treated this problem in great detail (Benson and Cohen, 1998; Benson, 1976) and has developed extensive tables of *group additivity values* constructed on the same principle as bond energies. For example, Benson, in seeking group additivity values for different kinds of CH_n groups defines primary P, secondary S, tertiary T, and quaternary Q carbons and then sets up the simultaneous equations to obtain energetic contributions for P, S, T, and Q.

$$\Delta_f H^{298}(\text{ethane}) = -83.81 = 2P$$
$$\Delta_f H^{298}(\text{propane}) = -104.7 = 2P + S$$
$$\Delta_f H^{298}(\text{isobutane}) = -134.2 = 3P + T$$
$$\Delta_f H^{298}(\text{neopentane}) = -168.1 = 4P + Q$$

The b vector in this equation set has been converted from kilocalories per mole (Benson and Cohen, 1998) to kilojoules per mol. Solve these simultaneous equations to obtain the energetic contributions for P, S, T, and Q.

Using these group values, predict $\Delta_f H^{298}$ isooctane (2,2,3-trimethylpentane) and tetramethylbutane. How do these values compare with the experimental values $(-224.1 \pm 1.3$ and $-225.9 \pm 1.9\,\text{kJ}\,\text{mol}^{-1}$, respectively)? What is the percent deviation from the experimental value in each case?

A variant on this procedure produces a first approximation to the molecular mechanics (MM) heat parameters (Chapters 4 and 5) for C—C and C—H. Instead of atomization energies, the *enthalpies of formation* of propane and butane $(-25.02$ and $-30.02\,\text{kcal}\,\text{mol}^{-1})$ are put directly into the b vector. The results (2.51 kcal mol^{-1} and $-3.76\,\text{kcal}\,\text{mol}^{-1})$ are not very good approximations to the heat parameters actually used (2.45 kcal mol^{-1} and $-4.59\,\text{kcal}\,\text{mol}^{-1})$ because of other factors to be taken up later, but the calculation illustrates the method and there is rough agreement.

Our results are in very good agreement with Benson's simpler *bond additivity values* (2.5 kcal mol^{-1} and $-3.75\,\text{kcal}\,\text{mol}^{-1}$; Benson and Cohen, 1998), as indeed they must be because they were obtained from the same set of experimental enthalpies of formation. Note that many applications in thermochemistry use energy units of kilocalories per mole, where 1.000 kcal $\text{mol}^{-1} = 4.184$ kJ mol^{-1}.

PROBLEMS

1. Carry out the multiplication

$$\begin{pmatrix} a_{11} & a_{12} & a_{13} \\ a_{21} & a_{22} & a_{23} \\ a_{31} & a_{32} & a_{33} \end{pmatrix} \begin{pmatrix} x_1 \\ x_2 \\ x_3 \end{pmatrix}$$

2. Solve, by hand, by *Mathcad*, and with your QBASIC programs

$$x + y = 4$$
$$4x - 3y = -1.5$$

3. Solve

$$x + 2y - z + t = 2$$
$$x - 2y + z - 3t = 6$$
$$2x + y + 2z + t = -4$$
$$3x + 3y + z - 2t = 10$$

4. Solve, using Cramer's rule,

$$x \sin\theta + y \cos\theta = x'$$
$$-\cos\theta + y \sin\theta = y'$$

5. Find the inverse of

$$\begin{pmatrix} -.5 & 3/2 & 0 \\ -3/2 & -.5 & 0 \\ 0 & 0 & 1 \end{pmatrix}$$

6. Find the determinant

$$\det \mathbf{A} = \begin{vmatrix} 2 & 1 & 0 \\ 1 & 0 & 1 \\ 3 & 3 & 2 \end{vmatrix}$$

7. Exchange any two columns of the determinant in the previous problem and evaluate the new determinant. Exchange any two rows of the determinant in the previous problem and evaluate the new determinant. Does the rule of Exercise 2-17 hold?

8. Find $\mathbf{A}^2$ given that

$$\mathbf{A} = \begin{pmatrix} 0 & 1 \\ 0 & 0 \end{pmatrix}$$

Does this answer seem surprising to you? Comment on it.

9. What is the atomization enthalpy of 2.00 mol of C and 3.00 mol of H_2? (See Computer Project 2-3.)

10. What is the atomization enthalpy of 1.00 mol of C_2H_6, ethane? The enthalpy of formation of ethane is $-83.8 \pm 0.4\,\text{kJ mol}^{-1}$.

11. If the C—H bond enthalpy is $413\,\text{kJ mol}^{-1}$, what is the C—C bond enthalpy?

12. Determine the molecular mechanics heat parameters for C—C and C—H using the enthalpies of formation of n-butane and n-pentane, which are -30.02 and $-35.11\,\text{kcal mol}^{-1}$ respectively.

13. Using the heat parameters from Problem 12, calculate $\Delta_f H^{298}$ of ethane. The experimental value is $-83.8 \pm 0.4\,\text{kJ mol}^{-1}$.

CHAPTER

3

Curve Fitting

A straight line drawn by eye through a scattered set of experimental points, presumed to represent a linear function, is not an acceptable representation of the data set. Many computer programs exist that fit analytical functions to data points by some statistical principle that avoids the subjectivity of visual curve fitting. Any set of experimental data points can be fit more or less well by an analytical function. One must select the function and then use routine curve fitting procedures to generate the analytical form of the function from the experimental observations. Statistical parameters generated in the fitting process tell us how good the fit is. Subjectivity has not been completely excluded from the process because we usually select the desired analytical function on subjective grounds like simplicity or conformity to some theory we want to test.

In this chapter, we shall use the principle of least squares to generate the equation of a unique curve for any given set of *x-y* pairs of data points. The curve so obtained is the best fit to the points subject to

1. The assumption of an *analytical form* (straight line, quadratic, etc.) and
2. The assumption that the deviations are *randomly distributed* about the analytical form.

We shall begin with the simplest case of a linear function passing through the origin to introduce the method and set up the ground rules. The more complicated

Computational Chemistry Using the PC, Third Edition, by Donald W. Rogers
ISBN 0-471-42800-0   Copyright © 2003 John Wiley & Sons, Inc.

case of a linear function not passing through the origin will be solved by a method that is general. The method will be extended to nonlinear functions and multivariate functions.

Information Loss

A data set $\{x, y\}$ can be represented in three ways: as a tabular collection of measurements, as a graph, or as an analytical function $y = f(x)$. In the process of reducing a tabular collection of results to its analytical form, some information is lost. Although $y = f(x)$ gives us the dependence of y on x, we no longer know where the particular measurement y_i at x_i is. That information has been lost. Often, selection of the form in which experimental results will be presented depends on how (or whether) information loss influences the conclusions we seek to reach.

The Method of Least Squares

We have already found that the probability function governing observation of a single event x_i from among a continuous random distribution of possible events x having a population mean μ and a population standard deviation σ is

$$p(x_i) \propto e^{-(x_i - \mu)^2 / 2\sigma^2} \tag{3-1}$$

The probability of observing a *distribution* of events requires that event x_1 (with probability p_1) occur, and x_2 (with probability p_2) occur, and so on. The probability of observing events, $(x_1, x_2, x_3, \ldots, x_n)$, is the simultaneous or sequential probability of observing all events in the distribution occurring once, that is, the product of the individual probabilities:

$$p(x_1 \text{ and } x_2 \text{ and } x_3 \text{ and} \ldots x_n) = p(x_1)p(x_2)p(x_3) \ldots p(x_n) = \prod_{i=1}^{n} p(x_i)$$
$$\prod_{i=1}^{n} p(x_i) \propto \prod_{i=1}^{n} e^{-(x_i - \mu)^2 / 2\sigma^2} \tag{3-2}$$

By the nature of exponential numbers, $e^a e^b = e^{a+b}$, so

$$\prod_{i=1}^{n} p(x_i) \propto e^{-\sum_{i=1}^{n} (x_i - \mu)^2 / 2\sigma^2} \tag{3-3}$$

Just as e^{-x} takes its maximum value when x is at a minimum, the right side of proportion (3-3) is a maximum when its exponent is a minimum. To minimize a fraction with a constant denominator, one minimizes the numerator

$$\sum_{i=1}^{n} (x_i - \mu)^2 = \text{a minimum} \tag{3-4}$$

In so doing, we obtain the condition of maximum probability (or, more properly, minimum probable prediction error) for the entire distribution of events, that is, the most probable distribution. The minimization condition [condition (3-4)] requires that the sum of *squares* of the differences between μ and all of the values x_i be simultaneously as small as possible. We cannot change the x_i, which are experimental measurements, so the problem becomes one of selecting the value of μ that best satisfies condition (3-4). It is reasonable to suppose that μ, subject to the minimization condition, will be the arithmetic mean, $\bar{x} = \left(\sum_{i=1}^{n} x_i\right)/n$, provided that the deviations are random, that is, that the distribution is Gaussian.

This method, because it involves minimizing the sum of squares of the deviations $x_i - \mu$, is called the *method of least squares*. We have encountered the principle before in our discussion of the most probable velocity of an individual particle (atom or molecule), given a Gaussian distribution of particle velocities. It is very powerful, and we shall use it in a number of different settings to obtain the best approximation to a data set of scalars (arithmetic mean), the best approximation to a straight line, and the best approximation to parabolic and higher-order data sets of two or more dimensions.

Exercise 3-1

For the simple data set $x_i = 2, 3, 7, 8, 10$ we have selected 5, 6, and 7 as possible values of μ. For which of these three is the sum of squared deviations from the data set a minimum?

Solution 3-1

The sum of squared deviations is least for $\mu = 6$. Conventional calculation of the arithmetic mean $\bar{x} = \left(\sum_{i=1}^{n} x_i\right)/n$ shows that it is also 6.

Least Squares Minimization

Clearly, proposing arbitrary candidates for μ and selecting the one with the smallest value of $\sum_{i=1}^{n} (x_i - \mu)^2$ to find $\bar{x}$ is not very efficient, nor can it be readily generalized. This is especially so because, even with a data set of integral numbers, the arithmetic mean does not have to be an integer.

A *systematic* method of arriving at the best value of μ is to find the minimum of $\sum_{i=1}^{n} (x_i - \mu)^2$ as a function of μ. This is the point at which the first derivative is zero

$$\frac{d}{d\mu} \sum_{i=1}^{n} (x_i - \mu)^2 = 0$$

The derivative of a sum is a sum of derivatives; hence,

$$\sum_{i=1}^{n} \frac{d}{d\mu} (x_i - \mu)^2 = -\sum_{i=1}^{n} 2(x_i - \mu) = 0$$

or

$$-\sum_{i=1}^{n} x_i + \sum_{i=1}^{n} \mu = 0$$

where μ is called a *minimization parameter* because the procedure amounts to selecting, from an infinite number of possible values, that μ for which the sum of squares of the deviations is a minimum. Because μ is a constant summed over n terms, $\sum_{i=1}^{n} \mu = n\mu$, whence

$$n\mu = \sum_{i=1}^{n} x_i$$

and we see that μ is the arithmetic mean

$$\mu = \frac{\sum_{i=1}^{n} x_i}{n} = \bar{x} \tag{3-5}$$

which is the conclusion we reached earlier. We shall now look at some problems to which the solution is not self-evident.

Linear Functions Passing Through the Origin

If the linear function through the origin $y = mx$ were obeyed with perfect precision by an experimental data set $\{x_i, y_i\}$, we would have

$$mx_i - y_i = 0$$

This is never the case for a real data set, which displays a deviation d_i for each data point owing to experimental error. For the real case,

$$mx_i - y_i = d_i \tag{3-6}$$

If the experimental error is random, the method of least squares applies to analysis of the set. Minimize the sum of squares of the deviations by differentiating with respect to m,

$$\frac{d}{dm}\sum_{i=1}^{n} d_i^2 = \frac{d}{dm}\sum_{i=1}^{n} (mx_i - y_i)^2 = 0 \tag{3-7}$$

which leads to

$$m = \frac{\sum_{i=1}^{n} x_i y_i}{\sum_{i=1}^{n} x_i^2} \tag{3-8}$$

The slope of the linear function is the minimization parameter. It is the only thing one can change to obtain a "better" fit to points for a line passing through the origin. (It is *NEVER* permissible to "adjust" experimental points.) The slope calculated by the least squares method is the "best" slope that can be obtained under the assumptions. Once one knows the slope of a linear function passing through the origin, one knows all that can be known about that function.

Exercise 3-2

Using a hand calculator, find the slope of the linear regression line that passes through the origin and best satisfies the points

x	1.0	2.0	3.0	4.0	5.0
y	1.1	1.9	2.9	4.0	4.8

Solution 3-2

Scanning the data set, it is evident that the slope should be slightly less than 1.0.

$$\sum_{1}^{5} x_i y_i = 53.6$$

$$\sum_{1}^{5} x_i^2 = 55.0$$

$$m = \frac{\sum_{i=1}^{n} x_i y_i}{\sum_{i=1}^{n} x_i^2} = 0.97$$

Linear Functions Not Passing Through the Origin

Deviations from a curve thought to be a straight line $y = mx + b$, not passing through the origin ($b \neq 0$), are

$$(mx_i + b) - y_i = d_i \tag{3-9}$$

Note that m and b do not have subscripts because there is only one slope and one intercept; they are the *minimization parameters* for the least squares function.

Now there are two minimization conditions

$$\frac{\partial}{\partial b} \sum_{i=1}^{n} d_i^2 = \frac{\partial}{\partial b} \sum_{i=1}^{n} (mx_i + b - y_i)^2 = 0 \tag{3-10a}$$

and

$$\frac{\partial}{\partial m} \sum_{i=1}^{n} d_i^2 = \frac{\partial}{\partial m} \sum_{i=1}^{n} (mx_i + b - y_i)^2 = 0 \tag{3-10b}$$

which must be satisfied simultaneously. Carrying out the differentiation, one obtains

$$\sum_{i=1}^{n} m x_i + \sum_{i=1}^{n} b - \sum_{i=1}^{n} y_i = 0 \tag{3-11a}$$

and

$$\sum_{i=1}^{n} m x_i^2 + \sum_{i=1}^{n} b x_i - \sum_{i=1}^{n} x_i y_i = 0 \tag{3-11b}$$

which are called the *normal equations*. Rewriting them as

$$nb + \sum_{i=1}^{n} x_i m = \sum_{i=1}^{n} y_i \tag{3-12a}$$

$$\sum_{i=1}^{n} x_i b + \sum_{i=1}^{n} x_i^2 m = \sum_{i=1}^{n} x_i y_i \tag{3-12b}$$

makes it clear that the intercept and slope are the two elements in the solution vector of a pair of simultaneous equations

$$\begin{pmatrix} n & \sum_{i=1}^{n} x_i \\ \sum_{i=1}^{n} x_i & \sum_{i=1}^{n} x_i^2 \end{pmatrix} \begin{pmatrix} b \\ m \end{pmatrix} = \begin{pmatrix} \sum_{i=1}^{n} y_i \\ \sum_{i=1}^{n} x_i y_i \end{pmatrix} \tag{3-13}$$

The coefficient matrix and nonhomogeneous vector can be made up simply by taking sums of the experimental results or the sums of squares or products of results, all of which are real numbers readily calculated from the data set.

Solving the normal equations by Cramer's rule leads to the solution set in determinantal form

$$b = \frac{D_b}{D} \tag{3-14a}$$

and

$$m = \frac{D_m}{D} \tag{3-14b}$$

or

$$b = \frac{\sum y_i \sum x_i^2 - \sum x_i y_i \sum y_i}{n \sum x_i^2 - \left(\sum x_i \right)^2} \tag{3-15a}$$

and

$$m = \frac{n\sum x_i y_i - \sum x_i \sum y_i}{n\sum x_i^2 - \left(\sum x_i\right)^2} \tag{3-15b}$$

where the limits on the sums, $i = 1 - n$, have been dropped to simplify the appearance of the equation. Equations (3-15) are in the form usually given in elementary treatments of least squares data fitting in analytical and physical chemistry laboratory texts.

Exercise 3-3

Find the slope and intercept of a straight line not passing through the origin of the data set

x	1.0	2.0	3.0	4.0	5.0
y	3.1	3.9	4.9	6.0	6.8

Solution 3-3

Using **Mathcad**, we regard the data set as two vectors $x = \{1.0, 2.0, 3.0, 4.0, 5.0\}$ and $y = \{3.1, 3.9, 4.9, 6.0, 6.8\}$. They are defined using the : keystroke for a 5×1 matrix

$$x := \begin{pmatrix} 1 \\ 2 \\ 3 \\ 4 \\ 5 \end{pmatrix} \qquad y := \begin{pmatrix} 3.1 \\ 3.9 \\ 4.9 \\ 6.0 \\ 6.8 \end{pmatrix}$$

$$\text{slope}(x, y) = 0.95$$
$$\text{intercept}(x, y) = 2.09$$

One might have expected the same slope for this function as for the function described in Exercise 3-2 because the y vector in this exercise is nothing but the y vector in Exercise 3-2 with 2.0 added to each element. Calculating the slope *and* intercept adds flexibility to the problem. Now the intercept that had been constrained to 0.0 in Exercise 3-2 is free to move a little, giving a better fit to the points. We find a slightly different slope and an intercept that is slightly different from the anticipated 2.0. The slope and intercept for this exercise should be reported as 0.95 and 2.1, retaining two significant figures.

If we go back and calculate the slope and intercept for the data set in Exercise 3-2 *without the constraint* that the line must pass through the origin, we get the solution vector {0.95, 0.09} for a line parallel to the line in Exercise 3-3 and 2.0 units distant from it, as expected.

Quadratic Functions

Many experimental functions approach linearity but are not really linear. (Many were historically thought to be linear until accurate experimental determinations showed some degree of nonlinearity.) Nearly linear behavior is often well represented by a quadratic equation

$$y = a + bx + cx^2 \tag{3-16}$$

The least squares derivation for quadratics is the same as it was for linear equations except that one more term is carried through the derivation and, of course, there are three normal equations rather than two. Random deviations from a quadratic are

$$\left(a + bx_i + cx_i^2\right) - y_i = d_i \tag{3-17}$$

The minimization conditions are

$$\frac{\partial}{\partial a} \sum_{i=1}^{n} d_i^2 = 0 \tag{3-18a}$$

$$\frac{\partial}{\partial b} \sum_{i=1}^{n} d_i^2 = 0 \tag{3-18b}$$

$$\frac{\partial}{\partial c} \sum_{i=1}^{n} d_i^2 = 0 \tag{3-18c}$$

which must be true simultaneously. Solution of these equations leads to the normal equations

$$na + b \sum x_i + c \sum x_i^2 = \sum y_i \tag{3-19a}$$

$$a \sum x_i + b \sum x_i^2 + c \sum x_i^3 = \sum x_i y_i \tag{3-19b}$$

$$a \sum x_i^2 + b \sum x_i^3 + c \sum x_i^4 = \sum x_i^2 y_i \tag{3-19c}$$

with the solution vector $\{a, b, c\}$ and the nonhomogeneous vector $\{\sum y_i, \sum x_i y_i, \sum x_i^2 y_i\}$. The matrix form of the set of normal equations is

$$\begin{pmatrix} a & \sum x_i & \sum x_i^2 \\ \sum x_i & \sum x_i^2 & \sum x_i^3 \\ \sum x_i^2 & \sum x_i^3 & \sum x_i^4 \end{pmatrix} \begin{pmatrix} a \\ b \\ c \end{pmatrix} = \begin{pmatrix} \sum y_i \\ \sum x_i y_i \\ \sum x_i^2 y_i \end{pmatrix} \tag{3-20}$$

or simply

$$\mathbf{Qs} = \mathbf{t} \tag{3-21}$$

where the meaning of matrix $\mathbf{Q}$ and vectors $\mathbf{s}$ and $\mathbf{t}$ are evident from Eq. (3-20). A general-purpose linear least squares program QLLSQ and a quadratic least squares program QQLSQ can be downloaded from *www.wiley.com/go/computational*. Others are available elsewhere (Carley and Morgan, 1989). The input format given in QQLSQ is used to solve Exercise 3-4.

Exercise 3-4

From the theory of the electrochemical cell, the potential in volts E of a silver-silver chloride-hydrogen cell is related to the molarity m of HCl by the equation

$$E + \frac{2RT}{F} \ln m = E^\circ + \frac{2.342\,RT}{F} m^{1/2} \tag{3-22}$$

where R is the gas constant, F is the Faraday constant (9.648×10^4 coulombs mol^{-1}), and T is 298.15 K. The silver-silver chloride half-cell potential E° is of critical importance in the theory of electrochemical cells and in the measurement of pH. (For a full treatment of this problem, see Mc Quarrie and Simon, 1999.) We can measure E at known values of m, and it would seem that simply solving Eq. (3-22) would lead to E°. So it would, except for the influence of nonideality on E. Interionic interference gives us an incorrect value of E° at any nonzero value of m. But if m is zero, there are no ions to give a voltage E. What do we do now?

The way out of this dilemma is to make measurements at several (nonideal) molarities m and *extrapolate* the results to a hypothetical value of E at $m = 0$. In so doing we have "extrapolated out" the nonideality because at $m = 0$ all solutions are ideal. Rather than ponder the philosophical meaning of a solution in which the solute is not there, it is better to concentrate on the *error* due to interionic interactions, which becomes smaller and smaller as the ions become more widely separated. At the *extrapolated* value of $m = 0$, ions have been moved to an infinite distance where they cannot interact.

Plotting the left side of Eq. (3-22) as a function of $m^{1/2}$ gives a curve with $\frac{2.342\,RT}{F}$ as the slope and E° as the intercept. Ionic interference causes this function to deviate from linearity at $m \neq 0$, but the limiting (ideal) slope and intercept are approached as $m \to 0$. Table 3-1 gives values of the left side of Eq. (3-22) as a function of $m^{1/2}$. The concentration axis is given as $m^{1/2}$ in the corresponding Fig. 3-1 because there are two ions present for each mole of a 1-1 electrolyte and the concentration variable for one ion is simply the square root of the concentration of both ions taken together.

The problem now is to find the best value of the intercept on the vertical axis. We can do this by fitting the experimental points to a parabola.

Solution 3-4

TableCurve gives $\{.2225, .05621, -.06226\}$ for a, b, and c of the quadratic fit, and program QQLSQ gives $\{0.2225, 5.621 \times 10^{-2},$ and $-6.226 \times 10{-2}$ (note that the order of terms is reversed in Output 3-4).

Table 3-1 The Left Side of Eq. 3-4 at Seven Values of $m^{1/2}$

$m^{1/2}$	$E + \frac{2RT}{F} \ln m$
.05670	.2256
.07496	.2263
.09559	.2273
.1158	.2282
.1601	.2300
.2322	.2322
.3519	.2346

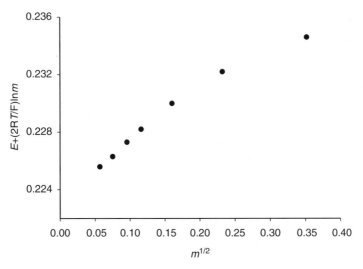

Figure 3-1   Voltage Measurements on a Silver-Silver Chloride, Hydrogen Cell at 298.15 K. The contribution of the Standard Hydrogen Electrode is taken as zero by convention.

Output 3-4

Program QQLSQ
How many data pairs? 7

THE FOLLOWING DATA PAIRS WERE USED:

X	Y	YEXPECTED
.0567	.2256	.2255161
.07496	.2263	.2263927
.09559	.2273	.2273332
.1158	.2282	.2282032
.1601	.23	.2299322
.2322	.2322	.2322238
.3519	.2346	.2345989

THE EQUATION OF THE PARABOLA IS:
$Y = -6.225745E\text{-}02X^{**}2 + (5.620676E\text{-}02)X + (.2225293)$
THE STANDARD DEVIATION OF Y VALUES IS 6.04448E-05

The two estimates for the first or a parameter of the parabolic fit are the intercepts on the voltage axis of Fig. 3-1, so both procedures arrive at a standard potential of the silver-silver chloride half-cell of 0.2225 V. The accepted modern value is 0.2223 V (Barrow, 1996).

Polynomials of Higher Degree

The form of the symmetric matrix of coefficients in Eq. 3-20 for the normal equations of the quadratic is very regular, suggesting a simple expansion to higher-degree equations. The coefficient matrix for a cubic fitting equation is a 4×4

matrix with x_i^6 in the 4, 4 position, the fourth-degree equation has x_i^8 in the 5, 5 position, and so on. It is routine (and tedious) to extend the method to higher degrees. Frequently, the experimental data are not sufficiently accurate to support higher-degree calculations and "differences" in fitting equations higher than the fourth degree are artifacts of the calculation rather than features of the data set.

Statistical Criteria for Curve Fitting

Frequently in curve fitting problems, one has two curves that fit the data set more or less well but they are so similar that it is difficult to decide which is the better curve fit. Commercially available curve fitting programs (e.g., *TableCurve*, **www.spss.com**) usually give many curves that fit the data (more or less well). The best of these are often very close in their goodness of fit. One needs an objective statistical criterion of this vague concept, "goodness of fit."

The *essential criterion* of goodness of fit is nothing more than the sum of squares of deviations as in Eq. (3-4), although this simple fact may be obscured by differences in notation and nomenclature. Starting with the simple case of comparing the fit of straight lines to the same data set, the deviation in minimization criterion (3-4) is referred to as the *residual*, $(\hat{y}_i - y_i)$, which is the difference between the ith experimental measurement y_i and the value $\hat{y}_i$ that y_i would have if the fit were perfect. The definition $(y_i - \hat{y}_i)$ is also used (Jurs, 1996) because the difference in sign is of no importance when we square the residual. Following Jurs's nomenclature and notation (Jurs, 1996), the sum of squares of residuals is thought of as the sum of squares due to *error*, hence the notation SSE

$$\text{SSE} = \sum_{i=1}^{n} (\hat{y}_i - y_i)^2 = \sum_{i=1}^{n} (y_i - \hat{y}_i)^2 \tag{3-23a}$$

The differences between corresponding elements in ordered sets (vectors) in, for example, columns 2 and 3 of output 3-4 give an ordered set of residuals. In this example, the residual for the first data point is $0.2256 - 0.2255161 = 8.39 \times 10^{-5}$.
    SSE is sometimes written

$$\text{SSE} = \sum_{i=1}^{n} w_i (\hat{y}_i - y_i)^2 \tag{3-23b}$$

where w_i is a weighting factor used to give data points a greater or lesser weight according to some system of estimated reliability. For example, if each data point were the arithmetic mean of experimental results, we might make $w_i = \sigma_0^2/\sigma_i^2$ where σ_0^2 is the maximum variance in the set ($w_0 = 1$) and all the other weighting factors are greater than 1 (Wentworth, 2000). For simplicity, let us take $w_i = 1$ for all data in what follows.

The sum of squares of differences between points on the *regression* line $\hat{y}_i$ at x_i and the arithmetic mean $\bar{y}$ is called SSR

$$SSR = \sum_{i=1}^{n} (\hat{y}_i - \bar{y})^2 \tag{3-24}$$

The *total* sum of squared deviations from the mean is

$$SST = \sum_{i=1}^{n} (y_i - \bar{y})^2 \tag{3-25}$$

which is made up of two parts, one contribution due to the regression line and one due to the residual or error

$$SST = SSR + SSE \tag{3-26}$$

Jurs defines a *regression coefficient* R as

$$R^2 = \frac{SSR}{SST} \tag{3-27}$$

If the fit is very good, the residuals will be small, SST→SSR as SSE→0, hence $R^2 \rightarrow 1.0$. If SSE is a substantial part of SST (as it is for a poor curve fit) $SSR < SST$ and $R^2 < 1.0$. In the limit as SSE becomes very large, $R^2 \rightarrow 0$.

To reconcile this notation with the output from *TableCurve*, note that

$$R^2 = \frac{SSR}{SST} = \frac{SST - SSE}{SST} = 1 - \frac{SSE}{SST} \tag{3-28}$$

Now, making only the change in notation of SSM for SST to indicate the total deviation from the mean and changing from upper to lower case r, we have

$$r^2 = 1 - \frac{SSE}{SSM} \tag{3-29}$$

which is the *coefficient of determination* as defined in *TableCurve* (TableCurve User's Manual, 1992). For an example of r^2 calculated by *TableCurve*, see Exercise 3-5.

Reliability of Fitted Parameters

For a specified mean and standard deviation the number of degrees of freedom for a one-dimensional distribution (see sections on the least squares method and least squares minimization) of n data is $(n - 1)$. This is because, given μ and σ, for $n > 1$ (say a half-dozen or more points), the first datum can have any value, the second datum can have any value, and so on, up to $n - 1$. When we come to find the last data point, we have no freedom. There is only one value that will make μ and σ come out right. If we select anything else, we violate the hypothesis of known μ and σ.

The situation is similar for a linear curve fit, except that now the data set is two-dimensional and the number of *degrees of freedom* is reduced to $(n-2)$. The analogs of the one-dimensional variance $\sigma^2 = (\sum_{i=1}^{n} d_i^2)/(n-1)$ and the standard deviation σ are the *mean square error*

$$\text{MSE} = \frac{\text{SSE}}{n-2} \tag{3-30}$$

and the standard deviation of the regression

$$s = \sqrt{\text{MSE}} \tag{3-31}$$

for the two-dimensional case.

If the data set is truly normal and the error in y is random about known values of x, residuals will be distributed about the regression line according to a normal or Gaussian distribution. If the distribution is anything else, one of the initial hypotheses has failed. Either the error distribution is not random about the straight line or $y = f(x)$ is not linear.

If the matrix form of the fitting procedure is used to solve for the intercept and slope of a straight line, Eq. (3-13)

$$\begin{pmatrix} n & \sum_{i=1}^{n} x_i \\ \sum_{i=1}^{n} x_i & \sum_{i=1}^{n} x_i^2 \end{pmatrix} \begin{pmatrix} b \\ m \end{pmatrix} = \begin{pmatrix} \sum_{i=1}^{n} y_i \\ \sum_{i=1}^{n} x_i y_i \end{pmatrix} \tag{3-32}$$

yields

$$\begin{pmatrix} b \\ m \end{pmatrix} = \begin{pmatrix} n & \sum_{i=1}^{n} x_i \\ \sum_{i=1}^{n} x_i & \sum_{i=1}^{n} x_i^2 \end{pmatrix}^{-1} \begin{pmatrix} \sum_{i=1}^{n} y_i \\ \sum_{i=1}^{n} x_i y_i \end{pmatrix} \tag{3-33}$$

The variance of the regression times the diagonal elements of the inverse coefficient matrix gives the variance of the intercept and slope,

$$s_b^2 = d_{11} s^2 \tag{3-34a}$$

and

$$s_m^2 = d_{22} s^2 \tag{3-34b}$$

which lead directly to the standard deviations of the intercept and slope, s_b and s_m. Their standard deviations are given in **TableCurve** under the heading Std Error in column 3, block 2 of the Numeric Summary output. We now have

$$y = a \pm \text{Std Error}_a + bx \pm \text{Std Error}_b \tag{3-35}$$

as the equation of our straight line fit. Further treatment, which we shall not go into here, involves finding the t value for each parameter and establishing the confidence limits at some confidence level, for example, 90%, 95%, or 99% (Rogers, 1983). These limits are given in block 2 of the *TableCurve* output. The confidence level can be specified in *TableCurve* by clicking the **intervals** button in the **Review Curve-Fit** window. The default value is 95%.

Exercise 3-5

Fit a linear equation to the following data set and give the uncertainties of the slope and intercept (Fig. 3-2).

x	0.400	0.800	1.20	1.60	2.00	2.40	2.80	3.20	3.60	4.00
y	2.79	3.82	4.75	4.85	5.55	6.71	7.81	8.12	9.31	9.65

Solution 3-5

Working the problem by the matrix method we get

$$\begin{pmatrix} b \\ m \end{pmatrix} = \begin{pmatrix} 10 & 22 \\ 22 & 61.6 \end{pmatrix}^{-1} \begin{pmatrix} 63.36 \\ 164.804 \end{pmatrix}$$
$$= \begin{pmatrix} 0.467 & -0.167 \\ -0.167 & 0.076 \end{pmatrix} \begin{pmatrix} 63.36 \\ 164.804 \end{pmatrix} = \begin{pmatrix} 2.10 \\ 1.92 \end{pmatrix}$$

Solved using the BASIC curve fitting program QLLSQ we get as a partial output block

10 POINTS, FIT WITH STD DEV OF THE REGRESSION .2842293

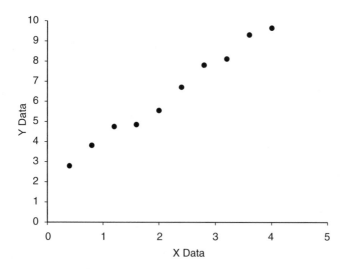

Figure 3-2   Data Set for Exercise 3-5.

SLOPE=1.925152 , Y INTERCEPT=2.100666

$$s_b = \sqrt{d_{11}s^2} = 0.194$$
$$s_m = \sqrt{d_{22}s^2} = 0.078$$

TableCurve gives

TableCurve Output 3-5
Rank 3 Eqn 1 $y = a + bx$

r2 Coef Det	DF Adj r2	Fit Std Err	F-value
0.9869616170	0.9832363647	0.2842291542	605.57301820

Parm	Value	Std Error	t-value	95% Confidence Limits		
a	2.100666667	0.194165477	10.81895043	1.651357072	2.549976262	
b	1.925151515	0.078231500	24.60839325	1.744119520	2.106183510	

The result of our analysis is that the data are fit by the straight line

$$y = 2.10 \pm 0.19 + 1.93 \pm 0.08\,x \tag{3-36}$$

Note that there are two equations with a higher rank than $y = a + bx$. They are the exponential and power equations $y = a + b\exp(-x/c)$ and $y = a + bx^c$. There is little to choose among the r^2 values of these three curve fits, 0.9871, 0.9871, and 0.9870, so we choose the simplest equation. Small differences in r^2 values should not be counted too heavily, and we should be wary of r^2 values that look impressive. Note that Fig. 3-2 shows substantial scatter despite the high r^2 value. Use common sense and an aesthetic of simplicity in choosing the best curve fit. As usual, the number of significant figures in the final equation (3-36) is determined by the number of significant figures going into the calculation (Fig. 3-2).

COMPUTER PROJECT 3-1 | *Linear Curve Fitting: KF Solvation*

Linear extrapolation of the experimental behavior of a real gas to zero pressure or a solute to infinite dilution is often used as a technique to "get rid" of molecular or ionic interactions so as to study some property of the molecule or ion to which these interferences are considered extraneous. Emsley (1971) studied the heat (enthalpy) of solutions of potassium fluoride KF and the monosolvated species KF·HOAc in glacial acetic acid at several concentrations. A known weight of the anhydrous salt KF was added to a known weight of glacial acetic acid in a Dewar flask fitted with a heating coil, a stirrer, and a sensitive thermometer. The temperature change on each addition was recorded. The heat capacity C of the flask and its contents was determined by supplying a known amount of electrical energy Q to the flask and noting the temperature rise ΔT in kelvins (K)

$$Q(\text{joules}) = C\Delta T \tag{3-37}$$

The experiment was repeated for the solvated salt KF·HOAc, where the molecule of solvation is acetic acid, HOAc. Some experimental results calculated from the original paper are shown in Table 3-2:

Table 3-2 Enthalpies of Solution $\Delta_{sol'n} H^{298}$ of KF and KF·HOAc in Glacial Acetic Acid at 298 K

KF: $C = 4.168$ kJ K^{-1}				
Molality	0.194	0.590	0.821	1.208
Temperature change, K	1.592	4.501	5.909	8.115
KF·HOAc: $C = 4.203$ kJ K^{-1}				
Molality	0.280	0.504	0.910	1.190
Temperature Change, K	−0.227	−0.432	−0.866	−1.189

Procedure. Calculate the heats of solution of the two species, KF and KF·HOAc, at each of the four given molalities from a knowledge of the heat capacity. Calculate the enthalpy of solution per mole of solute $\Delta_{sol'n} H^{298}$ at each concentration. Find the least squares curve fit and calculate the uncertainties $\pm \cdots$ for the function

$$\Delta_{sol'n} H^{298} = \Delta_{sol'n} H_0^{298} \pm \cdots_{\Delta_{sol'n}H_0^{298}} + (\text{SLOPE})M \pm \cdots_{\text{SLOPE}} \qquad (3\text{-}38)$$

where M is the molality and $\Delta_{sol'n} H_0^{298}$ is the enthalpy of solution at infinite dilution. Do this for each species, the anhydrous and the solvated fluoride. Record the slopes and intercepts of both functions, SLOPE1 for KF and SLOPE2 for KF·HOAc. Record the enthalpy changes and uncertainties, $\Delta_{sol'n} H_0^{298} \pm \cdots_{\Delta_{sol'n}H_0^{298}}$ (KF) and $\Delta_{sol'n} H_0^{298} \pm \cdots_{\Delta_{sol'n}H_0^{298}}$ (KF·HOAc). What are the units of SLOPE?

Read the article on the original research (Emsley, 1971) and include a commentary on these results in your report for this experiment. Emsley claims that the enthalpy of solution

$$\text{KF}(s) + \text{HOAc}(l) \rightarrow \text{KF} \cdot \text{OAcH}(\text{sol'n})$$

is $\Delta_{sol'n} H_0^{298} = -38.5 \text{ kJ mol}^{-1}$. What argument does he present to support this claim?

COMPUTER PROJECT 3-2 | *The Boltzmann Constant*

An interesting historical application of the Boltzmann equation involves examination of the number density of very small spherical globules of latex suspended in water. The particles are distributed in the potential gradient of the gravitational field. If an arbitrary point in the suspension is selected, the number of particles N at height h μm (1 μm $= 10^{-6}$ m) above the reference point can be counted with a magnifying lens. In one series of measurements, the number of particles per unit volume of the suspension as a function of h was as shown in Table 3-3.

The Boltzmann distribution gives

$$N = N_0 e^{\frac{-mgh}{kT}} \qquad (3\text{-}39)$$

Table 3-3   The Number of Latex Particles at Various Heights in μm Above a
Reference Point

$h/\mu m$	0	50	70	90	100	150	200
N	977	453	293	219	176	69	28

at constant temperature T, where m is the effective mass of the particle corrected for the buoyancy of the supporting medium.

Statistical analysis of the data set is best done by "linearizing" the function (Jurs, 1996), that is, by transforming it to a straight line of the form $y = a + bx$. In the case of the Boltzmann distribution, because $y = ae^{bx}$ leads to $\ln y = \ln a + bx$, we can take logarithms of both sides,

$$\ln N = \ln N_0 - \frac{mgh}{kT} \tag{3-40}$$

whereupon the slope of $\ln N$ vs. h is $-\frac{mgh}{kT}$, where g is the acceleration due to the gravitational field and k is a universal constant now called Boltzmann's constant and denoted k_B.

The supporting medium was water at 298 K ($\rho = 0.99727$), and the density of latex is 1.2049 g cm^{-3}. The latex particles had an average radius of 2.12×10^{-4} mm; hence, their effective mass corrected for buoyancy is their volume $\frac{4}{3}\pi r^3$ times the density difference $\Delta\rho$ between latex and the supporting medium, water

$$m = v\rho = (4/3)(2.12 \times 10^{-7}\,\text{m})^3(1.2049 - 0.99727)$$

This yields $m = 8.287 \times 10^{-18}$ kg.

Procedure.   Compute the slope of the function by a linear least squares procedure and obtain a value of Boltzmann's constant. How many particles do you expect to find 125 μm above the reference point? Take the uncertainty you have calculated for the slope, as the uncertainty in k_B. Is the modern value of $k_B = 1.381 \times 10^{-23}$ within these error limits?

Having determined k_B and knowing that the gas constant $R = 8.314\,\text{J K}^{-1}$ from macroscopic measurements on gases, determine Avogadro's number L from the relationship

$$R = k_B L \tag{3-41}$$

Calculate the % difference between L found by this method and the modern value of 6.022×10^{23}. Does this support the idea that the Boltzmann constant is the gas constant *per particle*?

COMPUTER PROJECT 3-3 | *The Ionization Energy of Hydrogen*
The ionization energy for hydrogen is the minimum amount of energy that is
required to bring about the reaction

$$H \rightarrow H^+ + e^-$$

The ionization energy for hydrogen (or other hydrogen-like systems) can be found
from the Rydberg equation

$$\bar{v} = \frac{1}{\lambda} = R\left(\frac{1}{n_1^2} - \frac{1}{n_2^2}\right) \tag{3-42}$$

along with an accurate set of spectral data. Eq. (3-42) leads us to believe that *both*
the slope *R and* the intercept *R* of the linear function

$$\bar{v} = R - R\left(\frac{1}{n_2^2}\right) \tag{3-43}$$

are equal to the ionization energy which has $n_2^2 = \infty$.

Procedure.   Use ***Mathcad***, QLLSQ, or ***TableCurve*** (or, preferably, all three) to
determine a value of the ionization energy of hydrogen from the wave numbers in
Table 3-4 taken from spectroscopic studies of the Lyman series of the hydrogen
spectrum where $n_1 = 1$.

Table 3-4   Spectral Wavenumbers $\bar{v}$ for the Lyman Series of Hydrogen

$n_2 = 2$	3	4	5	6	7
$n_2^2 = 4$	9	16	25	36	49
$\bar{v} = 82259$	97492	102824	105292	106632	107440

Note that we are interested in n_2, the atomic quantum number of the level *to which*
the electron "jumps" in a spectroscopic excitation. Use the results of this data
treatment to obtain a value of the Rydberg constant R. Compare the value you
obtain with an accepted value. Quote the source of the accepted value you use for
comparison in your report. What are the units of R? A conversion factor may be
necessary to obtain unit consistency. Express your value for the ionization energy
of H in units of hartrees (h), electron volts (eV), and kJ mol^{-1}. We will need it
later.

Reliability of Fitted Polynomial Parameters

The method of finding uncertainty limits for linear equations can be generalized
to higher-order polynomials. The matrix method for finding the minimization

parameters for the polynomial of next higher order (second order), $y = ax + bx^2$ is

$$\begin{pmatrix} a & \sum x_i & \sum x_i^2 \\ \sum x_i & \sum x_i^2 & \sum x_i^3 \\ \sum x_i^2 & \sum x_i^3 & \sum x_i^4 \end{pmatrix} \begin{pmatrix} a \\ b \\ c \end{pmatrix} = \begin{pmatrix} \sum y_i \\ \sum x_i y_i \\ \sum x_i^2 y_i \end{pmatrix} \tag{3-44}$$

with the solution vector

$$\begin{pmatrix} a \\ b \\ c \end{pmatrix} = \begin{pmatrix} a & \sum x_i & \sum x_i^2 \\ \sum x_i & \sum x_i^2 & \sum x_i^3 \\ \sum x_i^2 & \sum x_i^3 & \sum x_i^4 \end{pmatrix}^{-1} \begin{pmatrix} \sum y_i \\ \sum x_i y_i \\ \sum x_i^2 y_i \end{pmatrix} \tag{3-45}$$

The quadratic curve fit leads to a number of residuals equal to the number of points in the data set. The sum of squares of residuals gives SSE by Eqs. (3-23) and MSE by Eq. (3-30), except that now the number of degrees of freedom for n points is $n - 3$

$$MSE = \frac{SSE}{n - 3} \tag{3-46}$$

MSE gives the standard deviation of the regression, $s = \sqrt{MSE}$.

The uncertainties of the minimization parameters are calculated just as they were for the linear case except that now there are three of them

$$s_a = \sqrt{d_{11}s^2} \tag{3-47a}$$

$$s_b = \sqrt{d_{22}s^2} \tag{3-47b}$$

and

$$s_c = \sqrt{d_{33}s^2} \tag{3-47c}$$

In contrast to the linear case, there are three degrees of freedom, but there is still only one standard deviation of the regression, s. The reader has the opportunity to try out these ideas in Computer Project 3-4.

COMPUTER PROJECT 3-4 | *The Partial Molal Volume of ZnCl$_2$*

In general, the volume of a solution, say $ZnCl_2$ in water, is dependent on the number of moles n_i of each of the components. For a binary solution,

$$V = f(n_1, n_2) \tag{3-48}$$

The change in volume dV on adding a small amount dn_1 of water or dn_2 of $ZnCl_2$ is

$$dV = \left(\frac{\partial V}{\partial n_1}\right) dn_1 + \left(\frac{\partial V}{\partial n_2}\right) dn_2 \tag{3-49}$$

where we stipulate that P and T are constant for the process and we adopt the usual subscript convention, 1 for solvent and 2 for solute. If we specify 1 kg as the amount of water, n_2 is the *molality* of $ZnCl_2$. We expect that the volume of the solution will be greater than 1000 cm^3 by the volume taken up by the $ZnCl_2$. It may seem reasonable to take the volume of one mole of $ZnCl_2$ in the solid state V_m and add it to 1000 cm^3 to get the volume of a 1 molal solution. One-half the molar volume of solute would, by this scheme, lead to the volume of a 0.5 molar solution and so on. This does not work. The volume of 1000 g of water *in the solution* is not exactly 1000 cm^3, and it is dependent on the temperature. Nor are the volumes additive. Indeed, some solutes cause contraction of the solution to less than 1000 cm^3.

Interactions at the molecular or ionic level cause an expansion or contraction of the solution so that, in general

$$V \neq 1000 + V_m \qquad (3\text{-}50)$$

We define a *partial molar volume* $\bar{V}_i$ such that

$$V = n_1\bar{V}_1 + n_2\bar{V}_2 \qquad (3\text{-}51)$$

for a binary solution or, in general,

$$V = \sum_{i=1}^{N} n_i\bar{V}_i \qquad (3\text{-}52)$$

for a solution of N components.

It can be shown (Alberty, 1987) that

$$\bar{V}_i = \left(\frac{\partial V}{\partial n_i}\right)_{n_j} \qquad (3\text{-}53)$$

where the subscript n_j indicates that all components in the solution other than i are held constant. If the solution is a binary solution of n_2 moles of solute in 1 kg of water, $\bar{V}_2$ is the partial *molal* volume of component 2. A partial molal volume is a special case of the partial molar volume for 1 kg of solvent.

Procedure. A study on the partial molal volume of $ZnCl_2$ solutions gave the following data (Alberty, 1987)

g $ZnCl_2$ per kg of H_2O	20.00	60.00	100.00	140.00	180.00	200.00
density	1.0167	1.0532	1.0891	1.1275	1.1665	1.1866

Calculate the number of moles of $ZnCl_2$ per kilogram of water in each solution (the molality m). Calculate the volume V of solution containing 1 kg of water at each solute concentration. Plot V vs. m. Use program **Mathcad**, QQLSQ, or **TableCurve** (or all three, if available) to obtain a quadratic expression $V = a + bm + cm^2$. Obtain an expression for the slope dV/dm which is the same as $(\partial V/\partial n_2)_{n_1}$. This is the partial molal volume of $ZnCl_2$. It is a partial volume because V varies with both n_{ZnCl_2} and the number of moles of water n_1. What is the partial *molal* volume of $ZnCl_2$ in water at 1.00 molal concentration? What is the partial *molar* volume of water at this concentration?

PROBLEMS

1. Select several values for μ of the data set in Exercise 3-1 and calculate $(x_i - \mu)$ for each of them. Plot the curve of $\sum (x_i - \mu)^2$ as a function of the selected parameter μ and locate the minimum visually. Compare with Solution 3-1.

2. Obtain Eq. (3-8) from Eq. (3-6) through Eq. (3-7).

3. An excess of porphobilinogen in the urine is associated with hepatic disorders and lead poisoning. Porphobilinogen can be separated from other porphyrins by ion exchange chromatography and treated with p-dimethylaminobenzalde-hyde (PDMA) to produce a red compound that absorbs light strongly at 550 nm. A set of standard solutions was made up with concentrations of 50.0, 75.0, 100.0, 125.0, 150.0, 175.0, 200.0, 225.0, and 250.0 mg/100 mL of porphobi-linogen. Their absorbances A after treatment with PDMA were 0.039, 0.061, 0.087, 0.107, 0.119, 0.163, 0.179, 0.194, and 0.213. What is the spectro-photometric calibration curve $A = f$(concentration) for this method? What are the units of slope? Three urine specimens treated by this method yielded absorbances A of 0.180, 0.162, and 0.213. What were the porphobilinogen concentrations of these three samples?

4. Expand the three determinants D, D_b, and D_m for the least squares fit to a linear function not passing through the origin so as to obtain explicit algebraic expressions for b and m, the y-intercept and the slope of the best straight line representing the experimental data.

5. Set up the determinants D, D_a, D_b, and D_c for the least squares fit to a parabolic curve not passing through the origin.

6. Expand the four 3×3 determinants obtained in Problem 5.

7. Using the expanded determinants from Problem 6, write explicit algebraic expressions for the three minimization parameters a, b, and c for a parabolic curve fit.

8. Compare the solution for Problems 6 and 7 with the BASIC statements in Program QQLSQ.BAS or QLLSQ.tru. They should agree.

9. Emsley, in the same paper referred to in Computer Project 3-1, presents viscosity measurements η for solutions of KF in HOAc as a function of molality m with the following results

m	0.135	0.398	1.028	1.466	1.903	2.567	3.052	3.428	3.770	4.206	4.307
η	1.38	1.77	3.69	5.97	8.68	16.34	29.29	41.53	58.57	92.73	106.36

Carry out a statistical analysis of this data set including a fitting equation and all uncertainties.

10. Write the determinant for a sixth-degree curve fitting procedure.

11. The volume of $ZnCl_2$ solutions containing 1000 g of water varies according to the quadratic equation (Computer Project 3-4)

$$V = 999.71 + 21.148\,m + 4.471\,m^2$$

Find the partial molal volume of $ZnCl_2$ in these solutions at 0.5, 1.0, 1.5 and 2.0 molar concentrations.

Multivariate Least Squares Analysis

In multivariate least squares analysis, the dependent variable is a function of two or more independent variables. Because matrices are so conveniently handled by computer and because the mathematical formalism is simpler, multivariate analysis will be developed as a topic in matrix algebra rather than conventional algebra.

We have already seen the normal equations in matrix form. In the multivariate case, there are as many slope parameters as there are independent variables and there is one intercept. The simplest multivariate problem is that in which there are only two independent variables and the intercept is zero

$$y = m_1 x_1 + m_2 x_2 \tag{3-54}$$

To simplify the algebra, the error in the x variable will be considered negligible relative to the error in the y variable, although this is not a necessary condition. Equation (3-54) describes a plane passing through the origin. Let us restrict it to positive values of x_1, x_2, and y.

One measurement of the dependent variable yields y_1 for known values of the independent variables x_{11}, x_{12}

$$y_1 = m_1 x_{11} + m_2 x_{12} \tag{3-55a}$$

and a second yields

$$y_2 = m_1 x_{21} + m_2 x_{22} \tag{3-55b}$$

for new values of the independent variables x_{21}, x_{22}.

The mathematical requirements for unique determination of the two slopes m_1 and m_2 are satisfied by these two measurements, provided that the second equation is not a linear combination of the first. In practice, however, because of experimental error, this is a minimum requirement and may be expected to yield the least reliable solution set for the system, just as establishing the slope of a straight line through the origin by one experimental point may be expected to yield the least reliable slope, inferior in this respect to the slope obtained from 2, 3, or p experimental points. In univariate problems, accepted practice dictates that we obtain many experimental points and determine the "best" slope representing them by a suitable univariate regression procedure.

The analogous procedure for a multivariate problem is to obtain many experimental equations like Eqs. (3-55) and to extract the best slopes from them by regression. Optimal solution for n unknowns requires that the slope vector be obtained from p equations, where p is larger than n, preferably much larger. When there are more than the minimum number of equations from which the slope vector is to be extracted, we say that the equation set is an *overdetermined* set. Clearly, n equations can be selected from among the p available equations, but this is precisely what we do not wish to do because we must subjectively discard some of the experimental data that may have been gained at considerable expense in time and money.

Equation set (3-55) can be written in matrix form

$$\mathbf{y} = \mathbf{Xm} \tag{3-56}$$

where $\mathbf{X}$ is the matrix of (known) input variables

$$\mathbf{X} = \begin{pmatrix} x_{11} & x_{12} \\ x_{21} & x_{22} \end{pmatrix} \tag{3-57}$$

$\mathbf{y}$ is the nonhomogeneous vector of dependent experimental measurements, and $\mathbf{m}$ is the slope vector, that is, the solution vector of the regression problem.

The deviations or residuals of y_1 and y_2 from the regression plane passing through the origin are

$$d_1 = m_1 x_{11} + m_2 x_{12} - y_1 \tag{3-58a}$$
$$d_2 = m_1 x_{21} + m_2 x_{22} - y_2 \tag{3-58b}$$

which are minimized by setting $\frac{\partial}{\partial m_1} \sum d_i^2$ and $\frac{\partial}{\partial m_2} \sum d_i^2$ equal to zero as shown in the section on linear functions not passing through the origin. These minimizations yield

$$m_1 x_{11}^2 + m_2 x_{11} x_{12} + m_1 x_{21}^2 + m_2 x_{21} x_{22} = y_1 x_{11} + y_2 x_{21} \tag{3-59a}$$
$$m_1 x_{11} x_{12} + m_2 x_{12}^2 + m_1 x_{21} x_{22} + m_2 x_{22}^2 = y_1 x_{12} + y_2 x_{22} \tag{3-59b}$$

We are dealing with real numbers that commute; hence, it is evident that the right side of Equation set (3-59) is

$$\begin{pmatrix} x_{11} & x_{21} \\ x_{12} & x_{22} \end{pmatrix} \begin{pmatrix} y_1 \\ y_2 \end{pmatrix} \tag{3-60}$$

or

$$\mathbf{X}^{\mathbf{T}}\mathbf{y} \tag{3-61}$$

where the matrix $\mathbf{X^T}$ is the transpose of the input matrix $\mathbf{X}$.

The left side of the normal equations can be seen to be a product including $\mathbf{X}$, its transpose, and $\mathbf{m}$. Matrix multiplication shows that

$$\mathbf{X}^\mathbf{T}\mathbf{X}\mathbf{m} \qquad\qquad (3\text{-}62)$$

is the matrix representation of the left side of the normal equations (see Problems).

Two important facts emerge here. First, the method is general and can be worked out for the $n \times n$ case $(n > 2)$ as above but with added labor. Second, *the input matrix need not be square*. By the geometric nature of a rectangular matrix, it is always conformable for multiplication into its own inverse. The result is a square product matrix of the smaller of the two dimensions of the rectangular input matrix. Indeed, for the present treatment to be nontrivial, the input matrix must be rectangular; a square input matrix with $\mathbf{X}^{-1}$ defined ($\mathbf{X}$ nonsingular) represents the problem of n independent equations in n unknowns, which is the problem we said we do not want to solve. From this point on, envision $\mathbf{X}$ as a $p \times n$ matrix with $p > n$.

The normal equations are simultaneous equations

$$\mathbf{X}^\mathbf{T}\mathbf{X}\mathbf{m} = \mathbf{X}^\mathbf{T}\mathbf{y} \qquad\qquad (3\text{-}63)$$

in which $\mathbf{X}^\mathbf{T}\mathbf{y}$ yields a vector of the smaller dimension of $\mathbf{X}^\mathbf{T}$, the same dimension as the vector obtained as the product $\mathbf{X}^\mathbf{T}\mathbf{X}\mathbf{m}$. The normal equations are often written

$$\left(\mathbf{X}^\mathbf{T}\mathbf{X}\right)\mathbf{m} = \mathbf{q} \qquad\qquad (3\text{-}64)$$

where $\left(\mathbf{X}^\mathbf{T}\mathbf{X}\right)$ emerges as the coefficient matrix of a simple set of simultaneous equations, $\mathbf{m}$ is the solution vector, and $\mathbf{q}$ is the nonhomogeneous vector $\mathbf{X}^\mathbf{T}\mathbf{y}$. Solution follows by the usual method of inverting the coefficient matrix and premultiplying it into both sides

$$\left(\mathbf{X}^\mathbf{T}\mathbf{X}\right)^{-1}\left(\mathbf{X}^\mathbf{T}\mathbf{X}\right)\mathbf{m} = \left(\mathbf{X}^\mathbf{T}\mathbf{X}\right)^{-1}\mathbf{q}$$

or

$$\mathbf{m} = \left(\mathbf{X}^\mathbf{T}\mathbf{X}\right)^{-1}\mathbf{q} \qquad\qquad (3\text{-}65)$$

This equation permits us to generate an n-fold slope vector $\mathbf{m}$ from a rectangular matrix $\mathbf{X}$ and a dependent variable vector $\mathbf{y}$. The procedure is analogous to the univariate case of generating the slope of a calibration curve passing through the origin from p known values of x_i and the corresponding measured values $y_1, \ldots, y_i, \ldots, y_p$, with the intention of using m to determine unknown values of x (the concentration of an analyte perhaps) from future measurements of y.

In the multivariate case, one slope vector is not enough; a square slope *matrix* must be generated with dimensions equal to the number of independent unknowns x_i one wishes to determine. The slope matrix is

$$\mathbf{M} = \begin{pmatrix} m_{11} & m_{12} \\ m_{21} & m_{22} \end{pmatrix} \tag{3-66}$$

for a problem in two unknowns and larger for n unknowns. This is done by repeating the procedure just given for obtaining an m vector n times with different vectors $\mathbf{y_1}, \mathbf{y_2}, \ldots, \mathbf{y_p}$ to produce an $n \times n$ slope matrix $\mathbf{M}$. Once having $\mathbf{M}$, one can determine unknown values of $\mathbf{x}$

$$\mathbf{x} = \mathbf{M}^{-1}\mathbf{y} \tag{3-67}$$

The n-fold procedure $(n > 2)$ produces an n-dimensional hyperplane in $n + 1$ space. Lest this seem unnecessarily abstract, we may regard the $n \times n$ slope matrix as the matrix establishing a calibration surface from which we may determine n unknowns x_i by making n independent measurements y_i. As a final generalization, it should be noted that the calibration surface need not be planar. It might, for example, be a curved surface that can be represented by a family of quadratic equations.

An illustrative example generates a 2×2 calibration matrix from which we can determine the concentrations x_1 and x_2 of dichromate and permanganate ions simultaneously by making spectrophotometric measurements y_1 and y_2 at different wavelengths on an aqueous mixture of the unknowns. The advantage of this simple two-component analytical problem in 3-space is that one can envision the plane representing absorbance A as a linear function of two concentration variables $A = f(x_1, x_2)$.

Application: Simultaneous Analysis by Visual Spectrophotometry

Simultaneous $Cr_2O_7^{2-}$ and MnO_4^- determination (Ewing, 1985) is a multivariate spectrophotometric analysis that requires determination of a matrix of four calibration constants, one for each unknown at each of two wavelengths,

$$\begin{aligned} a_{11}x_1 + a_{12}x_2 &= A_1 \\ a_{21}x_1 + a_{22}x_2 &= A_2 \end{aligned} \tag{3-68}$$

The elements a_{ij} are *absorptivities* (or are proportional to absorptivities, depending on the concentration units and cell dimensions), $\mathbf{x} = \begin{pmatrix} x_1 \\ x_2 \end{pmatrix}$ is the unknown concentration vector, and $\mathbf{y} = \begin{pmatrix} A_1 \\ A_2 \end{pmatrix}$ is the *absorbance* vector, observed at wavelengths λ_1 and λ_2.

The absorbance A is proportional to x through Beer's law (see Computer Project 2-1). The analytical problem is to solve the matrix equation

$$\mathbf{Ax} = \mathbf{y} \tag{3-69}$$

for $\mathbf{x}$ from measured values of $\mathbf{y}$ once we have determined the matrix $\mathbf{A}$. Be careful to distinguish between $\mathbf{A}$ and A. (For accepted nomenclature, see Ewing, 1985.) The wavelengths λ_i should be chosen so as to make the $\mathbf{A}$ matrix as "nonsingular as possible," that is, wavelengths should be selected so that, insofar as possible, absorbance by one species dominates all the rest. The wavelengths selected for this problem are 440 nm for $Cr_2O_7^{2-}$ and 525 nm for MnO_4^-.

Dichromate-permanganate determination is an artificial problem because the matrix of coefficients can be obtained as the slopes of A vs. x from four univariate least squares regression treatments, one on solutions containing only $Cr_2O_7^{2-}$ at 440 nm, one on the same solution at 525 nm, and one on solutions containing only MnO_4^- at each of these two wavelengths. We did this for five concentrations of each absorbing species and obtained the matrix

$$\mathbf{A} = \begin{pmatrix} 4.39 \times 10^{-3} & 2.98 \times 10^{-3} \\ 3.85 \times 10^{-4} & 4.24 \times 10^{-2} \end{pmatrix} \qquad (3\text{-}70)$$

Elements in the slope matrix $\mathbf{A}$ are proportional to absorptivities and concentrations are in parts per million. We shall take this as the "true" slope matrix.

Part 1. Generate the slope matrix from Eq. (3-65)

To obtain this matrix by the multivariate method, we first generate two absorptivity vectors $\mathbf{a}_{1j}$ and $\mathbf{a}_{2j}$ from a known concentration matrix in parts per million

$$\mathbf{X} = \begin{pmatrix} 53.0 & 8.65 \\ 27.0 & 13.0 \\ 80.0 & 4.33 \\ 0.0 & 17.3 \\ 106 & 0.0 \end{pmatrix} \qquad (3\text{-}71)$$

and the measured absorbance vectors $\mathbf{y}_j$, one at the lower wavelength and one at the higher wavelength. From the measured absorbance vector $\mathbf{y} = \{0.251, 0.149, 0.361, 0.049, 0.456\}$ at 440 nm, we obtained the vector $\mathbf{a}_{1j} = \{4.32 \times 10^{-3}, 2.72 \times 10^{-3}\}$ rounded to the appropriate number of significant figures. This is the top row of the matrix $\mathbf{A}$.

Mathcad

$$\mathbf{X} := \begin{pmatrix} 53.0 & 8.65 \\ 27.0 & 13.0 \\ 80.0 & 4.33 \\ 0.0 & 17.3 \\ 106 & 0.0 \end{pmatrix} \qquad \mathbf{X^T} = \begin{pmatrix} 53 & 27 & 80 & 0 & 106 \\ 8.65 & 13 & 4.33 & 17.3 & 0 \end{pmatrix}$$

$$\mathbf{X^T \cdot X} = \begin{pmatrix} 2.117 \times 10^4 & 1.156 \times 10^3 \\ 1.156 \times 10^3 & 561.861 \end{pmatrix}$$

$$\mathbf{(X^T \cdot X)}^{-1} = \begin{pmatrix} 5.32 \times 10^{-5} & -1.094 \times 10^{-4} \\ -1.094 \times 10^{-4} & 2.005 \times 10^{-3} \end{pmatrix}$$

$$\mathbf{y} := \begin{pmatrix} .251 \\ .149 \\ .361 \\ .049 \\ .456 \end{pmatrix} \quad \mathbf{y1} := \begin{pmatrix} .401 \\ .568 \\ .209 \\ .740 \\ .042 \end{pmatrix} \quad \mathbf{(X^T \cdot X)}^{-1} \cdot \mathbf{X^T y} = \begin{pmatrix} 4.316 \times 10^{-3} \\ 2.723 \times 10^{-3} \end{pmatrix}$$

Repeating the process with a new measured absorbance vector {0.401, 0.568, 0.209, 0.740, 0.042} at 525 nm leads to the $\mathbf{a}_{2j}$ vector $\mathbf{y}_2 = \{3.85 \times 10^{-4}, 4.29 \times 10^{-2}\}$, the bottom row of the slope matrix. Together, they yield $\mathbf{A}$

$$\mathbf{A} = \begin{pmatrix} 4.32 \times 10^{-3} & 2.72 \times 10^{-3} \\ 3.85 \times 10^{-4} & 4.29 \times 10^{-2} \end{pmatrix} \tag{3-72}$$

Matrix (3-72) is essentially the same as matrix (3-70), but it is not exactly the same because it was obtained by the multivariate method from a different data set.

Part 2. Analyze unknown mixtures using the A matrix and two measurements of the absorbance, one at 440 nm and the other at 525 nm, as the y vector.

Having combined the two absorbance vectors into the absorbance matrix $\mathbf{A}$, we are in a position to use $\mathbf{A}$ to solve for unknown concentration vectors $\mathbf{x}$. Because $\mathbf{y} = \mathbf{Ax}$, it follows that

$$\mathbf{x} = \mathbf{A}^{-1}\mathbf{y}$$

Suppose that we have measured absorbance vectors $\mathbf{y}$ for three different solutions each containing both $Cr_2O_7^{2-}$ and MnO_4^-. Let us call them $\mathbf{ya}$, $\mathbf{yb}$, and $\mathbf{yc}$,

$$\mathbf{ya} := \begin{pmatrix} .331 \\ .401 \end{pmatrix} \quad \mathbf{yb} := \begin{pmatrix} .156 \\ .354 \end{pmatrix} \quad \mathbf{yc} := \begin{pmatrix} .177 \\ .723 \end{pmatrix}$$

where the measurement at 440 nm is the y_1 (top) element in each y vector and the 525 measurement is the y_2 (bottom) element in $\mathbf{y}$. The solution concentrations in parts per million follow easily.

Mathcad

$$\mathbf{A} := \begin{pmatrix} .004316 & .002723 \\ .000385 & .0429 \end{pmatrix} \quad \mathbf{ya} := \begin{pmatrix} .331 \\ .401 \end{pmatrix} \quad \mathbf{yb} := \begin{pmatrix} .156 \\ .354 \end{pmatrix} \quad \mathbf{yc} := \begin{pmatrix} .177 \\ .723 \end{pmatrix}$$

$$\mathbf{A}^{-1} \cdot \mathbf{ya} = \begin{pmatrix} 71.2 \\ 8.71 \end{pmatrix} \quad \mathbf{A}^{-1} \cdot \mathbf{yb} = \begin{pmatrix} 31.11 \\ 7.97 \end{pmatrix} \quad \mathbf{A}^{-1} \cdot \mathbf{yc} = \begin{pmatrix} 30.55 \\ 16.58 \end{pmatrix}$$

The fourth significant figures in the vectors $\mathbf{xb}$ and $\mathbf{xc}$ are artifacts of the calculation. The concentration vectors should be reported as $\mathbf{xa} = \{71.2, 8.71\}$, $\mathbf{xb} = \{31.1, 7.97\}$, and $\mathbf{xc} = \{30.5, 16.6\}$ ppm.

Error Analysis

Subtracting the slope matrix obtained by the multivariate least squares treatment from that obtained by univariate least squares slope matrix yields the error matrix

$$\begin{pmatrix} 0.07 \times 10^{-3} & 0.26 \times 10^{-3} \\ 0.0 & -0.50 \times 10^{-3} \end{pmatrix} \tag{3-73}$$

where the univariate results are taken as "true" values.

Normally, one does not have "true" values of the elements of the slope matrix $\mathbf{M}$ for comparison. It is always possible, however, to obtain $\hat{\mathbf{y}}$, the vector of predicted y values at each of the known x_i from any of the slope vectors $\mathbf{m}$ obtained by the multivariate procedure

$$\hat{\mathbf{y}} = \mathbf{Xm} \tag{3-74}$$

This permits error analysis of that vector. (Note that the order $\mathbf{Xm}$ is necessary for the matrix and vector to be conformable for multiplication.) Repeating the procedure for all m vectors leads to error analysis of the entire matrix $\mathbf{M}$.

We wish to carry out a procedure that is the multivariate analog to the analysis in the section on reliability of fitted parameters. A vector multiplied into its transpose gives a scalar that is the sum of squares of the elements in that vector. The $\hat{\mathbf{y}}$ vector leads to a vector of residuals

$$\mathbf{e} = \hat{\mathbf{y}} - \mathbf{y} = \mathbf{Xm} - \mathbf{y} \tag{3-75}$$

The product $\mathbf{e}^{\mathrm{T}}\mathbf{e}$ is the sum of squares of residuals from the vector of residuals. The variance is

$$s^2 = \frac{\mathbf{e}^{\mathrm{T}}\mathbf{e}}{p - n - 1} \tag{3-76}$$

where p is the number of measurements made to establish n components of the slope vector. Under the assumption that the residuals are normally distributed, the best estimator of the variance of the ith element in $\mathbf{m}$ is $s^2 d_{ii}$, where d_{ii} is the ith diagonal element of $\mathbf{D} = (\mathbf{X}^{\mathrm{T}}\mathbf{X})^{-1}$. These variances are analogous to those obtained by Eqs. (3-34).

For our sample data set of $Cr_2O_7^{2-}$ and MnO_4^- absorbances, we seek the first of the two $\hat{y}$ vectors of residuals at 440 nm.

$$\mathbf{X} := \begin{pmatrix} 53.0 & 8.65 \\ 27.0 & 13.0 \\ 80.0 & 4.33 \\ 0.0 & 17.3 \\ 106 & 0.0 \end{pmatrix} \qquad \mathbf{m1} := \begin{pmatrix} .00432 \\ .00272 \end{pmatrix}$$

$$\mathbf{X \cdot m1} = \begin{pmatrix} 0.252 \\ 0.152 \\ 0.357 \\ 0.047 \\ 0.458 \end{pmatrix}$$

This gives an error (residual) vector of

$$\mathbf{y} := \begin{pmatrix} .251 \\ .149 \\ .361 \\ .049 \\ .456 \end{pmatrix} \qquad \mathbf{y - X \cdot m1} = \begin{pmatrix} -1.488 \times 10^{-3} \\ -3 \times 10^{-3} \\ 3.622 \times 10^{-3} \\ 1.944 \times 10^{-3} \\ -1.92 \times 10^{-3} \end{pmatrix}$$

which leads to

$$\mathbf{SSE} := (\mathbf{y - X \cdot m1})^\mathbf{T} \cdot (\mathbf{y - X \cdot m1}) \qquad \mathbf{SSE} = (3.18 \times 10^{-5})$$

and an s^2 value of

$$s^2 = \frac{\mathbf{e}^\mathbf{T}\mathbf{e}}{p - n - 1} = \frac{\mathbf{SSE}}{5 - 2 - 1} = \frac{3.18 \times 10^{-5}}{2} = 1.6 \times 10^{-5}$$

The required inverse matrix $\mathbf{D} = (\mathbf{X}^\mathbf{T}\mathbf{X})^{-1}$ is

$$\mathbf{D} = \begin{pmatrix} 5.32 \times 10^{-5} & -1.09 \times 10^{-4} \\ -1.09 \times 10^{-4} & 2.01 \times 10^{-3} \end{pmatrix} \tag{3-77}$$

which gives the uncertainties in row 1 of the slope matrix as

$$\mathbf{A} = \begin{pmatrix} (4.32 \pm 0.03) \times 10^{-3} & (2.72 \pm 0.18) \times 10^{-3} \\ 3.85 \times 10^{-4} & 4.29 \times 10^{-2} \end{pmatrix} \tag{3-78}$$

The reader is asked to find the standard deviations of the slopes of row 2 of the $\mathbf{A}$ matrix in Problem 9 below.

Student's t statistics (Rogers, 1983) follow in the usual way as do the 95% confidence limits on the computed slopes, $\{(4.32 \pm 0.12) \times 10^{-3}, (2.72 \pm 0.77) \times 10^{-3}\}$ at 440 nm and $\{(3.85 \pm 2.7) \times 10^{-4}, (4.29 \pm 0.17) \times 10^{-2}\}$ at 525 nm. These are not the same as the standard deviations due to the t statistic. The relative uncertainty on element a_{21} is large because the parameter is an order of magnitude smaller than the other elements in the slope matrix.

COMPUTER PROJECT 3-5 | *Calibration Surfaces Not Passing*
                                 Through the Origin

Let the generalization of Eq. set (3-55)

$$y_i = \sum m_j x_{ij} \tag{3-79}$$

each contain a term, call it m_0, with the stipulation that x_{i0} is some constant value, take it to be 1.0 for simplicity. The normal equations and the solution for the **m** vectors follow just as they did in the previous section except that each equation in set (3-58) contains an additive constant m_0. The constant m_0, a minimization parameter along with the rest of the m_j, is the best estimator of the y intercept for a function not passing through the origin; it is the unique point at which the calibration surface cuts the y-axis.

To set up the problem for a microcomputer or **Mathcad**, one need only enter the input matrix with a 1.0 as each element of the 0th or leftmost column. Suitable modifications must be made in matrix and vector dimensions to accommodate matrices larger in one dimension than the **X** matrix of input data (3-56), and output vectors must be modified to contain one more minimization parameter than before, the intercept m_0.

Procedure. The method can be tested using the matrix of concentrations, in micromoles per liter (μmol L^{-1}), of tryptophan and tyrosine at 280 nm suitably modified to take into account constant absorption at 280 nm of some absorber that is neither tryptophan nor tyrosine

$$\mathbf{X} = \begin{pmatrix} 1.0 & 47.6 & 116 \\ 1.0 & 125 & 147 \\ 1.0 & 23.7 & 109 \\ 1.0 & 156 & 48.3 \\ 1.0 & 272 & 15.1 \end{pmatrix} \tag{3-80}$$

Columns 2 and 3 of matrix (3-80) are analogous to the concentration matrix (3-71). Using the absorbance vector $\{0.846, 1.121, 0.776, 0.599, 0.559\}$, compute the solution vector, $\mathbf{m}_{280}$. The first element of the solution vector is the intercept due to background absorption, and the second two elements are the absorbancies. What are the absorbancies of tyrosine and tryptophan at 280 nm by this method? Compare your results with the accepted values $a_{280}(\text{tyr}) = 1.28 \times 10^3$ and $a_{280}(\text{try}) = 5.69 \times 10^3$ L mol^{-1} cm^{-1} (Eisenberg and Crothers, 1979). What are the units of your results for tyrosine and tryptophan? What is the intercept of absorbance at 280 nm due to compounds that are neither tyrosine nor tryptophan?

COMPUTER PROJECT 3-6 | *Bond Energies of Hydrocarbons*

Determination of bond energies in hydrocarbons is a nontrivial example of multi-variate analysis because lone C—C and C—H bonds cannot be observed in stable hydrocarbons; they always appear in groups. One could take the C—H bond energy to be one-fourth of the energy of atomization of methane, subtract six times that value from the energy of atomization of ethane to get the C—C single bond energy, and proceed in a like way, using the atomization energies of an alkene and an alkyne to generate the carbon-carbon double and triple bonds. This strategy would be risky at best because the C—H bonds in a single molecule, methane, would be taken to represent all C—H bonds, ethane would be taken to represent all C—C bonds, and so on. It would be better to draw bond energies from a *basis set* of data for several, preferably many, molecules on the reasonable assumption that the mean result is more reliable than any single result from the set. Because the bonds cannot be observed singly, the problem is multivariate, and because we wish to generate a few bond energies from many experimental results, the input matrix will be overdetermined.

We will generate the energies for the carbon-hydrogen bond B_{CH} and the carbon-carbon single bond B_{CC} using the five linear alkanes from ethane through hexane as the five-member data base. The equation to be used is

$$hB_{CH} + sB_{CC} = H_a \tag{3-81}$$

where h is the number of C—H bonds in each hydrocarbon, s is the corresponding number of C—C single bonds, and H_a is the enthalpy of atomization. Enthalpies of atomization of carbon and hydrogen were taken as 716.7 and 218.0 kJ per mole of atoms produced (Lewis et al., 1961) and were combined with the appropriate enthalpy of formation (Cox and Pilcher, 1970; Pedley et al., 1986; **www.webbook. nist.gov**) to obtain the enthalpy of atomization of each hydrocarbon by the method shown in Fig. 3-3 (see also Computer Project 2-1). The enthalpy of formation of methane, $\Delta_f H^{298}(\text{methane}) = -74.5$ kJ mol^{-1}, is the enthalpy necessary to go from the elements in the standard state (0 by definition) to the molecule in the

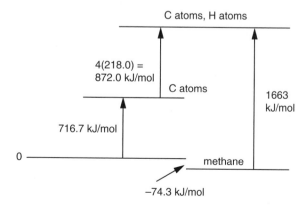

Figure 3-3   Enthalpy Diagram for the Atomization of Methane.

standard state at 298 K. In this illustration and in the computer project calculations that follow, discrepancies of ± 2 kJ mol^{-1} are not uncommon because of experimental uncertainty and differences among the various sources. We shall go into the difference between energy and enthalpy in this context in a later chapter.

Computation. Decide on an appropriate input matrix of bond numbers

$$\begin{pmatrix} h_1 & s_1 \\ h_2 & s_2 \\ \vdots & \vdots \\ h_n & s_n \end{pmatrix}$$

for ethane through hexane. The enthalpy of formation vector is $\{-20.24, -24.83, -30.36, -35.10, -39.92\}$ in the same order, where the units are kilocalories per mole. To convert from kilocalories per mole to kilojoules per mole, multiply by 4.184. Calculate the 5-fold enthalpy of atomization vector and the 2-fold vector of bond enthalpies. Obtain an error vector by comparing your result with the accepted values (Atkins, 1994) of 412 and 348 for the C—H and C—C bonds. respectively.

More than a reasonable number of significant figures is carried through the calculation to be rounded off at the end. When properly rounded, the uncertainty of the computed result should be reflected in the significant figures such that the rightmost digit is uncertain but no more than one uncertain digit is included in the final result. This is, of course, an approximate indicator of uncertainty; if a rigorous indicator is desired, the standard deviation, variance, or confidence level should be reported with the computed result.

COMPUTER PROJECT 3-7 | *Expanding the Basis Set*

Add ethylene, 1-propene, 1-butene, acetylene, and 1-propyne to the basis set. To do this, you must calculate five new atomization enthalpies from cycles similar to the one in Fig. 3-3. Also extend the input matrix to a 4×10 matrix. Generate the C—H, C—C, C=C, and C≡C bond energies. Comment on the magnitude of the bond enthalpies, particularly the enthalpies of the C—C single, double, and triple bonds. Is there any relationship between bond strength and bond energy for the three carbon-carbon bonds? Look up a set of accepted values for these bond energies and calculate a 4-fold error vector. Does the error for C—H and C—C get larger or smaller for the extended basis set as compared with the smaller basis set used in the first part of this experiment? Discuss this result.

PROBLEMS

1. Obtain the normal equations [Eq. set (3-63)] from the minimization conditions

$$\frac{\partial \sum d_i^2}{\partial m_1} = \frac{\partial \sum d_i^2}{\partial m_2} = 0$$

2. Multiply $\mathbf{X}^T\mathbf{y}$ from Eq. set (3-61) to show that it is equal to the right side of Eq. set (3-63).

3. Multiply $\mathbf{X}^T\mathbf{X}\mathbf{m}$ from Eq. set (3-61) to show that it is equal to the left side of Eq. set (3-63).

4. Can a rectangular matrix be both premultiplied and postmultiplied into its own transpose, or must multiplication be either pre- or post- for conformability? If multiplication must be either one or the other, which is it?

5. Show that $\mathbf{e}^T\mathbf{e}$ is the sum of squares of elements in the vector $\mathbf{e} = \{1, 2, 3\}$.

6. Does $\mathbf{e}^T\mathbf{e}$ commute with $\mathbf{e}\mathbf{e}^T$?

7. What is the average enthalpy of atomization of the four C—H bonds in methane? Compare this value with the accepted value of the C—H bond enthalpy.

8. Calculate the bond enthalpy of the C—C bond in ethane using only the enthalpies of atomization of methane and ethane. Compare this result with the accepted result.

9. Find the standard deviations of the slopes in matrix (3-78) for row 2, which refers to absorbances measured at 525 nm.

10. When 6 moles of the substrate analog PALA combine with the enzyme ACTase, two things happen at the same time. The enzyme T unfolds to a more active form R

$$T \rightarrow R$$

and 6 moles of PALA bind to the enzyme. The measured enthalpy of both reactions together is

$$\Delta H = 6\Delta_{PAL}H + \Delta_{T \rightarrow R}H = -209.2 \text{ kJ mol}^{-1}$$

(Klotz and Rosenberg, 2000).

   We would like to know the binding energy per mole of PALA and the enthalpy of the transformation $T \rightarrow R$, but we do not have enough information. Independent studies show that partial unfolding of ACTase occurs on binding of less PALA, in particular, 1.8 mol of PALA cause 43% unfolding and 4.8 mol cause 86%. The enthalpy changes are -63.2 and -184.5 kJ mol^{-1} respectively, leading to

$$\Delta H = 1.8\Delta_{PAL}H + 0.43\Delta_{T \rightarrow R}H = -63.2 \text{ kJ mol}^{-1}$$

and

$$\Delta H = 4.8\Delta_{PAL}H + 0.86\Delta_{T \rightarrow R}H = -184.5 \text{ kJ mol}^{-1}$$

Use all three equations to find the enthalpies of binding and of unfolding for this enzyme.

CHAPTER

4

Molecular Mechanics: Basic Theory

The first *molecular modeling* technique we shall look at is *molecular mechanics* (MM). MM is a very fast method of determining the geometry, molecular energies, vibrational spectra, and enthalpies of formation of stable ground-state molecules. Because of its speed, it is widely used on large molecules such as those of biological or pharmaceutical importance that are currently beyond the reach of more computer-intensive *molecular orbital* methods. MM is an empirical method, relying on a large number of parameters, drawn from experimental data, called, collectively, the *force field parameters*. The major drawback of MM is encountered when one or more of the parameters necessary to solve a problem is not known. Because they are parameterized using data from molecules in the ground state, neither MM nor semiempirical methods are not as useful for modeling transition-state chemistry as *ab initio* methods.

The Harmonic Oscillator

The harmonic oscillator (Fig. 4-1) is an idealized model of the simple mechanical system of a moving mass connected to a wall by a spring. Our interest is in very small masses (atoms). The harmonic oscillator might be used to model a hydrogen atom connected to a large molecule by a single bond. The large molecule is so

Computational Chemistry Using the PC, Third Edition, by Donald W. Rogers
ISBN 0-471-42800-0   Copyright © 2003 John Wiley & Sons, Inc.

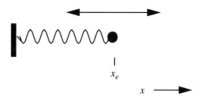

Figure 4-1   A Harmonic Oscillator in One
Dimension.

heavy that it can be considered a "wall" that is stationary relative to the quick
motions of the hydrogen atom.

If the spring follows Hooke's law, the force it exerts on the mass is directly
proportional and opposite to the excursion of the particle away from its equilibrium
point x_e. The particle of mass m is accelerated by the force $F = -kx$ of the spring.
By Newton's second law, $F = ma$, where a is the acceleration of the mass

$$F = ma = m\frac{d^2x}{dt^2} = -kx \tag{4-1}$$

where k is the *force constant* of the spring.

This is a differential equation that we shall see often. It has as one of its solutions

$$x(t) = A \cos \omega t \tag{4-2}$$

where ω is the *angular frequency* of oscillation expressed in radians

$$\omega = \sqrt{\frac{k}{m}} \tag{4-3}$$

The cycle of oscillation is 0 to 2π, precisely the circumference of a circle. After one
cycle of 2π radians is complete, another cycle begins, identical to the one before it.
The angular frequency in radians ω is related to the frequency expressed in units of
complete cycles per second v as $\omega = 2\pi v$, whence

$$v = \frac{1}{2\pi} \sqrt{\frac{k}{m}} \tag{4-4}$$

The modern unit, expressing frequency in cycles per second, is the *hertz* (Hz).

Note that we have taken a cosine rather than a sine function for our solution.
Substitution of either Eq. (4-2) or the equivalent sine function into Eq. (4-1) gives a
true statement (with certain restrictions on ω); therefore, both are solutions.
Moreover, the sum or difference

$$\cos \omega t \pm \sin \omega t \tag{4-5}$$

is a solution, as is

$$e^{-i\omega t} \tag{4-6}$$

It is a property of this family of differential equations that the sum or difference of two solutions is a solution and that a constant (including the constant $i = \sqrt{-1}$) times a solution is also a solution. This accounts for the acceptability of forms like $x(t) = A \cos \omega t$, where the constant A is an *amplitude factor* governing the maximum excursion of the mass away from its equilibrium position. The exponential form comes from Euler's equation

$$e^{-i\omega t} = \cos \omega t \pm i \sin \omega t \qquad (4\text{-}7)$$

a form that will be useful later.

The potential energy for a conservative system (system without frictional loss) is the negative integral of a displacement times the force overcome. In this case, the potential energy for a displacement x away from x_e, is

$$V = -\int_0^x -kx \, dx = \frac{kx^2}{2} \qquad (4\text{-}8)$$

where we have taken $V = 0$ at x_e.

The Two-Mass Problem

If we think about two masses connected by a spring, each vibrating with respect to a stationary center of mass x_c of the system, we should expect the situation to be very similar in form to one mass oscillating from a fixed point. Indeed it is, with only the substitution of the *reduced mass* μ for the mass m

$$\nu = \frac{1}{2\pi} \sqrt{\frac{k}{\mu}} \qquad (4\text{-}9)$$

where

$$\mu = \frac{m_1 m_2}{m_1 + m_2} \qquad (4\text{-}10)$$

for the two masses m_1 and m_2 as in Fig. 4-2.

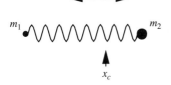

Figure 4-2 Two Masses Vibrating Harmonically with Respect to their Center of Mass. The center of mass may be stationary or moving with respect to an external coordinate system.

To an observer at the center of mass, the overall motion of the system (translation) is irrelevant. The only important motions are those motions relative to the center of mass. Distances from the center of mass to each particle are *internal coordinates* of the system, usually denoted r_1 and r_2 to emphasize that they are internal coordinates of a molecular system.

The harmonic oscillator of two masses is a model of a vibrating diatomic molecule. We ask the question, "What would the vibrational frequency be for H_2 if it were a harmonic oscillator?" The reduced mass of the hydrogen molecule is

$$\mu = \frac{m_1 m_2}{m_1 + m_2} = \frac{1}{2} = 0.5000 \text{ atomic mass units}$$

The atomic mass unit is 1.661×10^{-27} kg, so $\mu = 0.500$ atomic mass units $= 8.303 \times 10^{-28}$ kg.

The atomic harmonic oscillator follows the same frequency equation that the classical harmonic oscillator does. The difference is that the classical harmonic oscillator can have *any amplitude* of oscillation leading to a continuum of energy whereas the quantum harmonic oscillator can have only certain specific amplitudes of oscillation leading to a discrete set of allowed energy levels.

Let us "guess" that the force constant is about 500 Nm^{-1}. The vibrational frequency is

$$\nu = \frac{1}{2\pi} \sqrt{\frac{500}{8.303 \times 10^{-28}}} = 1.24 \times 10^{14} \text{ Hz} \qquad (4\text{-}11)$$

To get the frequency $\bar{\nu}$ in centimeters^{-1}, the nonstandard notation favored by spectroscopists, one divides the frequency in hertz by the speed of light in a vacuum, $c = 2.998 \times 10^{10}$ cm s^{-1}, to obtain a reciprocal wavelength, in this case, 4120 cm^{-1}. This relationship arises because the speed of any running wave is its frequency times its wavelength, $c = \nu\lambda$ in the case of electromagnetic radiation. The Raman spectral line for the fundamental vibration of H_2 is 4162 cm^{-1} ..., not a bad comparison for a simple model.

We are tempted to make some generalizations, for example, the guess 500 Nm^{-1} was pretty good for the H–H single bond. We might guess 500, 1000, and 1500 Nm^{-1} for the force constants of the C–C. C=C, and C≡C bonds on the grounds that double and triple bonds ought to be twice and three times as strong, respectively, as single bonds (see Computer Project 3-5). These guesses won't be bad either. We are led to the conclusion that the harmonic oscillator is a reasonably good approximation for the vibrational motion of at least some chemical bonds.

Of course, the "guesses" above aren't really guesses. They are predicated on many years of Raman and other spectroscopic experience and calculations that are the reverse of the calculation we described. In spectroscopic studies, one normally calculates the force constants from the stretching frequencies; in modeling, one seeks to find the stretching frequencies predicted by certain models. Moreover, the atomic weights themselves, and hence the reduced mass, are the results of over 200 years of chemical and physical determinations relative to the defined atomic weight of the 12 isotope of carbon as exactly 12.000... units.

These are all *empirical* measurements, so the model of the harmonic oscillator, which is purely theoretical, becomes semiempirical when experimental information is put into it to see how it compares with molecular vibration as determined spectroscopically. In what follows, we shall refer to *empirical* molecular models such as MM, which draw heavily on empirical information, *ab initio* molecular models such as advanced MO calculations, which one strives to derive purely from theory without any infusion of empirical data, and *semiempirical* models such as PM3, which are in between (see later chapters).

Polyatomic Molecules

Most of the molecules we shall be interested in are polyatomic. In polyatomic molecules, each atom is held in place by one *or more* chemical bonds. Each chemical bond may be modeled as a harmonic oscillator in a space defined by its potential energy as a function of the degree of stretching or compression of the bond along its axis (Fig. 4-3). The potential energy function $V = kx^2/2$ from Eq. (4-8), or $V = (k_i/2)(r_i - r_{i0})^2$ in terms of internal coordinates, is a parabola open upward in the V vs. r plane, where r_i replaces x as the extension of the ith chemical bond. The force constant k_i and the equilibrium bond distance r_{i0}, unique to each chemical bond, are typical *force field parameters*. Because there are many bonds, the potential energy-bond axis space is a many-dimensional space.

There are forces other than bond stretching forces acting within a typical polyatomic molecule. They include bending forces and interatomic repulsions. Each force adds a dimension to the space. Although the concept of a surface in a many-dimensional space is rather abstract, its application is simple. Each dimension has a potential energy equation that can be solved easily and rapidly by computer. The sum of potential energies from all sources within the molecule is the potential energy of the molecule relative to some arbitrary reference point. A

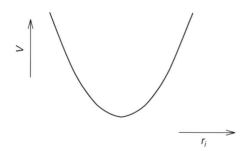

Figure 4-3 Potential Energy as a Function of Compression or Stretching of a One-Dimensional Harmonic Oscillator.

convenient potential energy reference point is the hypothetical molecule as it would be if no bond were either extended or compressed relative to its equilibrium bond distance and no other nonequilibrium energy interactions existed within the molecule.

We envision a potential energy surface with minima near the equilibrium positions of the atoms comprising the molecule. The MM model is intended to mimic the many-dimensional potential energy surface of real polyatomic molecules. (MM is little used for very small molecules like diatomics.) Once the potential energy surface has been established for an MM model by specifying the force constants for all forces operative within the molecule, the calculation can proceed.

The reason the equilibrium positions of the atoms are near but not at the minima in potential energy for each bond considered individually is that, in a polyatomic molecule, atomic positions are determined by a compromise among numerous forces. In general, atoms reside at positions leading to a minimum potential energy or equilibrium structure *for the molecule as a whole*. For example, the triatomic molecule A—B—C has a "natural" length for the bonds A—B and B—C and a "natural" angle for the angle ABC. If there is also an A—C bond, that bond will in general not be just the right length to fit into the space between A and C. It will be compressed or extended according to the forces tending to maintain the ABC bond angle. The ABC bond angle will also be distorted by the presence of the A—C bond.

The triangular molecule ABC will not display any of the "natural" bond lengths or angles; rather, it will have equilibrium bond lengths and angles that are fairly close to the natural lengths and angles. Each atom will be displaced some small distance away from its energy minimum, contributing a potential energy to the equilibrium structure. The sum of the potential energies brought about by displacements of all atoms from their natural bond lengths, angles, etc. is the *steric energy* of the molecule.

Molecular Mechanics

The problem of molecular mechanics is to find an unknown molecular geometry by minimizing all contributions to the steric energy. The strategy used to optimize a molecular geometry is to start with an approximate geometry and improve it incrementally by an iterative procedure. The input geometry of the molecule, which constitutes the major part of an MM *input file*, is specified by each of the three coordinates of each atom in Cartesian space. In the course of the program run, each atom is moved slightly.

How does one "move" an atom? The atomic coordinates are changed slightly from their initial values. One of two things can happen. Either the calculated total potential energy of the molecule goes up or it goes down. If it goes up, the move was in the wrong direction. The move was uphill on the potential energy surface, away from the equilibrium structure, and it is discarded. If the energy goes down, the move was in the right direction on the potential energy surface and the move is retained. This process is iterated many times. Once the equilibrium geometry of the molecule has been reached, the system must exit from the iterative loop.

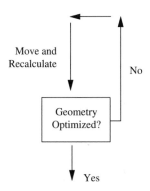

Concentrating on only one atom for the moment, there comes a time when the energy change is small because the atom is near the bottom of its parabolic potential energy function or *well*. When the atom is sufficiently near the bottom of the potential energy well that the change in *molecular* energy for a small change in position is within a predetermined limit, the potential energy is *minimized* and its position is said to be *optimized*.

One cannot simply optimize the position of each atom in sequence and say the job is done. Any change in an atomic position brings about a small change in the forces on all the other atoms. Optimization has to be repeated until the lowest molecular potential energy is found that satisfies all the forces on all the atoms. The final location of an atom will, in general, be at a position that is some small distance from the position it would have if it were not influenced by the other atoms in the molecule.

The equilibrium structure of a molecule can also be found by geometry minimization. To attain an efficient search of the potential energy surface, MM programs are written with a gradient calculation as part of the minimization routine. The gradient on a potential energy surface is essentially the slope. By selecting the direction of maximum steepness for the next change in x-, y-, and z-coordinates, the geometry is changed in the direction of maximum slope or "steepest descent." Iterations continue by a route of steepest descent, thus approaching the equilibrium geometry in the smallest number of iterations. An arbitrarily specified small gradient can be used as an exit criterion from the optimization loop. Other mathematical methods of finding the most advantageous path down a potential energy surface are also used in consideration of maximum speed and most effective use of computer resources (Grant and Richards, 1995).

A way of making the program efficient is to program the computer to make large changes in atomic coordinates when the gradient is large and small changes when the gradient is small. In this way, large "steps" are taken in the direction of the potential energy minimum when the atom is far from its equilibrium position but the steps become smaller as the atom approaches its equilibrium location. Large steps cover a lot of ground, to cut down on the number of steps, and small steps fine-tune the equilibrium structure. Dependence of step size on gradient also provides an exit from the iteration loop. When the average move of atoms has been brought to within an arbitrary limit on step size, the optimization satisfies a *geometric criterion* of equilibrium. The program exits the loop and prints out the results of the calculation.

If a molecule is strained, atoms may not be very close to the minimum of their individual potential energy wells when the best compromise geometry is reached. In such a case, the geometric criterion does not provide an exit from the loop. Programs are usually written so that they can automatically switch from a geometric minimization criterion to an energy minimization procedure.

Rather than continuing to deal with abstractions, let us plunge right in by carrying out a "bare bones" MM calculation. After the reader has a practical sense of how MM calculations work, we shall return to some of the topics referred to above in more detail and we shall introduce some others.

Ethylene: A Trial Run

A first illustration of an MM calculation is given by running the input file "**minimal**" to find the equilibrium geometry of ethylene, C_2H_4, using the program MM3. This input file has been stripped down so far that some chemists might not even recognize it as an MM input file, but the computer does. Using a graphical interface, it is common practice to carry out MM calculations without ever seeing an input file, but it is important to know that it is there. Computers do not process diagrams, they process numbers, strictly speaking, binary numbers. The structure of input files is an important and recurring theme throughout this book.

The first line contains the information that the number of atoms is (integer) 6 and that the output will be minimal, designated by the integer 4. The second line contains the information (also in integer format) that there will be 1 *connected atom list* and that there will be 4 *attached atoms*. Connected atoms can be thought of as making up the skeleton of the molecule, C=C in this case. Attached atoms are the four hydrogens attached to the C=C skeleton by C–H bonds. The third line is the actual connected atom list; atom 1 is connected to atom 2. The fourth line is the attached atom list; atom 1 is attached to atom 3, atom 1 to 4, 2 to 5, and 2 to 6. The block of information below line 4 is the geometry in the form of Cartesian coordinates in order of the atoms, for example, atom 1 is at $x = 2.$, $y = 3.$, and

$z = 0$.. Distance units are in angstroms. The periods are necessary because the computer system distinguishes between *floating point* numbers, which have a decimal point, and integers, which do not. The rightmost column of integers in the geometry block consists of atom identifiers, 2 for sp^2 carbon and 5 for hydrogen.

								6 4
1				4				
1	2							
1	3	1	4	2	5	2	6	
2.		3.		0.		2		
3.		3.		0.		2		
1.		4.		0.		5		
1.		2.		0.		5		
4.		4.		0.		5		
4.		2.		0.		5		

File 4-1a. A Minimal Input File for Ethylene. The format is for an MM3 calculation.

A very important aspect of File 4-1 is its strict format. If we look at the file once again with spaces indicated by a line over the file (not a part of a working file), we get File 4-1b.

5	10	15	20	25	30	35	40	45	50	55	60	65
												6 4
1						4						
1	2											
1	3	1	4	2	5	2	6					
2.		3.		0.		2						
3.		3.		0.		2						
1.		4.		0.		5						
1.		2.		0.		5						
4.		4.		0.		5						
4.		2.		0.		5						

File 4-1b. A Minimal File for Ethylene with Format Indicators. The italicized line of format indicators at the top shows intervals of five spaces each. It is illustrative only and is not part of the input file.

The format markers in File 4-1b are at intervals of five spaces each. Thus the entire file might be thought of as a 10×67 matrix with row 1 containing the integers 6 and 4 in columns 65 and 67 and zeros elsewhere. (In FORTRAN, a blank is read as a zero.) Row 5 has the floating point number 2. in columns 4 and 5. Both the 2 and the . (decimal point) occupy a column. Row 5 column 35 contains the integer 2 and so on.

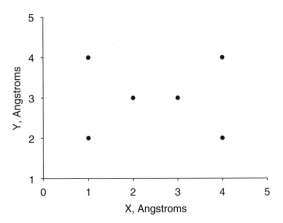

Figure 4-4 The Input Geometry for Ethylene.

The input geometry is shown in the x, y plane by Fig. 4-4 (all z-coordinates are
zero) with the carbon atoms assigned arbitrary y-coordinates 3 angstroms above the
x-axis and the hydrogens at approximate positions 2 and 4 angstroms above the x-
axis. The hydrogens are paired, 2 hydrogen atoms 1 angstrom to the left of one
carbon atom and 2 hydrogen atoms 1 angstrom to the right of the other.

When we graph the positions of all six atoms in the x, y plane, the approximate
nature of the input file is evident. Anyone who has used simple "ball and stick"
molecular models will see that the carbon atoms in Fig. 4-4 are too close together
and the entire molecule is compressed in the x-direction.

The Geo File

When the MM program is run, in this case MM3 from N. L. Allinger's group at the
University of Georgia, the final position of each atom is printed in two output files.
One output file is the *geo* file.

```
                                                        0   6 4  0 0   0 10.0
0    1        0.0000000        4      0        0    0    0    0    0        1    0
     1    2
     1    3    1    4    2    5    2    6
   1.77425   3.00009   0.00000 C   2(  1)
   3.11180   3.00003   0.00000 C   2(  2)
   1.20884   3.94637   0.00000 H   5(  3)
   1.20876   2.05386   0.00000 H   5(  4)
   3.67729   3.94627   0.00000 H   5(  5)
   3.67722   2.05375   0.00000 H   5(  6)
```

File 4-2 The Geo File for Ethylene. For exact formats, please see program
documentation (e.g., Tripos, 1992).

At first glance, the geo file looks different from the input file in many respects, but, remembering that FORTRAN is a strictly formatted language and that a blank is equal to zero in FORTRAN, we soon see that it is essentially the same except for some of the values in the geometry specification. The geo file is actually a legitimate input for a repeat calculation, although without some alteration of the first two control lines the output obtained from a second MM run using the geo file as input would merely be redundant with the first.

There are really only two elements in the geo file that we haven't seen already in the input file, the 10.0 in row 1 and a 1 in row 2. Neither is essential; the file will run without them. The first addition is a maximum time for the run, which is set at 10.0 minutes. Exceeding a time maximum usually indicates a fault in the input file that has sent the computer into an infinite loop. In practice, this safety check on the calculation should never be encountered. The 1 in row 2 is a "switch." A switch is a number that either turns a calculation on or turns it off. In this case, the original input file had a blank in row 2, column 75, indicating that a precise van der Waals energy need not be calculated until near the end of the computer run when the geometry is nearly at its final accuracy. Because the geo file is output with the correct geometry (within the accuracy of the model) the van der Waals calculation switch is *on* (integer 1 in position 2, 75). Although we haven't discussed van der Waals forces yet, the point here is that there are many features of an MM program that can be switched *on* or *off* by a properly formatted 0 or 1.

Format is the key to the remaining apparent discrepancies between the geo file and the minimal input file. The letters C or H identifying the six atoms in the molecule and the parenthesized 1 through 6, numbering them for convenience of the human reader, are not read by the computer because they are out of format. The computer can be programmed to read or to ignore any position in the input matrix. In this case, the alphabetic identifiers and sequential numbers are in positions that are ignored.

The only real difference between the input file and the geo file is that the x- and y-coordinates have changed. (The z-coordinates remain zero throughout the calculation.) Closer inspection shows that the changes in the y-coordinates are all very small. The only substantial changes during the MM program run came in the x-coordinates, as we might have anticipated by looking at Fig. 4-4. The distance between the carbon atoms has increased from 1.0 Å in the input file to about 1.34 Å in the geo file. The latter value is about what we would expect for a C=C bond. The x-distance between the H atoms has contracted from 3 Å to about 2.47 Å . This is a nonbonded interatomic distance (Fig. 4-5).

The Output File

The output file contains more information than the geo file and is documented to be easily understood by the human reader. The first block of information (PART 1) is an echo of the input file with some explanatory notation. The dielectric constant is arbitrarily set at the default value of 1.5. Default values are automatically used in

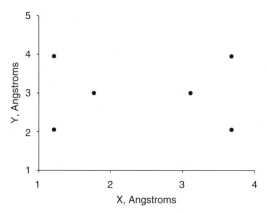

Figure 4-5   The Geometry of Ethylene with all Atoms at their Equilibrium Positions in the MM3 Force Field.

the program run when the operator declines or fails to specify a value in the input file. The second block of information in PART 1 is the INITIAL STERIC ENERGY. which we expect to be high because the molecule is not very close to its equilibrium geometry. The principal contribution to the initial steric energy is compression energy in the x-dimension brought about by the abnormally short C=C bond in the input. There is a significant contribution from the H—C—H bond angle, which is abnormally small in the input structure. The dipole moment of this symmetrical structure is zero.

PART 1

CHEMICAL FORMULA : C(2) H(4)
FORMULA WEIGHT :   28.032

DATE : 09/14/2001
TIME : 11:53:17

THE COORDINATES OF 6 ATOMS ARE READ IN.
CONFORMATIONAL ENERGY, PART 1: GEOMETRY AND STERIC ENERGY OF INITIAL CONFORMATION.
CONNECTED ATOMS
   1- 2-
ATTACHED ATOMS
   1- 3, 1- 4, 2- 5, 2- 6,

INITIAL ATOMIC COORDINATES

ATOM	X	Y	Z	TYPE
C(1)	2.00000	3.00000	0.00000	(2)
C(2)	3.00000	3.00000	0.00000	(2)
H(3)	1.00000	4.00000	0.00000	(5)
H(4)	1.00000	2.00000	0.00000	(5)
H(5)	4.00000	4.00000	0.00000	(5)
H(6)	4.00000	2.00000	0.00000	(5)

DIELECTRIC CONSTANT = 1.500
INITIAL STERIC ENERGY IS 251.2407 KCAL.

COMPRESSION	218.0511
BENDING	31.9368
BEND-BEND	−0.8293
STRETCH-BEND	−0.2189
VANDERWAALS	
1,4 ENERGY	−0.0439
OTHER	0.0000
TORSIONAL	0.0000
TORSION-STRETCH	0.0000
DIPOLE-DIPOLE	2.3450
CHARGE-DIPOLE	0.0000
CHARGE-CHARGE	0.0000
DIPOLE MOMENT =	0.000 D

CPU time for initial calculation is   3.13 seconds.

PART 2 tracks the progress of iterative geometry minimization until the average change in atomic position is less than the default limit of 1×10^{-5} Å. In this case, the geometry is optimized by the end of the second cycle. Note that both the average and maximum atomic movements diminish as the iterations proceed. The root mean square (RMS) gradient is given in units of energy per unit distance.

CONFORMATIONAL ENERGY, PART 2: GEOMETRY MINIMIZATION

GEOMETRY OPTIMIZATION IS CONTINUED UNTIL ATOM
MOVEMENT CONVERGES WITHIN FOLLOWING VALUES:
    AVERAGE MOVEMENT 0.00007 A
    MAXIMUM MOVEMENT 0.00073 A

<<<<<<<<<<<<<<<<<C Y C L E 1>>>>>>>>>>>>>>>>>>>
    (CH)-MOVEMENT = 1
ITER 1 AVG. MOVE = 0.07057 A   (MAX MOVE : ATOM 3 0.40396 A)
ITER 2 AVG. MOVE = 0.05886 A   (MAX MOVE : ATOM 3 0.43342 A)
ITER 3 AVG. MOVE = 0.01364 A   (MAX MOVE : ATOM 3 0.06326 A)
ITER 4 AVG. MOVE = 0.00272 A   (MAX MOVE : ATOM 3 0.01899 A)
ITER 5 AVG. MOVE = 0.00031 A   (MAX MOVE : ATOM 5 0.00196 A)
ITER 6 AVG. MOVE = 0.00004 A   (MAX MOVE : ATOM 3 0.00025 A)

<<<<<<<<<<<<<<<<<C Y C L E 2>>>>>>>>>>>>>>>>>>>

    (CH)-MOVEMENT = 0
ITER 7 AVG. MOVE = 0.00001 A   (MAX MOVE : ATOM 5 0.00003 A)

* * * * * * * * * GEOMETRY IS OPTIMIZED * * * * * * * * *

```
WITHIN : AVERAGE MOVEMENT 0.00001 A (LIMIT 0.00007 A)
    MAXIMUM MOVEMENT 0.00003 A (LIMIT 0.00073 A)
GRADIENT
    RMS GRADIENT                  0.000940 KCAL/MOL/A
    MAX GRADIENT
        X-DIRECTION : ATOM # 1    0.002595 KCAL/MOL/A
        Y-DIRECTION : ATOM # 2    0.001302 KCAL/MOL/A
        Z-DIRECTION : ATOM #***   0.000000 KCAL/MOL/A
    DELTA TIME = 0.00 SEC.   TOTAL CPU TIME = 3.18 SEC.
```

PART 3 is a repetition of PART 1 but with the final geometry (identical to the geo file). Both compression and bending contributions to the final energy of the ethylene molecule are small. Very small negative energies are sometimes encountered (e.g., the BEND-BEND energy) because steric energy is calculated relative to an arbitrary zero point. Dipole-dipole interactions are major contributors to the final steric energy. van der Waals repulsion and dipole-dipole interactions are discussed below. Although in this case relaxation of compression energy of the C=C bond and relaxation of the bending energy of the H–C–H group are the major contributors to the decrease in initial steric energy from $251.2 \, \text{kcal mol}^{-1}$ to the final steric energy of $2.6 \, \text{kcal mol}^{-1}$, one should beware of making fine geometric distinctions drawn from steric energies ascribed to different modes of motion. Force field parameters are to some degree composite, and a single steric energy may represent more than one mode of motion.

Once we have the atomic coordinates relative to any origin, they can be translated so that the origin is at the center of mass, permitting calculation of the moments of inertia about the x-, y-, and z-axes. A single mass rotating in a plane at a fixed distance r from a center of rotation has a moment of inertia $I = mr^2$. For a collection of masses m_i, each rotating at fixed r_i, the moments of inertia are additive

$$I_{\text{system}} = \sum m_i r_i^2 \tag{4-12}$$

In the case of a polyatomic molecule, rotation can occur in three dimensions about the molecular center of mass. Any possible mode of rotation can be expressed as projections on the three mutually perpendicular axes, x, y, and z; hence, three moments of inertia are necessary to give the resistance to angular acceleration by any torque (twisting force) in x, y, and z space. In the MM3 output file, they are denoted IX, IY, and IZ and are given in the nonstandard units of grams square centimeters.

```
        CHEMICAL FORMULA :  C( 2) H( 4)
        FORMULA WEIGHT :  28.032
                            DATE : 09/14/2001
                            TIME : 11:53:18
```

```
* M M 3 * * * * * * * * * * * * 1 9 9 2 *
*                                * PARAMETER RELIABILITY
*                                *
*                                * BLANK = GOOD
*                                *   B = FAIR
*                                *   C = TRIAL
*                                *   R = READ IN
*                                *   S = SUBSTITUTED
*                                *   U = PARAMETER NOT EXIST
*                                *
*                                * THIS IS SHOWN AT THE END
*                                * OF THE LINE.
* * * * * * * * * * * * * * * * * * * * *
```

CONFORMATIONAL ENERGY, PART 3: GEOMETRY AND STERIC
   ENERGY OF FINAL CONFORMATION.

CONNECTED ATOMS
   1- 2-
ATTACHED ATOMS
   1- 3, 1- 4, 2- 5, 2- 6,
FINAL ATOMIC COORDINATE

ATOM	X	Y	Z	TYPE
C(1)	1.77425	3.00009	0.00000	(2)
C(2)	3.11180	3.00003	0.00000	(2)
H(3)	1.20884	3.94637	0.00000	(5)
H(4)	1.20876	2.05386	0.00000	(5)
H(5)	3.67729	3.94627	0.00000	(5)
H(6)	3.67722	2.05375	0.00000	(5)

DIELECTRIC CONSTANT = 1.500

FINAL STERIC ENERGY IS   2.6017 KCAL.

COMPRESSION	0.0190
BENDING	0.0160
BEND-BEND	−0.0005
STRETCH-BEND	0.0020
VANDERWAALS	
1,4 ENERGY	0.4905
OTHER	0.0000
TORSIONAL	0.0000
TORSION-STRETCH	0.0000
DIPOLE-DIPOLE	2.0748
CHARGE-DIPOLE	0.0000
CHARGE-CHARGE	0.0000

- -

COORDINATES TRANSLATED TO NEW ORIGIN WHICH IS
   CENTER OF MASS

C(1)	−0.66877	0.00003	0.00000 (2)
C(2)	0.66877	−0.00003	0.00000 (2)
H(3)	−1.23419	0.94631	0.00000 (5)
H(4)	−1.23427	−0.94620	0.00000 (5)
H(5)	1.23426	0.94620	0.00000 (5)
H(6)	1.23419	−0.94631	0.00000 (5)

MOMENT OF INERTIA WITH THE PRINCIPAL AXES

(1) UNIT $= 10^{**}(-39)$ GM*CM**2
   IX $= 0.5994$ IY $= 2.8021$ IZ $= 3.4015$
(2) UNIT $=$ AU A**2
   IX $= 3.6103$ IY $= 16.8762$ IZ $= 20.4865$
DIPOLE MOMENT $= 0.000$ D
   COMPONENTS WITH PRINCIPAL AXES
      X $= 0.0000$ Y $= 0.0000$ Z $= 0.0000$
- -

End of
   Total cpu time is 3.18 seconds.
   This job completed at 11:53:18 (09/14/2001)

File 4-3 Conclusion of the Output File for Minimal Ethylene.

TINKER

For many individuals and chemistry departments the financial demands of commercial research level software packages are a burden. An extensive package of powerful MM and related programs called TINKER is available and can be downloaded from **dasher.wustl.edu/tinker/** as freeware. These programs, from J. W. Ponder's group at Washington University School of Medicine, St. Louis are not as easy to set up and operate as commercial programs, which are written to appeal to a wide audience of specialists and nonspecialists, but with an occasional hint from the local computer software guru, they can be up and running in a few hours. As with most molecular modelers, you will probably pick a few programs that do the job you want done and feel no guilt at all about ignoring the others. My favorite MM program is MM3. It is well to have a general idea what is out there, however, in case you run into a problem that your favorite program does not do or does poorly. The input file for a TINKER geometry minimization of ethylene is quite similar to **minimal.mm3**.

   The name of the TINKER input file in File 4-4 is **ethylene.xyz**, where the **.xyz** indicates that the geometry is given in Cartesian coordinates. (There are other coordinate systems for input files.) Comparison with the MM3 input file should enable you to describe the function of every integer and floating-point number in File 4-4.

6	Ethylene							
1	C	2.000000	3.000000	0.000000	2	2	3	4
2	C	3.000000	3.000000	0.000000	2	1	5	6
3	H	1.000000	4.000000	0.000000	5	1		
4	H	1.000000	2.000000	0.000000	5	1		
5	H	4.000000	4.000000	0.000000	5	2		
6	H	4.000000	2.090000	0.000000	5	2		
6	Ethylene							
1	C	1.831255	3.008674	0.000000	2	2	3	4
2	C	3.168740	3.021325	0.000000	2	1	5	6
3	H	1.256891	3.949547	0.000000	5	1		
4	H	1.274778	2.057116	0.000000	5	1		
5	H	3.725220	3.972888	0.000000	5	2		
6	H	3.743115	2.080450	0.000000	5	2		

Files 4-4a and b. The Initial (top) and Final (bottom) Geometries of Ethylene Calculated by TINKER Using the MM3 Force Field.

Version 3.9 of TINKER offers no fewer than 18 different force fields, some still under construction or in testing, and many oriented toward biochemical and medicinal applications, as appropriate for a system developed at a medical school. In running the programs, both MM3 and TINKER prompt the operator to provide information, including the force field desired, before the program run. Comparing the geometric output for MM3 and TINKER using the MM3 force field, we see that the specific values of the coordinates differ slightly (the entire molecule has moved during minimization) but the molecular geometry itself is the same for the two output files. The distance between carbon atoms is 1.338 ± 0.001 Å by both calculations.

Note the distinction between programs and force fields. It is possible to carry out a calculation using the MM3 force field and the MM3 program, or one can run the TINKER program with any one of its 18 resident force fields. It is also possible to modify a force field or to create one's own force field. This is a difficult advanced task and is not recommended. Good force fields are the result of decades of testing by competent scientific teams. Addition of a bad force field to the literature is a disservice to the science.

Within reasonable limits, different starting coordinates can be given for the initial geometry; optimization leads to the same *final* geometry. Changing the input bond distance of the C=C bond to 1.5 Å and the *y*-coordinates of two of the hydrogen atoms by 0.2 Å gives a final C=C bond length of 1.337 Å, in agreement with 1.338 Å obtained from the previous starting geometry. The slight difference of 0.001 Å = 1 mÅ is found because the geometry criterion for exit from the loop can never be zero but must be some small finite value. Thus two different calculations may arrive at slightly different locations very near the bottom of a potential energy well, both of which satisfy the geometric criterion for exit from the iterative loop.

COMPUTER PROJECT 4-1 | *The Geometry of Small Molecules*
We have just seen how to construct a TINKER input file for ethylene. We shall now construct several new models and study their geometries.

Procedure.   a) Using the procedure shown in constructing the ethylene input file, construct an approximate input file for H_2O. The atom type for oxygen is 6. The approximate input geometry can be taken as in File 4-5, where all of the z-coordinates are set at 0. Go to the tinker directory in the MS-DOS operating system and create an input file for your H_2O calculation. Be sure the extension of the input file is **.xyz**. Rename or edit the file as necessary.

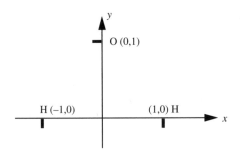

3	Water						
1	H	−1.	0.	0.	21	2	
2	O	0.	1.	0.	6	1	3
3	H	1.	0.	0.	21	2	

File 4-5. An Approximate Input File for the Water Molecule. A hydrogen attached to an oxygen has the special atom designator 21 as distinct from the designator 5 in hydrocarbons.

Run the file using program TINKER and force field MM3 to determine the H–O bond lengths and the H–O–H bond angle. The program runs on the command **minimize**. In responding to the prompt requesting the name of the input file, include the **.xyz** extension. Respond to the parameter request with **mm3**. Take the default gradient (hit Enter). The output file is stored under the name of your input file with a *tilde* 1, that is, the input file **h2o.xyz** produces the output file **h2o ~ 1.xyz**.

Search the literature for the experimental results for the H—O bond lengths and the H—O—H bond angle, and include a discussion of the comparison in your report.

Unlike program MM3, input format is not strict. The output file is formatted by TINKER, but the input file does not have to resemble it. After a successful run on H_2O, try cutting down on the number of spaces between elements in the input file until you have arrived at File 4-4. Do the more compact files run? Does File 4-4 run?

```
3  Water
1  H  −1.  0.  0.  21  2
2  O   0.  1.  0.   6  1  3
3  H   1.  0.  0.  21  2
```

File 4-6. TINKER Input File for the Water Molecule in Free Format

b) Construct a TINKER input file for ethane and determine its geometry. The numerical designator for an sp^3 carbon atom is 1, and the designator for hydrocarbon H is 5. You will have some nonzero z-coordinates.

c) Using a piece of graph paper, plot the approximate coordinates of ethylene from File 4-4b, the minimized or "optimized" structure of the model for ethylene in the MM3 force field. Replace one of the hydrogens in File 4-4b with a methyl group. Change the coordinates of File 4-4b by replacing atom 4 (hydrogen) with the four atoms of a methyl group at the approximate geometry that you have found from your graph paper sketch. Renumber the atoms 1 to 9 as necessary (File 4-7). You now have an approximate input geometry for propene. Rename your file **propene.xyz** and minimize to obtain the final geometry for the propene model. Many variations are possible but the following file runs.

```
9    Propene
1  C    1.831225    3.000000    0.000000     2   2   3   6
2  C    3.168775    3.000000    0.000000     2   1   4   5
3  H    1.265783    3.946259    0.000000     5   1
4  H    3.734217    3.946259    0.000000     5   2
5  H    3.734217    2.053741    0.000000     5   2
6  C    1.3    2.0    0.000000     1    1  7  8  9
7  H 0.2 2.0 0.0 5 6
8  H 0.5 1.5 1.0 5 6
9  H 1.2 1.2 0.0 5 6
```

File 4-7. TINKER Input File for Propene in Mixed Format. Mixed format can be used when one is modifying or editing an output file from a previous calculation.

Note that the strict format of the ethylene output file was not followed in adding new atoms. Be careful of your connected atom list to the right of the input file; it is a rich source of potential errors. Use your graph to keep the numbering straight.

Table 4-1   Bond Lengths for Carbon-Carbon Bonds in Alkanes and Alkenes

Bond	C—C	C=C	C—H
Bond length	—	—	—

d) Write out a table of bond lengths for the three bonds, C—C, C=C, and C—H that you have studied using the MM3 force field (Table 4-1).

The GUI Interface

Up to this point, we have used a numerical input file to stress the fact that computers work on numbers, not diagrams. MM3 and TINKER work from numerical input files that are similar but not identical. Both can be adapted to work under the command of a *graphical user interface*, GUI (pronounced "gooey"). Before going into more detail concerning MM, we shall solve a geometry optimization using the GUI of PCMODEL (Serena Software). The input is constructed by using a mouse to point and click on each atom of the connected atom list or skeleton of the molecule. This yields Fig. 4-6 (top).

After a rough estimate of the connected atom geometry has been entered, the hydrogen atoms can be added by exercising the **H/AD** option and the branch atom, by default a carbon atom at the top of Fig. 4-6, can be changed into an oxygen atom with the periodic table (**PT**) option. In this way, the skeleton at the top of Fig. 4-6 is converted to an approximate input structure at the bottom of the figure. The GUI translates the relative positions in the diagram into atomic coordinates to be input to the computer. The GUI also places the hydrogen atoms at default positions that, by the nature of the methyl group, for example, have nonzero z-coordinates (Fig. 4-7). As pictorial as all of this is, don't forget that the computer processes only numbers, and binary numbers at that.

The parameter sets for PCMODEL V 8.0 are MMX, a derivative of the MM2 parameter set of Allinger's group, MM3, MMFF94, AMBER, and Oplsaa. We shall continue using MM3. After minimization, the model, in this case that of propan-2-ol, has bond distances and angles at the equilibrium values as determined by the force field. Once one knows the Cartesian coordinates of all the atoms in the model, one knows everything that can possibly be known about its geometry including all

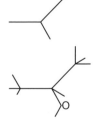

Figure 4-6   Connected Atom Skeleton for Propan-2-ol (Top).
Input Diagram with Hydrogen Atoms Added and the Oxygen Atom
Indicated (Bottom).

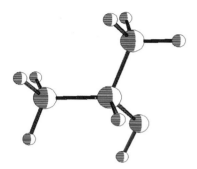

Figure 4-7 Output of the Optimized Geometry of Propan-2-ol Depicted as a Pluto Model. Pluto is one of several pictorial options provided in PCMODEL.

bond lengths, simple angles, dihedral angles, and nonbonded interatomic distances. Each of these features is automatically calculated by PCMODEL using its QUERY option. During minimization, the image of the molecule on the CRT screen may change a lot or a little depending on how much the input geometry is changed to obtain the final geometry. After minimization, two energies (enthalpies) are displayed prominently at the right of the screen, the enthalpy of formation and the strain energy. They are related to the steric energy and are discussed below.

Parameterization

There is at present no methodical way of obtaining precise values of parameters necessary to construct a useful MM force field. Bond lengths and bond angles can be determined from rotational spectroscopy, X-ray diffraction, neutron diffraction, or electron diffraction experiments, but, unfortunately, experimental values from these sources are not strictly comparable (Burkert and Allinger, 1982). This is not because of any fundamental flaw in either the methods or the experiments but because the methods measure slightly different things. For example, X rays are diffracted by the electrons surrounding a nucleus and electrons are diffracted by the nucleus itself. If the electron probability density is centered at the nucleus these results are essentially the same, but if the electron probability density is distorted away from the nucleus toward the bond the results will be different. An example of distortion away from a nucleus toward a bond is the distortion of electrons away from the proton in a C—H bond.

In general, we know bond lengths to within an uncertainty of 0.005 Å = 0.5 pm. Bond angles are reliably known only to one or two degrees, and there are many instances of more serious angle errors. In addition to experimental uncertainties and inaccuracies due to the model (lack of coincidence between model and molecule), some models present special problems unique to their geometry. For example, some force fields calculate the ammonia molecule, NH_3, to be planar when there is abundant experimental evidence that NH_3 is a trigonal pyramid.

H⁣ᐧᐧᐧN⁣ᐧᐧᐧH

H

Force constants can be calculated if a spectral line can be associated with a specific mode of motion. For example, if we take the C—H stretching frequency to be at 2900 cm^{-1} in the infra red spectrum of hydrocarbons, we have $k = 457$ N m^{-1} for the force constant of the C—H bond. The actual value of k taken from the MM3 force field differs somewhat from this value, being 4.74 mdyn/Å = 474 N m^{-1}. The frequency enters into this calculation as the square, making k very sensitive to v. Infrared C—H stretching frequencies are not identical from one molecule to the next, giving a range of values of v from which to calculate k. A choice of 2960 cm^{-1} (Barrow, 1999) leads to $k = 480$ N m^{-1}. General-purpose force fields like MM3 are parameterized to reproduce not only spectral vibrational frequencies but molecular geometry and energy as well. Selection of the stretching parameters is guided by peaks in the vibrational spectrum, but the final choice is intended to give the best results for all calculated values. In the absence of a methodically precise method of parameter generation, some degree of trial and error enters into the process.

Parameterization of organometallics and metal-ligand compounds is much more difficult than parameterization of organic compounds for several reasons (Jensen, 1999). Principal among these reasons are those related to the weakness of the coordinate covalent bonds relative to the covalent bond. Metal-ligand bonds are less rigid and the energy barriers between different structures are lower than those in organic compounds. Thus a metal might form a tetrahedral complex with one ligand and a square planar complex with another similar ligand. Relativistic effects are also more prominent in metal bonding than in the bonding of first-row elements.

The Energy Equation

The MM energy equation is a sum of sums that began in a simple form, for example,

$$V = \sum V_{\text{stretch}} + \sum V_{\text{bend}} + \sum V_{\text{torsion}} + \sum V_{\text{VDW}} \qquad (4\text{-}13)$$

for stretching, bending, torsional and van der Waals energies, respectively (Allinger, 1982), and has been made more elaborate as new terms were added for greater generality and accuracy. In writing an energy equation there are two ways to go. One can include only a few potential energy sums, for computational speed, or one can include many sums, for accuracy. The first option is taken when one seeks a force field that is applicable to large molecules as in enzyme and protein studies, and the second option is taken when one seeks accurate information on relatively small organic or inorganic molecules. Force fields applicable to polypeptides and proteins include AMBER of Kollman's group (Kollman, 1995) and CHARMM (Karplus, 1986). Both AMBER-95 and CHARMM22 are available in the TINKER force field collection. The most widely used of the force fields parameterized for accuracy on relatively small molecules is MM3.

It may seem strange that the number of sums in the energy equation is, within limits, arbitrary. If a mode of motion or intramolecular interaction exists in a given molecule, doesn't that demand a term in the energy equation? Not necessarily. Each energy equation has its own force-field parameter set. Obviously, increasing the number of sums in the energy equation by one requires a whole new set of parameters for that sum, but it also changes all the other parameters as well. Conversely, when we go from a more detailed parameter set like Allinger's MM3 or MM4 to a simpler set, we are "lumping together" two or more modes of motion or intramolecular interactions as one. Molecular mechanics being a purely empirical method, one can treat two modes of motion as one, hoping that the simpler parameter set will generate a potential energy close to what would be found by using a more complete energy equation with both modes included. Generally, this hope is not completely fulfilled and some accuracy is sacrificed in schemes that use simplified energy equations.

Another indication of the arbitrary nature of the MM force field is the united atom approach taken in parameterizing a number of force fields intended for large molecules, especially proteins. In this approximation, groups of atoms, say the $-CH_3$ group, are treated as single atom types and given parameters for the group. Speed is gained, but some accuracy is lost.

Sums in the Energy Equation: Modes of Motion

The four steric energy sums in Eq. (4-13) corresponding to stretching, bending, and torsional modes of motion and van der Waals intramolecular interaction appear to be about the smallest number one can use in an accurate MM geometry minimization.

The Stretching Mode

Taylor's expansion of a function at $x = a$ is usually written

$$f(x) = f(a) + f'(a)(x - a) + \frac{f''(a)}{2}(x - a)^2 + \cdots + \frac{f^n(a)}{n!}(x - a)^n + \cdots$$

$$(4\text{-}14)$$

If we carry out a Taylor's expansion of the potential energy about the equilibrium length of an isolated chemical bond, we get

$$V(r) = V_0 + \left(\frac{dV}{dr}\right)_{r_0}(r - r_0) + \frac{1}{2}\left(\frac{d^2V}{dr^2}\right)_{r_0}(r - r_0)^2 + \frac{1}{6}\left(\frac{d^3V}{dr^3}\right)_{r_0}(r - r_0)^3$$

$$+ \frac{1}{24}\left(\frac{d^4V}{dr^4}\right)_{r_0}(r - r_0)^4 + \cdots$$

$$(4\text{-}15)$$

The potential energy is never known in an absolute sense but is always measured relative to some arbitrary benchmark. Let us set the potential energy to zero at the equilibrium bond length, $V_0 = 0$, which is the bottom of the potential energy well.

Also, the first derivative at the minimum (or other extremum) of a function is zero, $(dV/dr)_{r_0} = 0$. The first nonzero term in Eq. (4-15) contains $\frac{1}{2}(d^2V/dr^2)_{r_0}$ which is precisely the force constant that multiplies $(r - r_0)^2$ to give what we called the harmonic oscillator equation

$$V_{\text{stretch}}(r) = \frac{k_{\text{stretch}}}{2}(r - r_0)^2 \tag{4-16}$$

in the section on the two-mass problem. The harmonic oscillator term is often called the *quadratic* term. It leads to the parabola in Fig. 4-3. The remaining terms in Eq. (4-15) involving the third and fourth derivatives are called the *cubic* and *quartic* terms.

Both the cubic and quartic terms widen the parabola of Fig. 4-3 at large values of r (to the right of the figure), giving a potential energy curve that is closer to what we suppose the real potential energy curve to be for extended bonds. Both of these terms have disadvantages aside from making the calculations longer and making the parameter set larger. The cubic term goes to $-\infty$ as the bond length becomes very large, and the quartic term goes to ∞. We know that the energy necessary to separate two bound atoms

$$A - B \rightarrow A + B$$

to a very large distance is neither $-\infty$ nor ∞. Rather, it is a finite value called the dissociation energy; therefore, neither the cubic nor the quartic term predicts the correct physical behavior at limiting values of r. Other methods are used to improve the curve fit at large values of r, but each carries with it some disadvantage (Jensen, 1999).

Fortunately for most ground-state calculations, and especially for equilibrium geometry calculations, the harmonic oscillator approximation holds and *anharmonic* extensions of Eq. (4-16) are not needed for bond stretching. The reason is that, under normal circumstances, the cost in energy of bond stretching is high and bonds stay within about ± 0.1 Å of their equilibrium values. Exceptions are highly strained molecules, excited-state calculations (for which MM may not be the method of choice anyway), and simulations.

The Bending Mode

Bond angle bending

is handled in much the same way as stretching. The potential energy of bending for an isolated bond angle can be expanded as a Taylor series about its equilibrium value, which is taken as zero, $V_{bend}(0) = 0$. Dropping the second term and terms higher than the third term leaves only the quadratic term expressing the excess potential energy over $V_{bend}(0)$

$$V_{bend}(\theta) = \frac{k_{bend}}{2}(\theta - \theta_0)^2 \qquad (4\text{-}17)$$

which is analogous to the harmonic oscillator expression for the stretching and compression energy, Eq. (4-16). The quadratic approximation for bending is, if anything, a better approximation to the actual energy curve than it is for stretching energy, and it is reliable for most angles within a range of about $\pm 30°$ of the equilibrium value.

When the three-atom bond angle is part of a molecular structure, its equilibrium angle is usually not at the equilibrium angle for an isolated bond because of the geometric compromise brought about by the rest of the molecule. This results in a nonzero contribution $V_{bend}(\theta) \neq 0$ to the steric energy for the equilibrium structure of the molecule as a whole. Like bond stretching, bond bending is an energetically costly deformation of the isolated bond angle; hence, at the equilibrium structure for the molecule as a whole, the bond is usually not deformed by more than a few degrees, well within the $\pm 30°$ limits of the quadratic function (4-17).

In those instances in which the bond angle is very strongly deformed, the usual expansion of the potential energy function to a cubic or quartic equation can improve the fit to the actual bending energy, but, in MM3, it has been found expedient to define entirely new atom types with entirely new parameter sets. For example, the C—C—C angles in cyclopropane (60°) or cyclobutene (90°) are distorted beyond the limit of the quadratic expressions for bending of sp^3 carbon atoms. Bending in cyclopropane is outside the range of even the cubic and quartic equations. Carbon atoms in cyclopropane and cyclobutene are given special atom type numbers (22 and 57, respectively, in the MM3 force field) and treated as though they have no relation to the sp^3 carbon atom, which, energetically, they do not. For this reason, there are more than a dozen atom type numbers for the single element carbon in MM3. Carbon is unique in having so many atom type numbers because carbon is unique in the number of chemical combinations it enters into. In general, force fields are constructed with an eye to achieving the greatest simplicity consistent with chemical reality and new atom types are admitted only grudgingly.

COMPUTER PROJECT 4-2 | *The MM3 Parameter Set*

One can start building up a list of MM3 parameters by use of the TINKER **analyze** command. Don't expect to build up the entire set, which occupies about 100 pages in the MM3 user's manual, but do obtain a few representative examples to get an idea of how a parameter set is constructed. From previous exercises and projects, you should have input and output geometries for an alkene, an alkane, and water. From these, the object is to determine the stretching and bending parameters for the C—C, C=C, C—H, and O—H bonds. The C—H bond parameters are not the same

Table 4-2   Some Stretching and Bending Parameters from the MM3 Parameter Set

	KS	KB	Length	Angle
C—C	4.4900	—	1.5247	—
C=C				
C=C—H				
C—C—H				
O—H	7.6300	0.6300		

Units are md/angstrom, angstrom/rad^2, angstrom, and degrees respectively. One millidyne (md) = 10^8 newtons (N).

for sp^2 and sp^3 carbon atoms. In MM3 the stretching parameter (Hooke's law constant k) is given the symbol KS and the bending constant is given the symbol KB. Fill out parameter Table 4-2.

Procedure.   One of the advantages of the TINKER system is that it is constructed with the same kind of input format for different modules of the program. This permits us to use either the input or the output file from one module, say the minimize module, as input to another module, in this case the analyze module. Start TINKER using the **analyze** command and respond to the prompt asking which program you want to analyze with **water.xyz** or **water ~ 1.xyz**, or whatever you have named your input or output file for determining the geometry of the H_2O molecule. Respond to the prompt for a parameter set with MM3. You will then be given several choices for your next step. Experiment with them. Some will give you information that you understand, and some will give you information that you may not understand. Each time you select a new response to this prompt, you will have to start the program again. Finally, settle on response **P** asking for the parameters. Specify atom numbers 1 and 2. You should get an output enabling you to fill in KS = 7.6300, KB = 0.6300 for the O—H bond stretch and bend constants with 0.9470; 105.0 for the equilibrium bond length (angstroms) and equilibrium angle (degrees). Continue in this way to fill out the entire table. For the carbon atom parameters, you may want to simplify your output by specifying only one atom, say atom 1. You will get parameters for all motions involving that atom.

As given, the bond angle is ambiguous because you cannot have an angle between only two atoms. Therefore, there must be a third atom involved in the last entry. In the case of water, it can only be H—O—H. Make it clear in your report what the third atom is, that is, give a bond angle for C—C—H and one for C=C—H, which will not be the same.

Discuss the quantitative differences among entries in your table. Why are some entries larger than other comparable entries? What are the torsional bending constants for C=C and C—C? Why do they assume the values you find? What is a "torsional angle," anyway?

Torsion.   Torsional deformation of an isolated equilibrium structure means twisting it so as to change the dihedral angle connecting two atoms. The dihedral angle

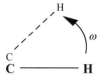

Figure 4-8 The H–C=C–H Torsional Bond Angle in Deformed Ethylene. The normal torsional angle is $\omega = 0°$.

of the structural unit H–C=C–H in ethylene, for example, is the angle made by the C–H bonds as seen looking down the C=C bond axis. In the equilibrium structure of ethylene the H–C=C–H dihedral angle is zero, but by applying a torsional force it can be changed to some angle ω (Fig. 4-8).

Torsional deformation of dihedral angles in molecules is different from bond stretching or bond bending in that it is periodic. After deformation through an angle ω of 2π, the original molecular structure is reproduced. For reasons of symmetry, the structure may be reproduced by deformations of less than 2π. For example, the structure of ethylene is reproduced at $\omega = \pi$ and that of ethane is reproduced at $\omega = \pi$ and $5\pi/3$. Ground-state ethane is in the staggered conformation, $\omega = \pi/3$. On rotation of one methyl group relative to the other, the molecule passes over three potential energy maxima to arrive at, sequentially, a staggered minimum at π and a staggered minimum at $5\pi/3$ before returning to its original conformation at $\omega = \pi/3$.

$V(\omega)$ is continuous and has continuous first derivatives over the interval $[0, 2\pi]$, which is the complete interval of ω. It is convenient to rename the interval $[-\pi, \pi]$ (which is the same as $[0, 2\pi]$) for the following discussion. Any continuous function can be represented over this interval by the *Fourier series*

$$f(x) = \tfrac{1}{2}a_0 + \sum_{n=1}^{\infty}(a_n \cos nx + b_n \sin nx) \tag{4-18}$$

where

$$a_n = \tfrac{1}{\pi}\int_{-\pi}^{\pi} f(x)\cos nx\, dx \tag{4-19a}$$

and

$$b_n = \tfrac{1}{\pi}\int_{-\pi}^{\pi} f(x)\sin nx\, dx \tag{4-19b}$$

$V(\omega)$ is an even function of ω over the interval $[-\pi, \pi]$. That is, if we calculate $V(\omega)$ at any ω and calculate $V(\omega)$ at $-\omega$, we get the same answer; the function is symmetrical about the central axis of the interval. Conversely, if we get $-V(\omega)$, the function is *odd* over the interval. The sine function is odd over the interval $[-\pi, \pi]$;

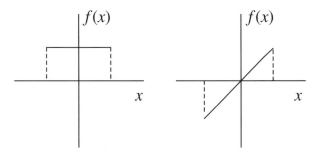

Figure 4-9 A Simple Even Function, $f(x) = $ const. and a Simple Odd Function, $f(x) = x$.

hence, $f(x) \sin nx$ for even $f(x)$ is the product of an even function times an odd function, which is an odd function (Fig. 4-9).

The integral of an odd function over a symmetrical interval is zero because every element on the left half of the interval is canceled by an equal and opposite element on the right. From this we know that all the constants $b_n = 0$ in Eq. (4-19b) and the *half-Fourier series*

$$f(x) = \tfrac{1}{2}a_0 + \sum_{n=1}^{\infty} a_n \cos nx \qquad (4\text{-}20)$$

expresses any symmetrical even function over the interval $[-\pi, \pi]$.

The torsional potential energy functions are symmetrical and even over the interval and can be written

$$V_{\text{tors}}(\omega) = \sum \frac{V_n}{n} \cos n\omega \qquad (4\text{-}21)$$

if we take $a_0 = 0$, which is equivalent to setting $V_{\text{tors}}(\omega) = 0$ as the base line of $V_{\text{tors}}(\omega)$. Within the series, one term dominates, for example, the $n = 2$ term for ethylene. Ethylene strongly resists torsional deformation to any angle other than 0 and π (180°) (Fig. 4-10).

The barriers in Fig. 4-10 are high because it is difficult to twist ethylene out of its normal planar *conformation*. The energy is the same at the midpoint and the end points in Fig. 4-10 because, on twisting an ethylene molecule 180° out of its normal conformation, one obtains a molecule that is indistinguishable from the original. The molecule has *2-fold torsional symmetry*.

Figure 4-10 The Potential Energy Form for Ethylene. The midpoint of the range of ω is 0° and the end points $\pm 180°$, that is, $[-\pi, \pi]$. The mid point and end points are identical by molecular symmetry.

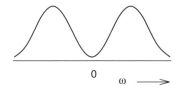

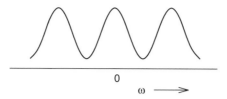

0

$\omega \longrightarrow$

Figure 4-11 The Potential Energy Form for Ethane. The midpoint of the range of ω is $\omega = 0°$ and the end points are $\pm 180°$. The end points and the minima are identical by molecular symmetry and correspond to the stable staggered form.

Ethane presents a different situation. It is easily twisted out of its low-energy stable form and it has 3-fold torsional symmetry; that is, by twisting one methyl group relative to the other by 120° or 240° (or −120°), one obtains conformers that are indistinguishable from the original. The $n = 3$ term of the Fourier series models ethane well. It has three potential energy minima for the staggered conformers separated by three potential energy barriers or maxima for the eclipsed conformers. The difference in height of the potential energy barriers between ethylene and ethane is modeled by selecting different values for the torsional constants with $V_3 < V_2$ (Fig. 4-11).

In the case of ethylene, because of 2-fold symmetry, odd terms drop out of the series, $V_3, V_5, \ldots = 0$. In the case of ethane, because of 3-fold symmetry, even terms drop out, $V_2, V_4, \ldots = 0$. Terms higher than three, even though permitted by symmetry, are usually quite small and force fields can often be limited to three torsional terms. Like cubic and quartic terms modifying the basic quadratic approximation for stretching and bending, terms in the Fourier expansion of $V_{\text{tors}}(\omega)$ beyond $n = 3$ have limited use in special cases, for example, in problems involving octahedrally bound complexes. In most cases we are left with the simple expression

$$V_{\text{tors}}(\omega) = \frac{V_1}{2}[1 + \cos \omega] + \frac{V_2}{2}[1 - \cos 2\omega] + \frac{V_3}{2}[1 + \cos 3\omega] \tag{4-22}$$

Although one of the three terms retained in the Fourier expansion is often sufficient for the potential energy function of molecules with symmetrical rotations about the central bond, systems of lower symmetry can be represented as well. Addition of an $n = 1$ term to the $n = 3$ term for ethane permits description of molecules that do not have 3-fold symmetry about the central bond. An example is *n*-butane. *n*-Butane has a low energy *anti* form separated by rotational barriers from two *gauche* forms that are somewhat higher in energy than the *anti* conformer but not as high as the barriers. This energetic situation is usually represented by a diagram similar to Fig. 4-12, where the *gauche* minima are at either side of the central maximum and the most stable conformer, the *anti* form, is at the arbitrary potential energy of zero, the end points $[-\pi, \pi]$ measured from $\omega = 0$ at the center of the diagram.

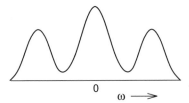

0
$\omega \longrightarrow$

Figure 4-12   The Potential Energy Form for *n*-Butane. The energy is expressed relative to the energy of the *anti* form which is at the end-points of the range of ω.

Addition of a $V_{n=1}$ term (Fig. 4-13) to the V_1, V_2, and V_3 term (Fig. 4-11) has the effect of raising the central maximum and the two symmetrical minima without changing the energy of the stable *anti* form, so producing a potential energy function that shows the qualitative form of Fig. 4-12. In this context, the $V_{n=1}$ term is called a *low-order torsional term.*

Figure 4-13   The Potential Energy Form Given by the $V_{n=1}$ Term in the Truncated Fourier Series (Eq. 4-21).

Experimental measurements have given the height of the energy minima of the *gauche* forms as 0.9 kcal mol^{-1}, the height of the central maximum as 4.5 kcal mol^{-1} above an arbitrary zero point established by the *anti* form, and the height of the two potential barriers separating the *gauche* from the *anti* form is 3.8 kcal mol^{-1} (Ege, 1998). By fitting these and other experimental results, empirical values of V_1, V_2, V_3, and the $V_{n=1}$ term can be determined.

The van der Waals Energy.   In the study of nonideal gases, we encounter attractive and repulsive forces between uncharged atoms and molecules among which no formal bonds exist. These forces were studied extensively by van der Waals in the nineteenth century. Even in particles that possess neither an ionic charge nor a permanent partial charge (permanent dipole), van der Waals forces exist. Indeed, van der Waals attractive forces must exist even in inert gases like helium, otherwise they could not be liquefied. Repulsive van der Waals energies are much larger than attractive forces over short distances.

In molecular mechanics, van der Waals forces are thought to influence atoms that are within the same molecule but are not connected by chemical bonds. They are sometimes called 1-4 interactions, implying that a chemical bond does not exist between atoms 1 and 4 but one does exist between atoms 2 and 3

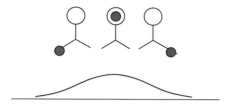

Figure 4-14 Excess Steric Energy of the Methyl-Methyl Conformation Relative to the Methyl-Hydrogen Conformation. The Staggered Forms are Ignored for Simplicity.

The origin of van der Waals repulsive forces is mutual interaction of electrons in atom 1 and those in atom 4.

If the methyls in *n*-butane are eclipsed, the hydrogens are also eclipsed. If the van der Waals repulsion energy in the *n*-butane conformation with methyl groups eclipsed is of a higher energy than the conformation with a methyl group eclipsing a hydrogen atom, we get a potential energy curve very like Fig. 4-13 by the mechanism shown in Fig. 4-14. Addition of this potential energy to the potential energy of the staggered and eclipsed forms of an ethane-like molecule as in Fig. 4-11 also gives the right qualitative form for the potential energy of *n*-butane shown in Fig. 4-12. Empirical selection of the correct torsional parameters V_1, V_2, V_3 and the $V_{n=1}$ term gives quantitative as well as qualitative agreement between the composite potential energy curve for butane (Fig. 4-12) and the actual butane molecule.

There are several serious problems in modeling the correct van der Waals potential function. First, it is by no means obvious that the methyl group exerts a greater van der Waals repulsion on another methyl group than it does on hydrogen. Neither methyl-methyl nor hydrogen-hydrogen repulsions can be studied in *n*-butane in the absence of the other. Despite what one might suppose from the relative sizes of methyl groups and hydrogen atoms, the van der Waals repulsion of a hydrogen atom is not negligible. Precisely because of its small size and consequent high charge density, hydrogen has a rather large effective atomic radius. Furthermore, it is not obvious whether the potential energy increase of the *gauche* forms of *n*-butane over the *anti* form is best modeled by inclusion of lower-order Fourier series terms or by an entirely independent van der Waals energy sum. Most recently, the latter solution has been preferred, but the form of the van der Waals potential energy function is difficult to extract from the empirical data available for the reasons just given.

Two potential energy expressions used for van der Waals interactions are the Lennard–Jones 6/12 potential function or some modification thereof,

$$V_{vdW} = \varepsilon \left\{ \left(\frac{r_0}{r}\right)^{12} - 2\left(\frac{r_0}{r}\right)^{6} \right\} \qquad (4\text{-}23)$$

where ε is an adjustable parameter governing the depth of the slight minimum in the potential energy curve and the Buckingham potential

$$V_{vdW} = Ae^{-Br} - \frac{C}{r^6} \qquad (4\text{-}24)$$

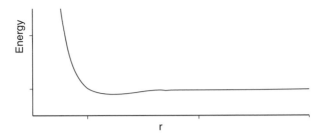

Figure 4-15 A van der Waals Potential Energy Function. The Energy minimum is shallow and the interatomic repulsion energy is steep near the van der Waals radius.

where A, B, and C are adjustable parameters. These equations are often used in their modified forms, for example by replacing r_0 in Eq. (4-23) by fitting parameters, so as to achieve more flexibility in fitting experimental data (Fig. 4-15).

Coulombic Terms. Coulombic energy of interaction arises from permanent dipoles within the molecule to be modeled, for example, the partial $+$ and $-$ charges within a carbonyl group

$$\delta+ \quad \delta-$$
$$C=O$$

Coulombic potential energy is calculated by modification and fitting of some form of Coulomb's equation

$$V = \frac{q_i q_j}{4\pi\varepsilon_0 \, r_{ij}} \tag{4-25}$$

where $4\pi\varepsilon_0$ is the vacuum permittivity (Atkins, 1994). In MM3, the energy of Coulombic interaction of dipoles is calculated by

$$V_\mu = \frac{\mu\mu'(\cos\chi - 3\cos\alpha\cos\beta)}{D\,r^3} \tag{4-26}$$

where μ and μ' are the dipole moments of interacting dipoles and D, the dipole moment, is given a default value of 1.5 (seen in the output file) (Fig. 4-16).

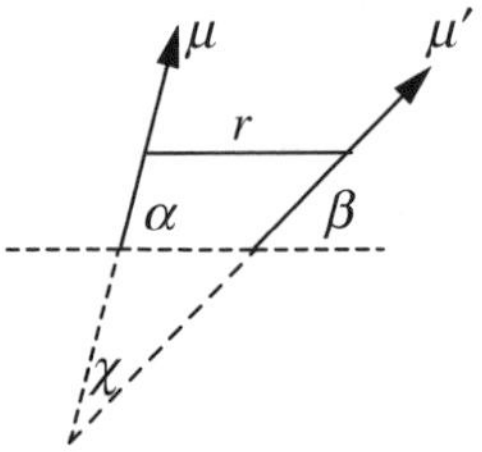

Figure 4-16 Orientation of Dipole Moment Vectors to give V_μ.

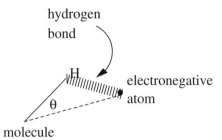

hydrogen
bond

H

θ

electronegative
atom

molecule

Figure 4-17 Hydrogen Bonding.

The angles α, β, and χ relate to the orientation of the dipole moment vectors. The geometry of interaction between two bonds is given in Fig. 4-16, where r is the distance between the centers of the bonds. It is noteworthy that only the bond moments need be read in for the calculation because all geometric features (angles, etc.) can be calculated from the atomic coordinates. A default value of 1.0 for dielectric constant of the medium would normally be expected for calculating structures of isolated molecules in a vacuum, but the actual default value has been increased 1.5 to account for some intramolecular dipole moment interaction. A dielectric constant other than the default value can be entered for calculations in which the presence of solvent molecules is assumed, but it is not a simple matter to know what the effective dipole moment of the solvent molecules actually is *in the immediate vicinity of the solute molecule.* It is probably wrong to assume that the effective dipole moment is the same as it is in the bulk pure solvent. The *molecular* dipole moment (File 4-3) is the vector sum of the individual dipole moments within the molecule.

The energy of intramolecular hydrogen bonding, if present, is calculated from a modified Buckingham potential [Eq. (4-25)] multiplied by a factor $\cos\theta\,(L/L_0)$ to take into account the angle θ between the actual hydrogen bond in the molecule and the ideal hydrogen bond angle ($0°$), along with the equilibrium length L of the covalent bond of hydrogen as compared to its normal bond length L_0 (Fig. 4-17).

In the case of ions, energies of charge-charge interaction are calculated directly from the Coulomb equation and charge-dipole interaction energies are calculated from a Coulomb equation with a geometric modification to account for the angle between the dipole moment and the charge (Tripos, 1992).

COMPUTER PROJECT 4-3 | *The Butane Conformational Mix*
We know from the discussion above that *n*-butane has an *anti* and two *gauche* forms (mirror images). These are called *stable conformers.* Molecules having a dihedral angle ω (see Fig. 4-18) that is not at a minimum of energy are unstable *configurations.* Although all three conformers, *anti* and two *gauche* forms, are stable in the energetic sense that they are at the minima of potential energy wells, there is enough thermal energy $k_B T$ that can be absorbed from the environment at any normal temperature $T \neq 0$ to drive any conformer up over the potential maximum in Fig. 4-18 and change (twist) it into any other conformer. Interconversion is

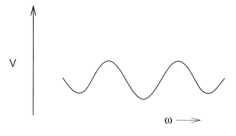

Figure 4-18 The Potential Energy for Rotation of *n*-Butane About its Central Bond Axis. The *anti* conformer in the center is slightly lower in energy than the two gauche conformers.

very rapid, and no conformer can be obtained in the absence of the others. The system exists as an equilibrium *conformational mix*.

We have it on good authority (Ege, 1998) that the *gauche* minimum on the potential energy coordinate is about 0.9 kcal mol^{-1} higher in energy than the *anti* conformation. This establishes a two-state energy system for the stable conformers, *gauche* and *anti* (Fig. 4-19).

We know from the section on molecular speads in Chapter 1 and Computer Project 3-2 that particles distribute themselves over an energy level spectrum in a very definite way, governed by the Boltzmann equation, Eq. (3-39), $N = N_0 e^{-(mgh/k_B T)}$ in the gravitational field. Equation (3-39) can be slightly modified by noting that *mgh* is the *potential energy* of an object of mass *m* in the potential field. If we use V to designate the potential energy not of gravity, but of twisting about the dihedral angle in *n*-butane, and if we use N to denote the number of *n*-butane molecules in the *gauche* condition relative to N_0 molecules as the *anti* conformer,

$$\frac{N}{N_0} = 2e^{-(V/k_B T)} \tag{4-27}$$

where we have arbitrarily set the potential energy of the *anti* conformer to 0 and we have included the number 2 to account for the two-fold degeneracy of the upper (*gauche*) energy level. (Each of the degenerate *gauche* levels can be considered to accommodate its own population of molecules, independent of the other.) Strictly speaking, the ratio N/N_0 is a ratio of probabilities, but we shall take 1 mole as the number of molecules so a calculated ratio of probabilities is a virtual certainty. There is, of course, no restriction that the number of conformers must be two; the method can be extended to many conformers as we would expect to do for more complicated molecules. The object of this computer project is to find precise values of *V*, the separation between or among conformers, and to use *V* to calculate the relative populations of the conformers in the equilibrium conformational mix.

Figure 4-19 Two-Level Energy Spectrum. The upper level is two-fold degenerate.

Procedure. Use the PCMODEL GUI to **draw** the *anti* carbon skeleton

and left click **H/AD** to attach hydrogen atoms. The approximate input model appears. Left click on **force field**, select MM3 and left click on **compute** → **minimize**. The final (optimized) model appears with the MM3 energy (the steric energy).

Do the same thing with the unstable configuration resulting from the carbon skeleton

It will relax to one of the *gauche* conformers. Use the difference in steric energies to calculate the equilibrium populations of the ground state and the two upper *gauche* conformations at 298.15 K. Calculate and report the relative populations at 200 and 500 K. Repeat the entire project using the MMFF94 force field of Halgren and Nachbar (Halgren and Nachbar, 1996) to be found using the force field option.

A Conformational Search-Global MM

Both PCMODEL (Serena Software) and MM3 (Tripos) have search routines that permit the user to enter one conformer of a molecule and find all others. (Strictly, no conformational search routine guarantees that *all* conformers have been found, but in simple cases the chances are pretty good.) There are several strategies for conformational searching, but a good one is to optimize the geometry for a plausible input structure, randomly change the input geometry, optimize again, and keep on doing this until one has reasonable assurance that all conformers have been found. Random changes in the input geometry are often called "kicks." This simplistic strategy is what anyone might think of, except that the computer implementation is a lot faster than you are. In taking many starting points, each with a different geometry and potential energy, one is said to have searched the potential surface for a specific mode of motion of the molecule. More than one mode of motion can be searched at the same time.

To conclude this computer project, we shall first search the potential surface for rotation of *n*-butane about its 2,3 C—C bond, for which we think we know the answer, then search the potential surface for 1-butene, for which we do not. In 1-butene, the double bond establishes a rigid plane but the methyl group can take up several different positions relative to it by rotation about the 2-3 single bond.

$$\overset{}{=}\!\!\diagup^{CH_2CH_3} \qquad \overset{CH_3}{=}\!\!\diagup^{CH_2} \quad \text{etc.}$$

Procedure. Open PCMODEL and work through this procedure step by step. Otherwise, these instructions will be cryptic.

***n*-Butane:** Minimize *anti* *n*-butane as in the MM3 procedure above. Go to **Compute** → **GMMX**. Search on **bonds**. Setup bonds, **select all** and hit **OK**. Enter **Job Name** 1-butene and **Run Gmmx**. You will see the model being "kicked" repeatedly. Left click outside of the **GMMX Run** box. You should see 3 # minimized and 3 # found. We already know that there are only three conformers, two of which are degenerate; hence, because $E_{current} \neq E_{minimum}$ we know all the energies there are. Subtracting the lower from the higher energy permits calculation of the population ratios of the conformational mix. If your energy difference is not the same as the one you got in the first part of this project, check to see that you are using the MM3 force field.

1-Butene: Now we shall look for the energies of the conformational mix of 1-butene. We know less about this molecule than we did about *n*-butane, so the procedure will be a little more complicated.

Minimize the *cis* rotamer in the MM3 force field starting from the skeleton

using **Add_B** to make your double bond. Go to **compute** → **GMMX**. Enter **Job Name** 1-butene and setup bonds, **select all** (you are searching only rotations about the 2,3 bond) and hit **OK**. **Run Gmmx**. Left click outside of the **GMMX Run** box. You should see 3 # minimized and 3 # found, but this time we do not know how many conformers there are. **Stop Job** and go to **File** → **open** 1-butene. You should see a structure list. Only three conformers are given, two of which are, again, degenerate. In fact, there are four conformers, but two have energies so high that they do not contribute significantly to the conformational mix. (Fig. 4-20). Calculate the population ratios at 200, 298.15, and 500 K in the MM3 and MMFF94 force fields for 1-butene just as you did for *n*-butane.

Figure 4-20   Rotamers (rotational conformers) of 1-Butene (Nevins, Chen, and Allinger, 1996).

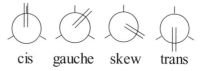

cis   gauche   skew   trans

Cross Terms

A simple example of a cross term is the stretch-bend interaction. If the angle ABC, having elastic AB and BC bonds, is closed, the bonds stretch because of repulsion between atoms A and C. The opposite is true if the angle is opened. Thus the stretch and bend of the system ABC are not independent; rather, they are coupled. The stretch-bend coupling term might take the bilinear form

$$V_{stretch-bend} = k_{SB}(r - r_0)(\theta - \theta_0) \tag{4-28}$$

where r is the bond length and θ is the simple angle. There are many possible cross terms

$$
\begin{aligned}
V_{\text{stretch}-\text{stretch}} &= k_{SS}(r - r_0)(r' - r'_0) \\
V_{\text{bend}-\text{bend}} &= k_{BB}(\theta - \theta_0)(\theta' - \theta'_0) \\
V_{\text{stretch}-\text{torsion}} &= k_{ST}(r - r_0)\cos n\omega
\end{aligned}
\tag{4-29}
$$

and others. It soon becomes evident that there are many terms that can be included in the energy equation, Eq. (4-13), and the question is how many to accept and how many to reject. Fewer terms are used in the energy equation for large molecules like proteins that tax computer resources. Many terms are used when one wants an accurate and detailed model of relatively small molecules or to calculate spectra, which demand a knowledge of not only the location of a potential well but also its depth and shape as well.

PROBLEMS

1. A 1.00-g mass connected to a fixed point by a spring oscillates at a frequency of 10.0 Hz. What is the Hooke's law force constant of the spring? Give units.
2. A slender vertical filament of negligible mass supports a 0.200-g mass at one end and is fixed at the other end. A force of 0.0800 N displaces the mass 0.0200 m. The mass executes simple harmonic motion as the filament bends. What is the bending constant KB of the filament? What is the frequency ν of the motion in Hz? What is the period t of oscillation?
3. The balance wheel of a chronometer is constructed so that its entire mass of 0.100 g may be considered to be concentrated in a ring of radius 0.600 cm. What is its moment of inertia?

4. The balance wheel in Problem 3 is driven by a coiled spring called a hairspring. The wheel executes simple harmonic angular motion between the two angular limits shown by the double arrow in Problem 3. Its oscillation over the marked excursion is complete every 0.500 s. What is the torsion constant κ of the spring?
5. Write the rotational analog of Hooke's law for the torque τ driving the oscillation in Problem 3. Write the rotational analog of Newton's second law. Combine the two laws to obtain the rotational analog of the Newton–Hooke equation, Eq. (4-1).
6. Calculate the reduced mass of $D^{35}Cl$ where D is the deuterium isotope of hydrogen (isotopic weight 2.014 atomic mass units) and ^{35}Cl is the 35 isotope of chlorine (isotopic weight 34.97 atomic mass units).

7. Gaseous $H^{35}Cl$ has a strong absorption band centered at about $\lambda = 3.40 \times 10^{-6}$ m in the infrared portion of the electromagnetic radiation spectrum. On the assumption that D bonds to Cl with the same strength that H does, predict the frequency of vibration in Hz and rad^{-1} of $D^{35}Cl$.

8. Under what circumstances would the maximum speed be numerically equal to the maximum excursion for a simple harmonic oscillator?

9. Given the O—H bond distance calculated from the MM3 parameter set for the water molecule as 0.947 Å = 94.7 pm and the H—O—H bond angle of 105°, what is the distance between the H nuclei in H_2O in the gas phase?

10. Given the bond distances and internuclear angle in Problem 9, what is the moment of inertia of the H_2O molecule about its principal axis through the oxygen atom (the y-axis in File 4-5)?

11. Calculate the moment of inertia about the x-axis of ethylene.

12. Convert the three moments of inertia in File 4-3 to MKS units.

13. What is the moment of inertia of acetylene about its C—C axis?

14. Using TINKER and the MM3 force field, determine the S—H bond length and the H—S—H bond angle in H_2S. Compare these values with what you found for the H_2O molecule. Compare them with the values you find in a general chemistry textbook. Using PCMODEL, with the MMFF94 force field, determine the S—H bond length and the H—S—H bond angle in H_2S. Compare these values with what you found for the H_2O molecule. Compare them with the values you find in a general chemistry textbook.

5

Molecular Mechanics II: Applications

The three main goals of a molecular mechanics program for small molecules are calculation of geometry, energy, and spectral absorbances due to vibrational excitation. Hagler (Hwang, Stockfish, and Hagler, 1994) has categorized force fields into class 1, intended to achieve the first of these increasingly demanding objectives; class 2, to achieve the first two; and class 3 to achieve all three objectives. Research and development on class 3 force fields is an active enterprise, as is extension of class 1 and class 2 force fields to less common molecules and larger, biologically important species. We have already introduced geometry determination in Chapter 4.

Coupling

We shall treat coupling of modes of motion in some detail because there are fundamental mechanical and mathematical topics involved that will be useful to us in both MM and quantum mechanical calculations. In the treatment of coupled harmonic oscillators, matrix diagonalization and normal coordinates are encountered in a simple form.

Computational Chemistry Using the PC, Third Edition, by Donald W. Rogers
ISBN 0-471-42800-0   Copyright © 2003 John Wiley & Sons, Inc.

a

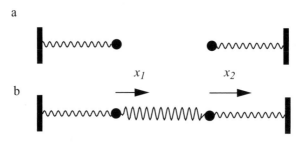

Figure 5-1 (a) Uncoupled and (b) Coupled Harmonic Oscillators.

Two harmonic oscillators consisting of identical masses driven by identical springs have identical frequencies of oscillation (Fig. 5-1a). If they are connected by a third spring as in Fig 5-1b, their motions are no longer independent. The tension on the spring accelerating one mass depends on the location of the other mass on the x-axis. The motions of m_1 and m_2 are said to be *coupled*. The excursions x_1 and x_2 in Fig. 5-1 represent displacements from an equilibrium configuration on the x-axis.

Suppose, for simplicity, that the masses in Fig. 5-1b are the same, $m_1 = m_2 = m$, and all three springs are the same, but velocities and displacements of the masses may not be the same. Let one mass be displaced by a distance x_1 from its equilibrium position while the other is displaced by a distance x_2. The only place the potential energy V

$$V = \int -F \, dx = \int kx_i \, dx_i = \tfrac{1}{2} kx_i^2 \tag{5-1}$$

can be stored is in the springs. V is the sum of the potential energies of the three springs, two lateral and one coupling spring

$$V = \tfrac{1}{2} kx_1^2 + \tfrac{1}{2} k(x_1 - x_2)^2 + \tfrac{1}{2} kx_2^2 \tag{5-2}$$

By hypothesis, the force constant of the coupling spring is the same as k for the other two springs, so the potential energy can be written

$$\begin{aligned} V &= \tfrac{1}{2} kx_1^2 + \tfrac{1}{2} k(x_1^2 - 2x_1x_2 + x_2^2) + \tfrac{1}{2} kx_2^2 \\ &= kx_1^2 - kx_1x_2 + kx_2^2 \end{aligned} \tag{5-3}$$

Note especially the center term in Eq. (5-3). Physical coupling of the masses leads to a term dependent on both x_1 and x_2. Equation (5-3) as written is not separable into an equation only in x_1 and an equation only in x_2.

From Newton's second law, noting that more than one force may influence each mass, $\sum f = ma$ where a is the acceleration $a = \frac{dv}{dt} = \frac{d^2x}{dt^2}$. For the coupled masses

$$m\ddot{x}_1 = ma_1 = \sum_i f_i$$

$$m\ddot{x}_2 = ma_2 = \sum_j f_j \tag{5-4}$$

We have used a common notation from mechanics in Eq. (5-4) by denoting velocity, the first time derivative of x, $\dot{x}$, and acceleration, the second time derivative, $\ddot{x}$. In a conservative system (one having no frictional loss), potential energy is dependent only on the location and the force on a particle $\frac{\partial V}{\partial x} = -f_x$; hence, by differentiating Eq. (5-3),

$$\sum_i f_i = -2kx_1 + kx_2$$

$$\sum_j f_j = -2kx_2 + kx_1 \tag{5-5}$$

which leads to

$$m\ddot{x}_1 = -2kx_1 + kx_2$$

$$m\ddot{x}_2 = -2kx_2 + kx_1 \tag{5-6}$$

and

$$\ddot{x}_1 + 2\frac{k}{m}x_1 - \frac{k}{m}x_2 = 0$$

$$\ddot{x}_2 + 2\frac{k}{m}x_2 - \frac{k}{m}x_1 = 0 \tag{5-7}$$

Let us take a pair of trial solutions on the reasonable guess that, however the coupling spring influences the motion, the masses will oscillate harmonically.

$$x_1 = A_1 \cos \omega t$$

$$x_2 = A_2 \cos \omega t \tag{5-8}$$

with an unknown angular frequency ω. Because the masses and the springs are the same, the system is symmetrical and we do not need to worry about different frequencies for m_1 and m_2. This leads to

$$\dot{x}_1 = \frac{d}{dt}A_1 \cos \omega t = -A_1\omega \sin \omega t$$

$$\dot{x}_2 = \frac{d}{dt}A_2 \cos \omega t = -A_2\omega \sin \omega t$$

$$\ddot{x}_1 = -\frac{d}{dt}A_1\omega \sin \omega t = -A_1\omega^2 \cos \omega t$$

$$\ddot{x}_2 = -\frac{d}{dt}A_2\omega \sin \omega t = -A_2\omega^2 \cos \omega t$$

and

$$-\omega^2 A_1 \cos \omega t + \frac{2k}{m} A_1 \cos \omega t - \frac{k}{m} A_2 \cos \omega t = 0$$
$$-\omega^2 A_2 \cos \omega t + \frac{2k}{m} A_2 \cos \omega t - \frac{k}{m} A_1 \cos \omega t = 0$$

$$(5\text{-}9)$$

for the equations of motion, Eqs. (5-6) or (5-7). Dividing by $\cos \omega t$, we get

$$\frac{2k}{m} A_1 - \omega^2 A_1 - \frac{k}{m} A_2 = 0$$
$$\frac{2k}{m} A_2 - \omega^2 A_2 - \frac{k}{m} A_1 = 0$$

$$(5\text{-}10)$$

This is a pair of simultaneous equations in A_1 and A_2 called the *secular equations*

$$\left(\frac{2k}{m} - \omega^2\right) A_1 - \frac{k}{m} A_2 = 0$$
$$-\frac{k}{m} A_1 + \left(\frac{2k}{m} - \omega^2\right) A_2 = 0$$

$$(5\text{-}11)$$

which has, as its secular determinantal equation

$$\begin{vmatrix} \left(\dfrac{2k}{m} - \omega^2\right) & -\dfrac{k}{m} \\ -\dfrac{k}{m} & \left(\dfrac{2k}{m} - \omega^2\right) \end{vmatrix} = 0$$

$$(5\text{-}12)$$

We recall that expansion of a 2×2 determinant follows the rule $\begin{vmatrix} a & b \\ c & d \end{vmatrix} = ad - bc$. Expansion of the determinant in Eq. (5-12) leads to the quadratic equation

$$\left(\frac{2k}{m} - \omega^2\right)^2 - \left(\frac{k}{m}\right)^2 = 0$$

$$(5\text{-}13)$$

which has two solutions for the square of the frequency of oscillation

$$\frac{2k}{m} - \omega^2 = \mp \frac{k}{m}$$

and

$$\omega^2 = \frac{2k}{m} \pm \frac{k}{m}$$

$$(5\text{-}14)$$

The two solutions for ω^2 are

$$\omega^2 = \frac{3k}{m}, \frac{k}{m} \tag{5-15}$$

Two solutions for ω^2 lead to four solutions for ω

$$\omega_1 = \sqrt{\frac{3k}{m}}, \quad \omega_2 = -\sqrt{\frac{3k}{m}}, \quad \omega_3 = \sqrt{\frac{k}{m}}, \quad \omega_4 = -\sqrt{\frac{k}{m}} \tag{5-16}$$

The general solutions for x_1 and x_2 are *superpositions*, that is, linear combinations of all of the solutions we have found

$$x_1 = A_1 \cos \omega_1 t + A_1' \cos \omega_2 t + A_1'' \cos \omega_3 t + A_1''' \cos \omega_4 t$$
$$x_2 = A_2 \cos \omega_1 t + A_2' \cos \omega_2 t + A_2'' \cos \omega_3 t + A_2''' \cos \omega_4 t \tag{5-17}$$

but the amplitude constants in Eq. (5-17) are not all independent. We can get rid of some of them.

Going back to the secular Eqs. (5-10) or (5-11),

$$\frac{2k}{m}A_1 - \omega^2 A_1 = \frac{k}{m}A_2$$
$$\frac{2k}{m}A_2 - \omega^2 A_2 = \frac{k}{m}A_1 \tag{5-18}$$

substituting $\omega^2 = k/m$,

$$\frac{2k}{m}A_1 - \frac{k}{m}A_1 = \frac{k}{m}A_2$$
$$\frac{k}{m}A_1 = \frac{k}{m}A_2 \tag{5-19}$$

which can be true only if $A_1 = A_2$. Substituting $\omega^2 = 3\,k/m$ gives $A_1'' = -A_2''$. The same calculation for all eight amplitudes yields $A_1 = A_2, A_1' = A_2', A_1'' = -A_2''$, and $A_1''' = -A_2'''$. These simplifications lead to

$$x_1 = A_1 \cos \omega_1 t + A_1' \cos \omega_2 t + A_1'' \cos \omega_3 t + A_1''' \cos \omega_4 t$$
$$x_2 = A_1 \cos \omega_1 t + A_1' \cos \omega_2 t - A_1'' \cos \omega_3 t - A_1''' \cos \omega_4 t \tag{5-20}$$

There are now four constants rather than eight. We expect four constants from two second-order differential equations. Dropping the unnecessary subscript 1 and replacing the cumbersome "prime notation",

$$x_1 = A \cos \omega_1 t + B \cos \omega_2 t + C \cos \omega_3 t + D \cos \omega_4 t$$
$$x_2 = A \cos \omega_1 t + B \cos \omega_2 t - C \cos \omega_3 t - D \cos \omega_4 t \tag{5-21}$$

Normal Coordinates

The procedure we followed in the previous section was to take a pair of *coupled equations*, Eqs. (5-6) or (5-17) and express their solutions as a sum and difference, that is, as *linear combinations*. (Don't forget that the sum or difference of solutions of a linear homogeneous differential equation with constant coefficients is also a solution of the equation.) This recasts the original equations in the form of *uncoupled equations*. To show this, take the sum and difference of Eqs. (5-21),

$$X_1 = x_1 + x_2$$
$$X_2 = x_1 - x_2$$

(5-22)

we arrive at

$$X_1 = 2(A \cos \omega_1 t + B \cos \omega_2 t)$$
$$X_2 = 2(C \cos \omega_3 t + D \cos \omega_4 t)$$

(5-23)

By the same double differentiation that gave us Eqs. (5-10),

$$\ddot{X}_1 = -\omega_1^2 X_1$$

(5-24a)

and

$$\ddot{X}_2 = -\omega_3^2 X_2$$

(5-24b)

which are just the equations we would get for uncoupled oscillators (Fig. 5-1a), except that the *coordinates* have undergone the linear transformations Eqs. (5-22). We should not be surprised that a transformation of the coordinate system leaves the solutions unchanged in Eqs. (5-22), leading to Eqs. (5-24). These new coordinates, X_1 and X_2, are called the *normal coordinates*.

Normal Modes of Motion

Let us take a 1.00-kg oscillator and couple it with an identical oscillator by means of a coupling spring of $k = 1.00 \, \text{Nm}^{-1}$. The force constants of the lateral springs are also $1.00 \, \text{N m}^{-1}$. Now the positive, real solutions for the frequencies are, from Eq. (5-15)

$$\omega_1, \omega_3 = \sqrt{\frac{k}{m}}, \sqrt{\frac{3k}{m}} = \sqrt{1}, \sqrt{3}$$

(Negative frequencies are physically meaningless.) Does this mean that one mass oscillates at $1.00 \, \text{rad s}^{-1}$ and the other at $\sqrt{3} = 1.73 \, \text{rad s}^{-1}$? Not exactly. Behavior depends on the initial conditions. In the special case that both masses start from rest

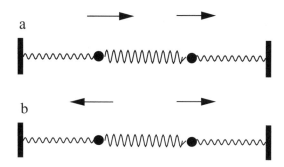

Figure 5-2 Synchronous and Antisynchronous Modes of Motion in a Bound, Two-Mass Harmonic Oscillator.

at the same displacement in the same direction, they will execute synchronous motion as in Fig. 5-2a, at $1.00 \, \text{rad s}^{-1}$. If the initial displacements are equal and opposite (starting from rest also), they will execute antisynchronous motion at a frequency of $\sqrt{3} = 1.73 \, \text{rad s}^{-1}$.

Two degrees of freedom lead to two modes of motion. These two modes of motion, synchronous and antisynchronous, are the *normal modes of motion* for this system. If only synchronous motion is excited, the antisynchronous mode will never contribute to the motion. The same is true for the *pure* antisynchronous mode (Fig. 5-2b); there will never be a synchronous contribution. Under these conditions, but only under these conditions, energy does not pass from one mass to the other.

In general, energy does pass from one normal mode to the other. If only one mass is displaced to amplitude A and released, it will excite motion in the second mass through the coupling spring until the second mass is oscillating with amplitude A. But, in the process, the first mass gradually loses all its energy until it stops. (Total energy is conserved.)

When the first mass has stopped, the situation is reversed. The second mass, oscillating at amplitude A, excites motion in the first mass, gradually losing its own energy, until it has excited the first mass back to amplitude A. This energy exchange, back and forth, goes on for ever (in the absence of friction). The *envelope* of amplitude of either mass in this exchange of energy is sinusoidal, with a frequency less than that of the individual masses. The frequency of transfer from one mode to the other is called the *beat frequency*.

If the masses are displaced in an arbitrary way or arbitrary initial velocities are given to them, the motion is *asynchronous*, a complex mixture of synchronous and antisynchronous motion. But the point here is that even this complex motion can be broken down into two normal modes. In this example, the synchronous mode of motion has a lower frequency than the antisynchronous mode. This is generally true; in systems with many modes of motion, the mode of motion with the highest symmetry has the lowest frequency.

An Introduction to Matrix Formalism for Two Masses

The general case of two masses bound by two lateral springs k_1 and k_2, and coupled by a coupling spring, k_c in Fig. 5-2b, has

$$m_1\ddot{x}_1 = \sum_i f_i = -k_1 x_1 - k_c(x_1 - x_2) = -(k_1 + k_c)x_1 + k_c x_2$$

$$m_2\ddot{x}_2 = \sum_j f_j = -k_2 x_2 - k_c(x_2 - x_1) = k_c x_1 - (k_2 + k_c)x_2$$

$$(5\text{-}25)$$

Eqs. (5-25) can be written in matrix form

$$\begin{pmatrix} m_1 & 0 \\ 0 & m_2 \end{pmatrix} \begin{pmatrix} \ddot{x}_1 \\ \ddot{x}_2 \end{pmatrix} = \begin{pmatrix} -(k_1 + k_c) & k_c \\ k_c & -(k_2 + k_c) \end{pmatrix} \begin{pmatrix} x_1 \\ x_2 \end{pmatrix} \tag{5-26}$$

or

$$\mathbf{M}\begin{pmatrix} \ddot{x}_1 \\ \ddot{x}_2 \end{pmatrix} = \mathbf{K}\begin{pmatrix} x_1 \\ x_2 \end{pmatrix} \tag{5-27}$$

The rules of matrix-vector multiplication show that the matrix form is the same as the algebraic form, Eq. (5-25)

$$\mathbf{M}\ddot{\mathbf{x}} = \mathbf{K}\mathbf{x} = \begin{pmatrix} -(k_1 + k_c) & k_c \\ k_c & -(k_2 + k_c) \end{pmatrix} \begin{pmatrix} x_1 \\ x_2 \end{pmatrix} = \begin{matrix} -(k_1 + k_c)x_1 + k_c x_2 \\ k_c x_1 - (k_2 + k_c)x_2 \end{matrix}$$

$$(5\text{-}28)$$

where $\mathbf{M}$ and $\mathbf{K}$ are matrices and lower-case bold $\ddot{\mathbf{x}}$ and $\mathbf{x}$ designate vectors

$$\ddot{\mathbf{x}} = \begin{pmatrix} \ddot{x}_1 \\ \ddot{x}_2 \end{pmatrix}$$

and

$$\mathbf{x} = \begin{pmatrix} x_1 \\ x_2 \end{pmatrix}$$

Interest now centers on the matrix $\mathbf{K}$. Equation (5-26) is general, but to introduce the method, let all springs have the same force constant. We have the matrix arising from the equations of motion,

$$\mathbf{K} = \begin{pmatrix} -2k & k \\ k & -2k \end{pmatrix} = k\begin{pmatrix} -2 & 1 \\ 1 & -2 \end{pmatrix} \tag{5-29}$$

which is the problem for symmetric and antisymmetric vibration that we have already solved.

If the coupling spring is missing [$k_c = 0$ in Eq. (5-28)], we get a unique *diagonal* matrix

$$\mathbf{K} = \begin{pmatrix} -k & 0 \\ 0 & -k \end{pmatrix} = k \begin{pmatrix} -1 & 0 \\ 0 & -1 \end{pmatrix} \tag{5-30}$$

leading to a pair of *uncoupled* equations,

$$m_1 \ddot{x}_1 = -kx_1$$
$$m_2 \ddot{x}_2 = -kx_2$$

Because the masses are the same, these equations describe independent oscillators with identical frequencies, as in Fig. 5-1a.

These matrices, for coupled and uncoupled oscillators,

$$\begin{pmatrix} -2 & 1 \\ 1 & -2 \end{pmatrix} \tag{5-31a}$$

and

$$\begin{pmatrix} -1 & 0 \\ 0 & -1 \end{pmatrix}, \tag{5-31b}$$

have the roots

$$(-1, -3); \quad \text{coupled} \qquad \text{and} \qquad (-1, -1); \quad \text{uncoupled}$$

Exercise 5-1

Find the roots of

$$\begin{pmatrix} -2 & 1 \\ 1 & -2 \end{pmatrix}$$

Solution 5.4.1

Set

$$\begin{vmatrix} -2 - z & 1 \\ 1 & -2 - z \end{vmatrix} = 0$$

where z is the root. This leads to the quadratic equation

$$(-2 - z)^2 - 1 = 0$$
$$(-2 - z) = \pm 1$$
$$z = -3, -1$$

Comparison with Eq. (5-15) enables us to identify the roots of the K matrix as $z = -\omega^2$.

Taking the roots from Exercise 5-1, we can write them as the elements of a *diagonal* matrix,

$$\begin{pmatrix} m_1 & 0 \\ 0 & m_2 \end{pmatrix} \begin{pmatrix} \ddot{X}_1 \\ \ddot{X}_2 \end{pmatrix} = m \begin{pmatrix} 1 & 0 \\ 0 & 1 \end{pmatrix} \begin{pmatrix} \ddot{X}_1 \\ \ddot{X}_2 \end{pmatrix} = m \begin{pmatrix} \ddot{X}_1 \\ \ddot{X}_2 \end{pmatrix} = -k \begin{pmatrix} 1 & 0 \\ 0 & 3 \end{pmatrix} \begin{pmatrix} X_1 \\ X_2 \end{pmatrix}$$

(5-32)

where the simplifying assumption $m_1 = m_2$ permits factoring m from the diagonal matrix **M**. The ordinary algebraic form of Eq. (5-32) is

$$m\ddot{X}_1 = -kX_1$$
$$m\ddot{X}_2 = -3kX_2$$

(5-33)

By comparison to Eqs. (5-23), it is evident that

$$\omega_1^2 = -\frac{k}{m}$$

(5-34a)

and

$$\omega_2^2 = -\frac{3k}{m}$$

(5-34b)

as found in Eq. (5-15). We have uncoupled the two original equations by diagonalizing the K matrix. Diagonalization reorients the coordinates to coincide with the vibrational modes, that is, it converts arbitrary coordinates to normal coordinates, X_i. We have determined the symmetric and antisymmetric frequencies by a matrix formalism that is readily generalizable to more complicated systems. Solving a determinantal equation and finding the roots of the corresponding matrix Eq. (5-32) amount to two ways of doing the same thing; we are solving simultaneous equations of the form Eqs. (5-11).

The Hessian Matrix

A single harmonic oscillator constrained to the x-axis has one force constant k that—stretching a point—we might think of k as a 1×1 force constant matrix. Two oscillators that interact with one another lead to a 2×2 force constant matrix

$$\begin{pmatrix} k_{11} & k_{12} \\ k_{21} & k_{22} \end{pmatrix}$$

(5-35)

where the lateral spring 1 controlling mass 1 has a force constant k_{11}, the other lateral spring has a force constant k_{22}, and the connecting spring has a force constant $k_{12} = k_{21}$.

In a molecule, these latter two force constants are equal to one another because the coupling of atom 1 for atom 2 is the same as the coupling of atom 2 for atom 1.

The force constants k_{12} and k_{21} are the *off-diagonal elements* of the matrix. If they are zero, the oscillators are uncoupled, but even if they are not zero, the K matrix takes the simple form of a symmetrical matrix because $k_{12} = k_{21}$. The matrix is symmetrical even though k_{11} may not be equal to k_{22}.

The force constants are second derivatives of the potential energy with respect to infinitesimal displacements of mass 1 and mass 2.

$$\begin{pmatrix} \dfrac{\partial^2 V}{\partial x_1^2} & \dfrac{\partial^2 V}{\partial x_1 \partial x_2} \\[2ex] \dfrac{\partial^2 V}{\partial x_2 \partial x_1} & \dfrac{\partial^2 V}{\partial x_2^2} \end{pmatrix} \tag{5-36}$$

This kind of matrix is called a *Hessian matrix*. The derivatives give the curvature of $V(x_1, x_2)$ in a two-dimensional space because there are two masses, even though both masses are constrained to move on the x-axis. As we have already seen, these derivatives are part of the Taylor series expansion

$$f(x) = f(a) + f'(a)(x - a) + \frac{f''(a)}{2}(x - a)^2 + \cdots + \frac{f^n(a)}{n!}(x - a)^n + \cdots$$

in a one-dimensional x-space. If we drop all terms except the quadratic in the harmonic approximation and expand the function about the equilibrium atomic position $a = x_0$,

$$V(x) = \tfrac{1}{2}\left(\frac{d^2 V}{dx^2}\right)_{x_0}(x - x_0)^2$$

as in Eq. (4-15) for one mass. For two masses in a one-dimensional space

$$dV = \tfrac{1}{2}\sum_{i,j}^{2}\left(\frac{d^2 V}{dq_i q_j}\right)_0 q_i q_j \tag{5-37}$$

where q_i and q_j are *mass-weighted* generalized displacement coordinates

$$q_1 = \sqrt{m_1}\,\Delta x_1, \quad q_2 = \sqrt{m_1}\,\Delta y_1, \quad q_3 = \sqrt{m_1}\,\Delta z_1, \quad q_4 = \sqrt{m_2}\,\Delta x_2, \ldots \tag{5-38}$$

Mass weighting the generalized displacement coordinates q_i and q_j retains the form of Eq. (5-37) even when the actual masses are not unit masses.

If there were three masses moving on the x-axis and interacting with one another, the Hessian matrix would be 3×3

$$\begin{pmatrix} k_{11} & k_{12} & k_{13} \\ k_{21} & k_{22} & k_{23} \\ k_{31} & k_{32} & k_{33} \end{pmatrix} \tag{5-39}$$

Two coupled masses oscillating in a plane have four degrees of freedom, x_1, y_1, x_2, and y_2 and so on (Fig. 5-3).

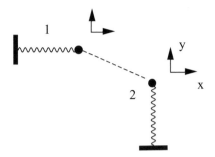

Figure 5-3 Coupled Harmonic Oscillators in the x-y Plane.

The Hessian matrix is

$$\begin{pmatrix} \dfrac{\partial^2 V}{\partial x_1^2} & \dfrac{\partial^2 V}{\partial y_1 \partial x_1} & \dfrac{\partial^2 V}{\partial x_2 \partial x_1} & \dfrac{\partial^2 V}{\partial y_2 \partial x_1} \\[2ex] \dfrac{\partial^2 V}{\partial x_1 \partial y_1} & \dfrac{\partial^2 V}{\partial y_1^2} & \dfrac{\partial^2 V}{\partial x_2 \partial y_1} & \dfrac{\partial^2 V}{\partial y_2 \partial y_1} \\[2ex] \dfrac{\partial^2 V}{\partial x_1 \partial x_2} & \dfrac{\partial^2 V}{\partial y_1 \partial x_2} & \dfrac{\partial^2 V}{\partial x_2^2} & \dfrac{\partial^2 V}{\partial y_2 \partial x_2} \\[2ex] \dfrac{\partial^2 V}{\partial x_1 \partial y_2} & \dfrac{\partial^2 V}{\partial y_1 \partial y_2} & \dfrac{\partial^2 V}{\partial x_2 \partial y_2} & \dfrac{\partial^2 V}{\partial y_2^2} \end{pmatrix} \qquad (5\text{-}40)$$

A Hessian matrix for a molecule containing n atoms is more complicated. In the most general case, it is a very large $3n \times 3n$ matrix brought about because each atom moves in Cartesian 3-space. This large matrix can be constructed by starting out with elements for atom 1 in the first 3 rows (x, y, and z) and the first 3 columns of the matrix. This results in the 3×3 submatrix at the top left of the Hessian matrix, which gives the potential energy increase for atom 1 moving in its own 3-space, uncoupled to any other atom. The 3×3 submatrix immediately to the right, occupying rows 1 through 3 and columns 4 through 6, describes the energy of interaction (coupling) of atoms 1 and 2.

$$\begin{pmatrix} \dfrac{\partial^2 V}{\partial x_1^2} & \dfrac{\partial^2 V}{\partial y_1 \partial x_1} & \dfrac{\partial^2 V}{\partial z_1 \partial x_1} & \dfrac{\partial^2 V}{\partial x_2 \partial x_1} & \dfrac{\partial^2 V}{\partial y_2 \partial x_1} & \dfrac{\partial^2 V}{\partial z_2 \partial x_1} & \\[2ex] \dfrac{\partial^2 V}{\partial x_1 \partial y_1} & \dfrac{\partial^2 V}{\partial y_1^2} & \dfrac{\partial^2 V}{\partial z_1 \partial y_1} & \dfrac{\partial^2 V}{\partial x_2 \partial y_1} & \dfrac{\partial^2 V}{\partial y_2 \partial y_1} & \dfrac{\partial^2 V}{\partial z_2 \partial y_1} & \cdots \text{etc.} \\[2ex] \dfrac{\partial^2 V}{\partial x_1 \partial z_1} & \dfrac{\partial^2 V}{\partial y_1 \partial z_1} & \dfrac{\partial^2 V}{\partial z_1^2} & \dfrac{\partial^2 V}{\partial x_2 \partial z_1} & \dfrac{\partial^2 V}{\partial y_2 \partial z_1} & \dfrac{\partial^2 V}{\partial z_2 \partial z_1} & \\[2ex] \dfrac{\partial^2 V}{\partial x_1 \partial x_2} & \dfrac{\partial^2 V}{\partial y_1 \partial x_2} & \dfrac{\partial^2 V}{\partial z_1 \partial x_2} & \dfrac{\partial^2 V}{\partial x_2^2} & \dfrac{\partial^2 V}{\partial y_2 \partial x_2} & \dfrac{\partial^2 V}{\partial z_2 \partial x_2} & \\[2ex] \dfrac{\partial^2 V}{\partial x_1 \partial y_2} & \dfrac{\partial^2 V}{\partial y_1 \partial y_2} & \dfrac{\partial^2 V}{\partial z_1 \partial y_2} & \dfrac{\partial^2 V}{\partial x_2 \partial y_2} & \dfrac{\partial^2 V}{\partial y_2^2} & \dfrac{\partial^2 V}{\partial z_2 \partial y_2} & \\[2ex] \dfrac{\partial^2 V}{\partial x_1 \partial z_2} & \dfrac{\partial^2 V}{\partial y_1 \partial z_2} & \dfrac{\partial^2 V}{\partial z_1 \partial z_2} & \dfrac{\partial^2 V}{\partial x_2 \partial z_2} & \dfrac{\partial^2 V}{\partial y_2 \partial z_2} & \dfrac{\partial^2 V}{\partial z_2^2} & \\[2ex] & \vdots & & & & & \ddots \\[1ex] & \text{etc.} & & & & \text{etc.} \end{pmatrix} \qquad (5\text{-}41)$$

Diagonally below and to the right of the matrix containing only 1,1 subscripts is a 3×3 matrix containing only 2, 2 subscripts. It describes atom 2 independent of coupling. Continuing down the diagonal, one encounters a submatrix for atoms 3, 4, and so on. These are called *block matrices*. If all coupling submatrices are set equal to zero, only the block diagonal matrix remains.

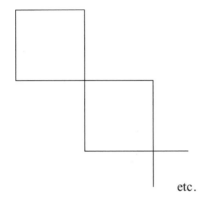

etc.

Diagonalization of a block diagonal matrix is much easier and faster than diagonalization of the full matrix for structures significantly different from the equilibrium geometry because much of the work has already been done by setting the off-diagonal blocks to zero. This works because the bonding forces are much greater than the coupling forces. The coupling forces are not zero, however, and cannot be completely ignored in an accurate calculation. MM programs may be specifically written to begin by optimizing the block diagonal matrix and to switch over to a full matrix optimization when the geometry has been brought near the equilibrium geometry as determined by the size of the steps taken toward the equilibrium geometry.

Why So Much Fuss About Coupling?

The subject of force-coupled harmonic oscillation, leading inevitably to normal modes of motion and to the Hessian matrix, has been developed in far greater detail than other topics in molecular mechanics because the mathematical formalism is basic to almost everything we do in molecular computational chemistry, extending well beyond the classical mechanics of atomic vibrations. Virtually every mathematical technique described in this book uses some kind of minimization: minimization of the error in statistics, minimization of classical mechanical energy in molecular mechanics, or minimization of the electronic energy in molecular orbital calculations. Less frequently, in the study of reactive intermediates or excited species, a "saddle point" is sought. The term *optimization* can be used to include techniques that seek saddle points as well as minima. The term *stationary point* is used to denote the result of a generalized optimization procedure.

Generalizing the Newton–Raphson method of optimization (Chapter 1) to a surface in many dimensions, the function to be optimized is expanded about the many-dimensional position vector of a point $\mathbf{x}_0$

$$f(\mathbf{x}) = f(\mathbf{x}_0) + \mathbf{g}^T(\mathbf{x} - \mathbf{x}_0) + \tfrac{1}{2}(\mathbf{x} - \mathbf{x}_0)^T \mathbf{H}(\mathbf{x} - \mathbf{x}_0) \tag{5-42}$$

where $\mathbf{g}$ is the gradient vector corresponding to the first derivative of a one-dimensional expansion and $\mathbf{H}$ is the Hessian matrix that corresponds to the second derivative. For a small gradient, this leads to (Jensen, 1999)

$$(\mathbf{x} - \mathbf{x}_0) = -\mathbf{g}\mathbf{H}^{-1} \tag{5-43}$$

If the coordinate system has been transformed to the normal coordinate system by a unitary transformation $\mathbf{U}$, the Hessian is diagonal and

$$\Delta x' = \sum_i \Delta x_i \tag{5-44}$$

or

$$\Delta x' = \frac{f_i}{\varepsilon_i} \tag{5-45}$$

where f_i is the projection of the gradient along the Hessian eigenvector with the eigenvalue ε_i.

Having filled in some of the mathematical foundations of optimization procedures, we shall return to the practical calculation of quantities of everyday use to the chemist.

The Enthalpy of Formation

We shall modify the minimal ethylene file **minimal.mm3** by placing an identifying name (Ethylene) starting in the first column of the first row of the input file and by placing a switch 1 in column 65 of the second line. This switch causes the *enthalpy of formation* option to be activated. An additional block of energy and enthalpy information, shown in File Segment 5-1, is generated by the MM3 calculation and added to the output file. The term "heat of formation," more properly "enthalpy of formation," means the enthalpy change $\Delta_f H^{298}$ brought about by the reaction forming a molecule from its elements in the standard state. Standard temperature is usually taken as 298.15 K. For example, $\Delta_f H^{298}$(ethene) is the enthalpy of the reaction

$$2C(gr) + 2H_2(g) = C_2H_4(g)$$

at $p = 1.000$ atm and $T = 298.15$ K where gr indicates graphitic carbon. The experimental value for this reaction ($\Delta_f H^{298}$(ethene) $= 52.5 \pm 0.4\,\text{kJ mol}^{-1} = 12.55 \pm 0.08\,\text{kcal mol}^{-1}$; Pedley et al., 1986) is obtained indirectly from heats (energies) of combustion of the reactants and products.

```
HEAT OF FORMATION AND STRAIN ENERGY CALCULATIONS
   (UNIT=KCAL/MOLE)
   (# = TRIPLE BOND)
BOND ENTHALPY (BE) AND STRAINLESS BOND ENTHALPY (SBE)
CONSTANTS AND SUMS
   BOND OR STRUCTURE       NO      —NORMAL—        –STRAINLESS–
   C=C SP2-SP2              1     26.430   26.43   24.503     24.50
   C–H OLEFINIC             4     -4.590  –18.36   -3.460    –13.84
                                     - - - - - - -        - - - - - - -
                                       BE = 8.07          SBE = 10.66
PARTITION FUNCTION CONTRIBUTION (PFC)
   CONFORMATIONAL POPULATION INCREMENT (POP)                    0.00
   TORSIONAL CONTRIBUTION (TOR)                                 0.00
   TRANSLATION/ROTATION TERM (T/R)                              2.40
                                                          PFC = 2.40
                                                          - - - - - - -
   HEAT OF FORMATION (HF0) = E + BE + PFC                      13.07
   STRAINLESS HEAT OF FORMATION FOR SIGMA SYSTEM (HFS)
HFS = SBE + T/R + ESCF – ECPI                                  13.06
   INHERENT SIGMA STRAIN (SI) = E + BE–SBE                      0.01
   SIGMA STRAIN ENERGY (S)=POP+TOR+SI                           0.01
- - - - - - - - - - - - - - - - - - - - - - - - - - - - - - - - - - - - - - - - - -
End of Ethylene
   Total cpu time is    2.20 seconds.
```

File Segment 5-1   Heat of Formation and Strain Energy Output.

At the top of File Segment 5-1 is a heat of formation information block. Two sums are listed: One is a sum of normal bond enthalpies for ethylene, and the other is a sum selected from a parameter set of strainless bonds. Both sets of bond enthalpies have been empirically chosen. A group of molecules selected as "normal" generates one parameter set, and a group supposed to be "strainless" is selected to generate a second set of strainless bond enthalpies designated SBE in File Segment 5-1. The subject of parameterization has been treated in detail in Chapter 4. See Computer Projects 3-6 and 3-7 for the specific problem of bond enthalpies.

After initial estimates of BE and SBE have been made, application to a larger set of test molecules suggests adjustments to bring about the smallest discrepancy between the experimental and calculated values of $\Delta_f H^{298}$ (or other selected criteria) for the entire set. The success or failure of bond energy assignment is determined, of course, not so much by one's ability to reproduce known $\Delta_f H^{298}$ values within the data set from which BEs are drawn but by the more important practical goal of correctly predicting unknown $\Delta_f H^{298}$ values from outside the test data set, leading to useful experiments and industrial applications.

Once the BEs and SBEs have been decided upon, the normal functioning of the MM program causes each bond to be multiplied by the number of times it appears in the computed molecule to find its contribution to the total bond enthalpy. In ethylene, $26.43 + 4(-4.59) = 8.07 \text{ kcal mol}^{-1}$. In File Segment 5-1, this *sum* is denoted BE. This whole procedure is essentially a conventional bond energy calculation.

The difference between an MM calculation of the enthalpy of formation and a bond energy scheme comes in the steric energy, which was shown in File 4-3. The sum of compression, bending, etc. energies is the steric energy, $E = 2.60 \text{ kcal mol}^{-1}$ in File 4-3. This is added to BE, as is the *partition function energy* contribution (see below), $PCE = 2.40 \text{ kcal mol}^{-1}$, to yield

$$8.07 + 2.60 + 2.40 = 13.07 \text{ kcal mol}^{-1}$$

for the enthalpy of formation $\Delta_f H^{298}$ of ethylene. The same thing is done with the strainless bond energies except that there is no steric energy E to add. In ethylene, the strain energy is essentially zero, therefore normal $\Delta_f H^{298}$ is essentially the same as strainless $\Delta_f H^{298}$. This is not true in general.

The Partition Function Contribution.   If we ignore vibrational motion for the moment, all translational and rotational modes of motion are fully excited, leading to an energy contribution of $\frac{1}{2}RT$ per degree of freedom per mole or $3RT$. Taking the ideal gas approximation for granted in calculating molar enthalpy (at this level, intermolecular interactions are assumed to be zero), H is RT higher than the energy. The total is $4RT$. This accounts for the partition function contribution $PFC = 2.40 \text{ kcal mol}^{-1}$.

The vibrational energy over all chemical bonds, i

$$U = \sum h\nu_i \left\{ \frac{1}{2} + \frac{1}{e^{h\nu_i/kT} - 1} \right\} \tag{5-46}$$

would normally be added to this sum. This term is difficult to evaluate, and, depending on the way in which the bond enthalpy parameters are obtained, it may be an overcalculation anyway. If bond energies are taken from $\Delta_f H^{298}$, as they are for MM3, they already contain the thermal energy of vibration.

Finally, the system "knows" to print End of Ethylene at the end of the program because we included the name Ethylene in the first line of the input file.

Enthalpy of Reaction

If one knows the enthalpy of formation at 298 K of all the constituents of a chemical reaction, one knows the enthalpy of reaction, $\Delta_r H^{298}$. The calculation rests on the law that thermodynamic functions, of which enthalpy is one, sum to zero around any cyclical path, formally,

$$\Delta_r H^{298} = \sum \Delta_f H^{298} (\text{products}) - \sum \Delta_f H^{298} (\text{reactants}) \qquad (5\text{-}47)$$

For example, in the hydrogenation of ethylene

$$CH_2{=}CH_2 + H_2 \rightarrow CH_3{-}CH_3$$

a diagram with enthalpy as the vertical axis can be drawn as in Fig. 5-4.

Because there is no natural zero point for the enthalpy in classical thermo-dynamics, we are free to define one, but only one, level in the enthalpy level diagram. The enthalpy levels in Fig. 5-4 are relative to a defined level of zero for the elements. The cyclical path includes $\Delta_f H^{298}$ values that are up for a positive ΔH^{298} and down for a negative ΔH^{298}. The horizontal lines represent the standard states of ethylene, elements, and ethane. The rightmost arrow represents the transition from the ethylene standard state to the ethane standard state $\Delta_r H^{298}$, the contribution from elemental $H_2(g)$ being zero. Thus the sum of the arrow lengths for hydro-genation of ethylene has a magnitude of $12.52 + 19.75 = 32.27\,\text{kcal mol}^{-1}$ in agreement with Eq. (5-47) and the direction is down, so $\Delta_r H^{298} = -32.27\,\text{kcal mol}^{-1}$ by this calculation. This reaction was studied experimentally with great care many years ago by Kistiakowsky (Kistiakowsky and Nickle, 1951), who found that $\Delta_r H^{298} = -32.60 \pm 0.05\,\text{kcal mol}^{-1}$. One should not expect this extraordinary precision and agreement between calculations and experimental measurements in general. Uncertainties of experiments and calculations are rarely better than $0.5\,\text{kcal mol}^{-1}$; hence, the uncertainty in a calculated $\Delta_f H^{298}$ will not normally be better than the square root of the sum of squares of uncertainties in $\Delta_f H^{298}$, that is, not better than $\pm 0.7\,\text{kcal mol}^{-1}$ for a simple A $\rightarrow$ B reaction.

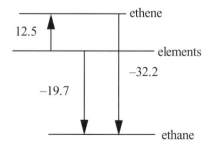

Figure 5-4 Enthalpy Diagram for Hydrogena-tion of Ethylene. The $\Delta_f H^{298}$ values were calculated with MM4 (Allinger et al., 1996; Nevins et al., 1996).

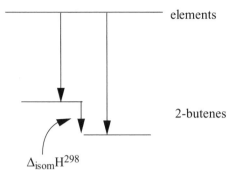

Figure 5-5 Enthalpy Changes in
Isomerization of 2-Butene.

$\Delta_{isom}H^{298}$

COMPUTER PROJECT 5-1 | *The Enthalpy of Isomerization of
cis- and trans-2-Butene*

Isomerizations are even simpler than hydrogenations inasmuch as they involve only
the transformation of one isomer into the other. An example is the transformation of
cis-2-butene to *trans*-2-butene

$cis \rightarrow trans$

The goal of this project is to determine the enthalpies of formation of *cis*- and *trans*-
2-butene and to calculate the enthalpy of isomerization $\Delta_{isom}H^{298}$ between them.
Remarkably accurate experimental results have been obtained, also by Kistiakowsky (Conant and Kistiakowsky, 1937), which permit indirect calculation of the
enthalpy change of this isomerization $\Delta_{isom}H^{298}$ The enthalpy diagram for
isomerization of 2-butene is given in Fig. 5-5. We shall use TINKER and
PCMODEL to find the length of each arrow in Fig. 5-5. We shall also use the
PCMODEL GUI, a powerful tool for file construction.

Procedure.   Using graph paper with Fig. 4-5 as a guide, construct the approximate
carbon atom skeletons of *cis*- and *trans*-2-butene,

$$\begin{array}{ccc} & \text{C=C} & \\ \text{C} & & \text{C} \end{array}$$

and

$$\begin{array}{cc} & \text{C} \\ \text{C=C} & \\ \text{C} & \end{array}$$

Because of its free format, the input file for TINKER is easier to construct by hand
than the input file for Program MM3. Place the hydrogen atoms at plausible
distances from the carbon atoms to produce two input files for determining the

steric energies of the two butene isomers using the MM3 force field. Store the input files in the TINKER directory as **.xyz** files, for example, **cbu2.xyz** for *cis*-2-butene.

Carry out this computation using 1) Program TINKER (force field MM3) and 2) Program MM3 (force field MM3).

1) TINKER. Go to your DOS operating system, TINKER directory. Execute the command **minimize**. Respond to the choices offered:

Enter Cartesian Coordinate File Name: **cbu2.xyz**
Enter Potential Parameter File Name: **mm3**
Enter RMS Gradient per Atom Criterion [0.01]: **Enter** (this is the default option)

Take the "Final Function Value" of TINKER as the steric energy for this calculation. Compare the results with each other and with a standard value from a good elementary organic chemistry text (e.g., Ege, 1994). Calculate $\Delta_{isom}H^{298}$ for the reaction *cis* → *trans* and compare it with a standard text and with Kistiakowsky's original value. Kistiakowsky's original work was carried out at 355 K, but the temperature difference between 298 K and 355 K cancels for this isomerization.

The geometric output can be found as you would find any other file in DOS

```
D:\tinker>dir cbu*.*
Directory of D:\tinker
CBU2   XYZ    287  02-16-02 2:04p cbu2.xyz
CBU2~1 XYZ    815  10-25-02 6:48a cbu2.xyz_2
  2 file(s)      1,102 bytesD:\tinker>dir cbu2*.*
```
You can see the optimized geometry using the edit command.

```
D:\tinker>edit cbu2~1.xyz
```

2) The PCMODEL Interface. Using the **draw** option of the PCMODEL interface, construct the carbon atom skeleton of *cis*-2-butene. Use **Add_B** to add a bond in the 2-3 position, thereby making it a double bond. Click on **H/AD** and observe that the correct number of hydrogen atoms appears at plausible distances and angles to the carbon atoms. Hydrogen atoms at either end may not be visible because of the hydrogens in front of them. Later, we shall learn how to rotate the molecular image in three-dimensional space so as to make all atoms visible. This is not the optimized structure of *cis*-2-butene; it is the input or starting structure before minimization.

If you are running an updated version (V 8.0) of PC Model, click on **force field→mm3**. Omit this step for older versions. Click on Analyze (or **compute** depending on the version of PCMODEL) to obtain a menu of options. Select **minimize**. The geometry changes can be seen on the screen and a sequence of numbers appears in the right panel of the CRT screen, ending in Hf, the enthalpy of formation. This is the PCMODEL-MM3 calculated value of $\Delta_f H^{298}$ for *cis*-2-butene. Repeat the entire procedure for *trans*-2-butene and calculate the required $\Delta_{isom}H^{298}$.

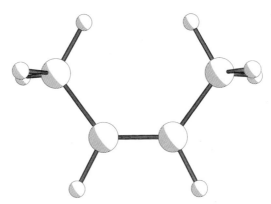

Figure 5-6 Pluto drawing of *cis*-2-Butene PCMODEL v8.0. The double bond is not rendered.

Minimize *cis*-2-butene. From the **View** menu, select **Pluto**. Your screen image should resemble Fig. 5-6. Go to **File** and **Print** your pluto drawing. Repeat with *trans*-2-butene. Include the structure drawings with your report. You can create pluto drawings for any of the stick figures in future computer projects. There are other graphical options.

Enthalpy of Reaction at Temperatures $\neq$ 298 K

If one knows $\Delta_r H^{298}$ *and the heat capacities* of all the constituents in a chemical reaction as a function of temperature, one can calculate $\Delta_r H^{298}$ at any temperature. For a simple A $\rightarrow$ B reaction, suppose that the heat capacity C_P of A is larger than the heat capacity for B. The enthalpy of A rises more steeply with temperature increase than that of B by the definition of heat capacity

$$C_P = \left(\frac{\partial H}{\partial T}\right)_P \tag{5-48}$$

Consequently, for any exothermic reaction, $\Delta_r H^{298}$ increases with temperature as well (Fig. 5-7).

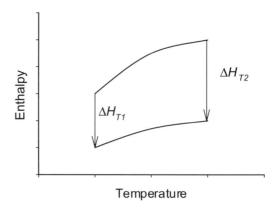

Figure 5-7 Enthalpy as a Function of Temperature for the Exothermic Reaction A $\rightarrow$ B.

More complicated reactions and heat capacity functions of the form $C_P = a + bT + cT^2 + \cdots$ are treated in thermodynamics textbooks (e.g., Klotz and Rosenberg, 2000). Unfortunately, experimental values of heat capacities are not usually available over a wide temperature range and they present some computational problems as well [see Eq. (5-46)].

Population Energy Increments

There are two additive terms to the energy, POP and TORS, that have not been mentioned yet because they are zero in minimal ethylene. The POP term comes from higher-energy conformers. If the energy at the global minimum is not too far removed from one or more higher conformational minima, molecules will be distributed over the conformers according to the Boltzmann distribution

$$\frac{N_i}{N_0} = e^{-[(E_i-E_0)/k_{\mathrm{B}}T]} \tag{5-49}$$

where $\frac{N_i}{N_0}$ is the ratio of molecules in the ith high-energy conformational state to molecules in the ground state (see Computer Project 4-2). $E_i - E_0$ is the energy difference between the energetic conformer and the ground state. If there is degeneracy of any energy level, it is counted into the conformational mix. If there are several conformers not too far from the ground state, they will be simultaneously populated and all must be taken into account. The energy of the compound under investigation, as measured by experimental means, is that of the weighted average determined by the conformational mix at any selected temperature.

Exercise 5-2

A molecule has three nondegenerate conformers (Fig. 5-8). One is 450×10^{-23} J above the ground state, and the second high-energy conformer is 900×10^{-23} J above the ground state. What are the percentages of each of the three conformers relative to the total number of molecules in a sample of the normal conformational mixture at 300 K?

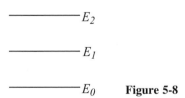

E_2

E_1

E_0 **Figure 5-8**

Solution 5-2

The exponent in the Boltzmann expression for the conformer at E_1 is

$$-\frac{(E_1 - E_0)}{k_{\mathrm{B}}T} = -\frac{450 \times 10^{-23}}{1.381 \times 10^{-23}(300)} = -1.086$$

where k_B is Boltzmann's constant. The exponent in the Boltzmann expression for the highest conformer is just twice this value

$$-\frac{(E_2 - E_0)}{k_B T} = -\frac{900 \times 10^{-23}}{1.381 \times 10^{-23}(300)} = -2.172$$

The ratio of conformers is

$$\frac{N_1}{N_0} = e^{-((E_1 - E_0)/k_B T)} = e^{-1.086} = 0.3376$$

between the middle conformer and the ground state and it is

$$\frac{N_2}{N_0} = e^{-((E_2 - E_0)/k_B T)} = e^{-2.172} = 0.1139$$

between the highest energy conformer and the ground state.
The total number of molecules is

$$N = N_0 + 0.3376N_0 + 0.1139N_0 = 1.452N_0$$

so

$$N_0 = \frac{N}{1.452} = 0.6889 = 68.9\ \% \text{ of the total}$$
$$N_1 = 0.3376\,N_0 = 23.3\ \%$$
$$N_2 = 0.1139\,N_0 = 7.6\ \%$$

Equal spacing between energy levels is not unusual. In the case of the harmonic oscillator, it is the rule.

Exercise 5-3

Suppose that the computed ground-state enthalpy of formation of a molecule is $\Delta_f H^{298} = -130.0\,\text{kJ mol}^{-1}$ but that it has two higher-energy conformers as described in Exercise 5-2. What is the expected experimental $\Delta_f H^{298}$ of the equilibrium conformational mixture?

Solution 5-3

The energies of the two higher-energy conformers described in Exercise 5-2 are

$$450 \times 10^{-23}\text{J}(6.022 \times 10^{23}\text{mol}^{-1}) = 2.71\,\text{kJ mol}^{-1}$$

and

$$900 \times 10^{-23}\text{J}(6.022 \times 10^{23}\text{mol}^{-1}) = 5.42\,\text{kJ mol}^{-1}$$

higher than the ground state, where the conversion factor is Avogadro's number. Knowing the % composition of the conformational mixture, we have

$$0.2325(2.71 \, \text{kJ mol}^{-1}) = 0.63 \, \text{kJ mol}^{-1}$$

and

$$0.0784(5.42 \, \text{kJ mol}^{-1}) = 0.43 \, \text{kJ mol}^{-1}$$

as the molar contributions from the higher-energy conformers to the conformational mixture. The total contribution is 1.06 kJ mol^{-1}, so we expect an experimental measurement of $\Delta_f H^{298}$ to yield $-130.0 + 1.06 = -128.9 \, \text{kJ mol}^{-1}$.

The discrepancy between calculation and experiment in Exercise 5-3 is within the uncertainty limits of many thermochemical measurements and would probably not be noticed in a single experiment. A systematic error of this magnitude would certainly be detected by a careful examination of many experimental results such as those carried out by Allinger (1989).

The term "*higher*-energy conformer" has been used in Exercise 5-3 to denote a conformer that has a slightly higher energy than the ground state, say 10 kJ mol^{-1} or less, as distinct from conformers that are much higher in energy than the ground state, say many tens or hundreds of kilojoules per mole. Truly high-energy conformers need not be considered when calculating the energy or enthalpy of a conformational mix because, even though each one contributes much to the mixture, there are so few of them (by Boltzmann's distribution) that the total contribution is nil. Notice that in Solution 5-3, the conformer at E_2 makes a smaller total contribution to $\Delta_f H^{298}$ than the conformer at E_1 even though it has a higher energy because there are fewer molecules at E_2 than at E_1.

Torsional Modes of Motion

Many flexible molecules have a low-frequency torsional (twisting) mode of motion. The resulting energy levels are closely spaced, not far removed from the ground state, and are appreciably populated at room temperature. There is a torsional contribution to the energy of flexible molecules owing to the cumulative contribution of the several upper torsional states, even though the contribution from each state is small. In rigid molecules, this low-frequency contribution is zero.

Rather than calculate the enthalpy contribution of the torsional states individually, an empirical sum that is an integral multiple of 0.42 kcal mol^{-1} per torsional degree of freedom is assigned to flexible molecules in MM3. Torsional motion of a methyl group is not added to a calculated $\Delta_f H^{298}$ because it is included in the methyl parameterization.

In propane, there are two low-frequency torsional motions at the C—C bonds

$$H_3C - CH_2 - CH_3$$

but both have been accounted for in the empirical bond energy scheme by the methyl group increments; therefore, no TORS correction is added to the BE sum to obtain $\Delta_f H^{298}$.

In n-butane,

$$H_3C - CH_2 - CH_2 - CH_3$$

the central C—C bond contributes a torsional energy that is not carried into $\Delta_f H^{298}$ by a contiguous methyl group. Hence, in n-butane, 1 unit of torsional energy for the 2-3 bond is added to the $\Delta_f H^{298}$ sum. In n-pentane, 2 units are added, one for each internal C—C bond. In isobutane, there is no internal C—C bond contribution, isopentane has 1, and so on. *Ethane is an exceptional case.* There is one torsional motion at the C—C bond, but there are *two* methyl groups so the molecule has been overcorrected. One torsional energy parameter must be *subtracted* to obtain the correct $\Delta_f H^{298}$.

The value of the torsional energy increment has been variously estimated, but TORS $= 0.42$ kcal mol^{-1} was settled on for the bond contribution method in MM3. In the full statistical method (see below), low-frequency torsional motion should be calculated along with all the others so the empirical TORS increment should be zero. In fact, TORS is not zero (Allinger, 1996). It appears that the TORS increment is a repository for an energy error or errors in the method that are as yet unknown.

COMPUTER PROJECT 5-2 | *The Heat of Hydrogenation of Ethylene*
In this experiment, we shall run two MM3 files, one for ethylene and one for ethane. Each run gives a value for the ground-state enthalpy of formation without torsional corrections of the target compound. After correcting ethane for the torsional increment of two methyl groups and only one torsional motion, calculate the "heat" (strictly, enthalpy) of hydrogenation $\Delta_h H^{298}$ for the reaction

$$CH_2 = CH_2 + H_2 \rightarrow CH_3 - CH_3$$

Procedure.   1) **MM3.** Atom types for carbon are 1 and 2 for sp^3 and sp^2, respectively, and hydrogen is type 5 as previously seen. Be sure the switch is on in column 65 or you will not get an enthalpy of formation. If you develop your ethane input file from the output file of ethylene (which is probably the best way to do it if you are running without a GUI) remember that you can use a nonzero z-coordinate to take into account the nonplanar nature of the ethane molecule. One should be aware of the three-dimensional nature of molecules in drawing an input geometry because, starting from a planar input file, it is possible for atoms that should move out of the input plane to become frozen in it, leading to the wrong geometry and energy (see false minima below). An example is tetrahedral methane, which can become frozen in a false planar configuration.

$$
\begin{array}{c}
H \\
| \\
H-C-H \\
| \\
H
\end{array}
$$

Activate MM3 with the command **mm3**. Answer questions: file? **ethene.mm3**, parameter file? **Enter** (default) line number **1**, option **2**. The default parameter set is the MM3 parameter set; don't change it. The line number starts the system reading on the first line of your input file, and option 2 is the block diagonal followed by full matrix minimization mentioned at the end of the section on the Hessian matrix. You will see intermediate atomic coordinates as the system minimizes the geometry, followed by a final steric energy. End with **0**, output **Enter**, coordinates **Enter**.

Your output is stored as TAPE4.MM3, and your final geometry is TAPE9.MM3. Read TAPE4.MM3 (**cat** TAPE4.MM3 in UNIX). If you are using a UNIX or LINUX system, this is case sensitive, that is, tape4.mm3 won't work because it is lower case. Once you get into Program MM3, procedures are identical from one system to another but file handling and editing are another matter. They are usually system specific, and you may need the help of your local system guru to establish a routine for your computer.

2) PCMODEL. **Draw** the carbon skeleton (a single horizontal line for ethylene) and add a bond using **Add_B**. Add hydrogens **H/AD** select **Force field MM3** and **Compute→minimize**. Record Hf, which is $\Delta_f H^{298}$ uncorrected for torsional energy for the molecule. Use the **Edit→erase** option to clear the screen and repeat the entire procedure for ethane omitting only the **Add_B** step because we want a single bond between the carbon atoms.

3) Output. Upon successful execution, you will obtain an output file from which you can follow the geometry change during iterative minimization. As the atoms approach their respective potential energy minima, they are moved less and less until the criterion of minimum geometry change is met (see also PART 2, File 4-3).

It is wise to keep an archive of all input and output files using **Save**. At the end of the output, various energies are calculated along with the "heat" of formation that we seek in order to obtain $\Delta_{hyd} H^{298}$. In calculating $\Delta_{hyd} H^{298}$, remember that the enthalpy of formation of an element in the standard state is zero.

Pi Electron Calculations

It has long been known that the enthalpy of hydrogenation of benzene ($49.8 \, \text{kcal mol}^{-1}$; Conant and Kistiakowsky, 1937) is not the same as three times the enthalpy of hydrogenation of cyclohexene ($3 \times 28.6 \, \text{kcal mol}^{-1}$). Evidently, the double bonds that we write in the Kekule structure of benzene

are very different from the double bond in cyclohexene. We ascribe a *resonance stabilization* to benzene and similar molecules (toluene, naphthalene, etc.) to account for the difference. Molecules possessing a benzenoid resonance stabilization energy are said to be *aromatic*. Stabilization is observed to a much lesser degree in a molecule like 1,3-butadiene when its enthalpy of hydrogenation is compared with two times the enthalpy of hydrogenation of 1-butene. Double bonds occupying alternant positions in a molecular structure are called *conjugated* double bonds. In the case of conjugated double bonds, we call the energy discrepancy a *conjugation energy*. Both resonance energy and conjugation energy are quantum mechanical in origin. Neither is treated by pure molecular mechanical calculations because MM is a classical mechanical theory as distinct from a quantum mechanical theory.

Each C=C bond in a conjugated or aromatic system has two electrons, designated π electrons, which are largely responsible for its quantum mechanical features. One approach to the mechanics of a system of alternant double bonds is to carry out a relatively simple quantum mechanical treatment of the π system called a *valence electron self consistent field* (VESCF) calculation. Self-consistent field calculations bring about a change in carbon-carbon bond orders that are no longer simply single or double. Force constants depend on bond order; hence, they are changed by the VESCF calculation. The VESCF calculation is followed by a MM minimization, which causes the atom coordinates to change, necessitating a new VESCF optimization, which necessitates a new MM calculation, and so on. This alternation between classical and quantum mechanical calculations is cut off at some point when repeated calculations do not produce a significant change in the energy. The procedure is not as difficult as it sounds, and with a fast workstation or microcomputer conjugated and aromatic systems can be treated rather easily.

Exercise 5-4

Using graph paper, construct a simple hexagon of carbon atoms and place hydrogen atoms outside the hexagon about 1 Å from the carbon atoms. The x, y, z-coordinates of all 12 atoms plus atom type designators constitute a minimal MM3 starting geometry for benzene.

Solution 5-4

One of many possible solutions is

−0.8	1.	0.	2
0.8	1.	0.	2
1.7	0.	0.	2
.8	−1.	0.	2
−0.8	−1.	0.	2
−1.7	−0.	0.	2
−1.5	2.	0.	5
1.5	2.	0.	5
3.	0.	0.	5
1.5	−2.	0.	5
−1.5	−2.	0.	5
−3.	−0.	0.	5

Exercise 5-5

Add the necessary control lines to obtain the full MM3 minimal input file and run the file under the MM3 force field to obtain the enthalpy of formation of the aromatic molecule benzene.

Solution 5-5

```
minbenz                                        1    12
T T T T T
    1                       6                        1
    1    2    3    4    5    6    1
    1    7    2    8    3    9    4    10   5    11   6    12
```

Placing the 5-line control block above the geometry specification block of Exercise 5-4 gives the complete minimal input file for benzene, which we can call **minbenz.mm3** (or anything else you like with the extension .mm3). Aside from the geometry block, there are two important differences between **minbenz.mm3** and the file **minimal.mm3** for ethylene in File 4-1a. One is the switch in column 61 of the first line, the other is the set of switches TTTTT that constitutes the entire second line. The first switch tells the system that there are alternant sp^2 carbon atoms in the system, and the series of T designators responds to the logical statement "atoms 1 through 6 are alternant sp^2 carbon atoms." The input line says that this statement is "True for each of the first 6 carbon atoms." Remember that in digital logic, T for True is 1 and False is 0, so we are justified in calling these letters logical switches. Other differences between **minbenz.mm3** and **mini-mal.mm3** are the absence of a 4 in column 67 of line 1, which changes the amount of output, and the enthalpy of formation switch in column 65 of line 3 in the **minbenz.mm3**. The computed enthalpy of formation of benzene found by execution of this file in Program MM3 is $20.36 \, \text{kcal mol}^{-1}$.

COMPUTER PROJECT 5-3 | *The Resonance Energy of Benzene*
Compute the resonance energy of benzene.

Procedure. We already have the computed $\Delta_f H^{298}$ of benzene from the previous paragraph. After adding four hydrogen atoms in the appropriate places of the sketch you have made for benzene and modifying the minimal input file for the new molecule, compute the $\Delta_f H^{298}$ of cyclohexene. Don't forget to change the control block of your input file by changing the added atoms list, the total atoms entry, and so on. Repeat the procedure with cyclohexane.

Alternatively, any or all three files for benzene, cyclohexene, and cyclohexane can be generated using the **draw** option of PCMODEL. Either way, the cyclohex-ene file is

```
Cyclohexene                                         0    16   3    0  0
    0    1    0.00000        10   0    0    0    0    0    0    1
         1    2    3    4    5    6    1    0    0    0    0    0  0  0
         1    7    2    8    3    9    3    10   4    11   4    12   5   13  6  14
         6    15   6    16
```

−0.05627	1.36350	0.99040	2
−1.21221	0.74441	0.70886	2
−1.29552	−0.51251	etc.	1
0.00069	etc.		1
etc.			1
			etc.

File 5-2.   An Input File For Cyclohexene.

If one remembers that zeros and blanks are the same, this is essentially a minimal file for cyclohexene. Once having obtained $\Delta_f H^{298}$(benzene) $= 20.36\,\text{kcal mol}^{-1}$, $\Delta_f H^{298}$ (cyclohexene), and $\Delta_f H^{298}$ (cyclohexane), you have enough information to obtain $\Delta_{hyd} H^{298}$ (benzene), $\Delta_{hyd} H^{298}$ (cyclohexene), and the resonance energy of benzene. Construct an enthalpy-level diagram showing how the resonance energy of benzene is obtained from the enthalpies of hydrogenation you have calculated.

Strain Energy

A useful computed property of some molecules is the *strain energy*, included in the MM3 output, which is the difference between the energy (enthalpy) of a target molecule and the energy of the same molecule calculated with a parameter set derived from reference molecules that are supposed to be strain free. (The rigorous thermodynamic distinction between energy and enthalpy drops out in the comparison.) Note that the steric energy is not the strain energy because most normal molecules have a nonzero steric energy but not all molecules have strain. Cyclohexane has a steric energy of $7.72\,\text{kcal mol}^{-1}$ but a strain energy (SE) of only $1.4\,\text{kcal mol}^{-1}$ in the MM3 force field. Relative values of strain energies are useful in qualitative chemical thinking, but absolute values are questionable because of the arbitrary selection of a set of strain-free reference molecules.

False Minima

In trying alternate ways of building up input files, one is bound to discover paths that lead to different enthalpies of formation for the same molecule. In general, MM minimizations seek the extremum closest to the starting geometry. Not all extrema are minima, nor is any one minimum necessarily the *global minimum*. Molecular geometries may come to rest at a local minimum, which is a conformer of the minimum geometry or (rarely) a saddle point on the potential energy surface. A conformer is distinguished from an *isomer* by the fact that one can go from one conformer to another without breaking chemical bonds, for example, by rotation of the 2-3 bond in butane whereas a change from one isomer to another requires breaking chemical bonds, for example, the isomerization from *cis*- to *trans*-2-butene.

False minima may entrap the unwary; a structure may be mistaken for the ground state that does not represent the most stable conformer. If so, the calculated enthalpy of formation will be higher than the thermodynamic enthalpy of formation. The error may be small, as in the case of a *gauche* conformer of butane mistaken for the ground-state *anti* conformer, or it may be tens or hundreds of kilocalories per mole resulting from a geometry that is a far cry from the desired equilibrium geometry.

There is no known computational method for finding the absolute molecular energy minimum. In the case of small molecules, one can determine the enthalpy of formation for all of the limited number of minima that can exist simply by starting from many different geometries and convincing oneself that all reasonable alternatives have been tried. Depending on molecular rigidity and symmetry, the number of conformational choices may increase rapidly for larger molecules.

The opposite approach to judicious choice uses a program that perturbs the starting geometry in a random way and then allows it to relax (Saunders, 1987). After many random "kicks" of this kind, the output files are examined for the lowest energy. Entry into Program MM3 is just as it was for the single minimizations leading to $\Delta_f H^{298}$. Activating the random or stochastic search routine in MM3 is by choosing option 8 or 9 from the routine menu presented at the start of the MM3 run. The system needs to know how many random kicks are requested as well as the kick size. This information is provided by adding a line at the bottom of the input file with an integer right justified in a field of 5 followed by a floating point number anywhere in the next field of 10 columns. Using 0 as a place marker for the integer input, the line 00005 2.0 as the very last line of a file like File 5-2 would specify five kicks of size 2.0 Å.

One sees much more output as the computer "kicks" or "pushes" the molecular geometry, carries out a new minimization, and stores the energy on each kick. You are free to specify a different number of kicks or a different kick size. The kick size of 2.0 Å has been found by experience to be a pretty good compromise between kicking the molecule so hard that it ends up in a bizarre geometry and not kicking it hard enough to go over conformational energy barriers.

The stored output of a stochastic search, STOCHASTIC.MEM, for the various local minima of *n*-pentane using Program MM3 is accessed by the following four lines

select
    MM3 COORDINATE FILE (INPUT FILE)? **NPENTA.MM3**
    LINE NUMBER = **1**
    SEARCH OUT PUT FILE [STOCHASTIC.MEM]? **STOCHASTIC.MEM**

in which lines 2, 3, and 4 are responses to program prompts (be careful of case in line 2; it may be **npenta.mm3**). The computer responds with File 5-3.

npenta
    4 CONFORMATIONS FOUND

	ITERATION FIRST FOUND	ENERGY	# OF TIMES FOUND	NEGATIVE FREQUENCY
1	1	4.1418	8	0

2	2	4.9948	21	0
3	3	5.7613	13	0
4	4	7.4594	8	0

File 5-3   Output From a Stochastic Search of the *n*-Pentane Molecule in MM3. The input structure was subjected to 50 "kicks."

The search reported in File 5-3 there found four energetically distinguishable conformers. Steric energies 2, 3, and 4 are of racemic pairs. They are degenerate. Strictly speaking, energy 4 corresponds to a pair of structural conformers that are somewhat twisted relative to one another; hence, though they are degenerate, they are not exactly a racemic pair (Mencarelli, 1995).

The energy differences among conformers relative to the ground state are 0.0, 0.85, 1.62, and 3.32 kcal mol^{-1}. The relative populations of the states, judged by the number of times they were found in a random search or 50 trials, are 0.16, 0.21, 0.15, and 0.08 when degeneracy is taken into account. In the limit of very many runs, a Boltzmann distribution would lead us to expect a ground state that is much more populous than the output indicates, but this sample is much too small for a statistical law to be valid.

The last column in File 5-3 shows that no imaginary frequencies are found in this example. In general, imaginary frequencies are found when optimization settles on a saddle point for a transition from one conformer to another. Because the force constant on a saddle point is negative, it has an imaginary root, leading to an imaginary frequency. Searching the potential energy surface of propene sometimes reveals a saddle point with the double bond eclipsed by one hydrogen of the methyl group as it rotates from one staggered conformation to another.

Dihedral Driver

A dihedral or "twist" angle A-B-C-D that A makes with D can be driven around the B-C axis in arbitrarily chosen angular increments. At each step during the drive, an energy is recorded so that the locus of energies as a function of drive angle gives a profile of the potential energy hill a molecule goes over on being driven away from a minimum. For example, a switch 1 in column 80 of line 2 of an input deck with a line like

    1000100002000030000400000018 0.000.0010.0

as the last line of an input deck, rotates **1** the dihedral angle connecting atoms 1 through 4, **0001000020000300004**, (skip 5 spaces, **00000**), from the *anti* angle **0180.** to the eclipsed angle **000.0** in increments of **010.0** degrees. (If you know some FORTRAN this incantation will make sense to you.) The zeroes are not necessary parts of the input file; they serve as place keepers. If a zero is omitted, however, a blank must take its place because this file is in strict MM3 format.

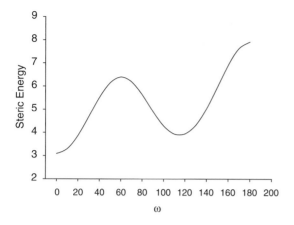

Figure 5-9 Steric Energy as Determined using the Dihedral Driver Option of MM3. Compare with Figure 4-12.

Exercise 5-6

Using Program MM3, the MM3 force field, SigmaPlot, and the comments in this section on dihedral driver, plot the steric energy of *n*-butane as a function of the dihedral angle from the *anti* conformation to the eclipsed conformation. Do not use a full matrix option as your optimization method; use the block diagonal option 1. Out of curiosity, you might want to try the other options to see what happens, just don't regard the results as your solution. The answer to any "What would happen if I..." question is, "Try it."

Solution 5-6

Your plot should resemble Fig. 5-9. Note that the angles of the dihedral driver rotate through negative values but it doesn't make any difference because the plot comes out the same, except for the change of $\omega = (-180 + \text{dihedral angle})$. The steric energies are most conveniently read from the **geo** file, which is TAPE9.MM3. PCMODEL contains a convenient dihedral driver and graphing combination (see PCMODEL user's manual).

One precaution is that, especially with congested molecules, these potential energy loci should not be taken too literally because rotated atoms or groups (within the model) can "stick" during rotation, then suddenly "snap into place", giving a potential energy discontinuity that has no counterpart in the real molecule.

Full Statistical Method

Very early force fields were used in an attempt to calculate structures, enthalpies of formation, and vibrational spectra, but it was soon found that accuracy suffered severely in either the structure-energy calculations or the vibrational spectra. Force constants were, on the whole, not transferable from one field to another. The result was that early force fields evolved so as to calculate *either* structure and energy or spectra, but not both.

To determine geometry, energy, and vibrational spectral frequencies, one must be able to determine the location, depth, and shape of the potential well corresponding to each bond in the model. Off-diagonal elements correspond to interactions like stretch-bend interactions that are reflected in the shape of the well. Early structure-energy force fields contained few off-diagonal elements, largely because of machine limitations. Machine limitations have become less severe during the later evolution of MM methods so that contemporary force fields can be used to calculate geometry and $\Delta_f H$ and also give good results for vibrational spectral lines. MM3 is one example of a newer, more versatile MM force field. The partial MM3 output file in Fig. 5-10 shows good agreement between calculated and experimental values.

no	Frequency	Symmetry	A(i)
1.	3721.5	(B2)	vs
2.	3666.3	(A1)	vs
3.	1593.7	(A1)	vs

Figure 5-10 Partial MM3 Output as Related to the Vibrational Spectrum of H_2O. The experimental values of the two stretching and one bending frequencies of water are 3756, 3657, and 1595 cm^{-1}. The IR intensities are all very strong (vs).

Entropy and Heat Capacity

A force field that can produce vibrational spectra has a second advantage in that the $\Delta_f H$ calculations can be put on a much more satisfactory theoretical base by calculating an enthalpy of formation at 0 K as in *ab initio* procedures and then adding various thermal energies by more rigorous means than simply lumping them in with empirical bond enthalpy contributions to $\Delta_f H^{298}$. The stronger the theoretical base, the less likely is an unwelcome surprise in the output.

Calculations by the more rigorous procedure yield, in MM3, a sum of (a) bond energies, (b) steric energy, (c) vibrational *zero point* and thermal energies, and (d) structural features POP and TORS. Energies (a), (b), and (d) are calculated as before. Bond energy parameters appear to be quite different from those of the default MM3 calculations carried out so far because zero point and thermal energies are not included in the parameters but are added later.

Energies (c) are calculated from the harmonic oscillator model (**Polyatomic Molecules**, Chapter 4). The quantized harmonic oscillator has a ladder of equidistant energy levels within a parabolic potential energy well. With a force field that provides good vibrational spectra, we can calculate a molecular energy at 0 K by summing bond energies from the constituent atoms and add energies (b) and (d) and then add (c), a half-quantum of energy called the zero point energy, *and a vibrational contribution* to the thermal energy that is calculated from rigorous statistical mechanical principles. The zero point and vibrational energy

	ENERGY (Kcal/mol)	ENTHALPY (Kcal/mol)	ENTROPY (eu)	FREE ENERGY (Kcal/mol)	HEAT CAPACITY (cal/mol/deg)
Translational	.889	1.481	34.593	−8.833	4.967
Rotational	.889	.889	10.375	−2.205	2.980
Vibrational	12.842%	12.842%	.008*	12.839%	.054*
Potential	.000#	.000#	.000	.000#	.000
Mixing	.000	.000	.000	.000	.000
Total	14.619	15.211	44.976	1.801	8.002

Figure 5-11 Partial MM3 Output Showing Entropy and Heat Capacities for Water. Experimental values are 45.13 cal K^{-1} mol^{-1} for the absolute entropy and 8.03 cal K^{-1} mol^{-1} for the heat capacity.

contributions are known as accurately as the vibrational spacings are known, that is, rather well if enough accurate off-diagonal elements are involved in the Hessian matrix (Fig. 5-11).

Fig. 5-11 shows that, for water, entropy and heat capacity are summations in which two terms dominate, the translational energy of motion of molecules treated as ideal gas particles, and rotational, energy of spin about axes having nonzero moments of inertia terms (see Problems).

Free Energy and Equilibrium

Having calculated the standard values $\Delta_f H°$ and $S°$ for the participants in a chemical reaction, the obvious next step is to calculate the standard Gibbs free energy change of reaction $\Delta_r G°$ and the equilibrium constant from

$$\Delta_r H° = \sum \Delta_f H°_{prod} - \sum \Delta_f H°_{react} \tag{5-50}$$

$$\Delta_r S° = \sum S°_{prod} - \sum S°_{react} \tag{5-51}$$

$$\Delta_r G° = \Delta_r H° - T\Delta_r S° \tag{5-52}$$

and

$$\Delta_r G° = -RT \ln K_{eq} \tag{5-53}$$

or, writing Eq. (5-53) in an equivalent form

$$K_{eq} = e^{-\Delta_r G°/RT} \tag{5-54}$$

Equation (5-54) makes clear a difficulty that will bedevil us throughout computational chemistry: *Although the accuracy of computational chemistry is extremely high, the demands placed on our results may be even higher.* In the present case, the equilibrium constant is dependent on the *exponential* of the standard free energy change; hence, a small error in any of the computations going into $\Delta_r G^\circ$ is reflected in a large error in K_{eq}. This inescapable exponential relationship must be borne in mind when evaluating calculated equilibrium constants and many other computed results in a practical or industrial context. However, lest we complain that nature has been unfair in placing such exacting demands on us for the accuracy of our calculations, let us be grateful that she permits us to do them at all.

Exercise 5-7: Calculation of K_{eq} at 298 K

Because of the severe demands placed on us for accuracy if we are to calculate an equilibrium constant, let us choose a simple reaction, the isomerization of but-2-ene.

cis-but-2-ene $\rightarrow$ $trans$-but-2-ene

Using MM3, calculate $\Delta_f H^\circ$ and S° leading to $\Delta_r G^\circ$ and K_{eq}. This reaction has been the subject of computational studies (Kar, Lenz, and Vaughan, 1994) and experimental studies by Akimoto et al. (Akimoto, Sprung, and Pitts, 1972) and by Kapeijn et al. (Kapeijn, van der Steen, and Mol, 1983). Quantum mechanical systems, including the quantum harmonic oscillator, will be treated in more detail in later chapters.

Solution 5-7

We found $K_{eq} = 5.33$ (**mm3 cbu2.mm3**, option **4**, TEMP, VIB, VIB, **Enter** (default); read the output file using **more TAPE4.MM3**, hit spacebar to advance page by page to STATISTICAL THERMODYNAMICS). Kar, Lenz, and Vaughan calculated $K_{eq} = 4.03$. Kapeijn, van der Steen, and Mol found $K_{eq} = 3.12$, and an experimental study by Akimoto, Sprung, and Pitts produced $K_{eq} = 2.98$ at 298 K.

COMPUTER PROJECT 5-4 | *More Complicated Systems*

Minimal input files for cyclopentene and cyclopentane can be constructed from a pentagon drawn on graph paper the way minimal ethylene was drawn (Fig. 4-4). For more complicated molecules, however, the **draw** function of PCMODEL or some similar file constructing program becomes less a convenience and more a necessity.

a) **Cyclopentene.** Calculate the enthalpy of hydrogenation of cyclopentene to cyclopentane and compare the result with $\Delta_{hyd} H^{298}$ (ethylene) found in Computer Project 5-2. The result for hydrogenation of cyclopentene differs from that for ethene by an amount that is well outside the expected error for MM calculations. Suggest a reason for this.

b) **Dimethylcyclopentene.** The hydrogenation product of dimethylcyclopentene

cis   *trans*

can exist in two forms, called *cis* or *trans*, according to whether the reaction product has both methyl groups on the same side of the plane of the cyclopentane ring or on opposite sides of the ring. (Most of the hydrogen atoms have been left out of the equation for simplicity.) The object of this part of the project is to compute the $\Delta_{hyd}H^{298}$ of the reactions leading to the *cis* and *trans* isomers and to compare them with the experimental value to see which of the isomers is actually formed.

Procedure. PCMODEL has a tool called **Rings**. Use this tool to establish a base molecule, either cyclopentene or cyclopentane. The **Build** tool converts a hydrogen atom into a methyl group, enabling you to build more complex molecules from simpler ones. Use **Build** to convert your base molecules to 1,2-dimethylcyclo-pentene, *cis*-dimethylcyclopentane, and *trans*-dimethylcyclopentane. Determine $\Delta_f H^{298}$ for all three molecules, and from these values determine which of the two reaction paths

cis

or

trans

is followed. [In the actual experiment (Allinger et al., 1982), both are formed but one predominates.] The experimental value is $\Delta_{hyd}H^{298}$ (1,2-dimethylcyclopen-tene) $= -22.5 \pm 0.2 \, \text{kcal mol}^{-1}$. This result is unusual, not only because it is a thermochemical analysis of parallel hydrogenation reactions, but because the enthalpy of hydrogenation is one of the smallest ever measured for a hydrocarbon. The reactant molecule is stabilized by methyl groups in the 1 and 2 positions, and the product, particularly the *cis* isomer, is destabilized by crowding. Both of these factors make hydrogenation less exothermic.

Among the unusual features of this reaction is that it is a surface-catalyzed reaction (Pd or Pt) but it gives a mixture of isomeric products. Suggest a mechanism by which this might occur.

c) Bicyclo[3.3.0]octane. If two cyclopentane molecules are fused at one bond the product is bicyclo[3.3.0]octane. Bicyclo[3.3.0]octane can exist as *cis* and *trans* isomers according to whether the two hydrogen atoms at the ring fusion are on the same side of the molecular plane or on opposite sides of it. Moreover, in the *cis* isomer, the pentagons themselves are bent slightly out of the plane of the molecule such that, viewed edgewise, three conformers are possible

The object of this part of the project is to determine the energy (enthalpy) levels in each the three conformers and so to determine the composition of the equilibrium conformational mixture. That having been done for the *cis* isomer, the procedure is repeated for the *trans* isomer.

Procedure.   Carry out a stochastic search on the *cis* isomer of bicyclo[3.3.0]octane to determine the relative energies of these three conformers. You need only the steric energies obtainable from STOCHASTIC.MEM for these relative energies because the summation of the bond energies is the same for all three conformers. Giving the lowest conformer an arbitrary zero energy, calculate the relative populations of the three states. You will need to match each energy with its structure.

  Carry out the same stochastic search over the conformational space of the *trans* isomer. The result of this search may surprise you at first, but there is a simple explanation, which you should include in your report.

PROBLEMS

1. A 10.0-g mass connected by a spring to a stationary point executes exactly 4 complete cycles of harmonic oscillation in 1.00 s. What are the period of oscillation, the frequency, and the angular frequency? What is the force constant of the spring?
2. The electric field of electromagnetic radiation completes 4.00×10^{13} complete cycles in 1.00 s. What are the period and frequency of the oscillation, and what is its wavelength? What is the frequency in units of cm^{-1}?
3. The hydrogen atom attached to an alkane molecule vibrates along the bond axis at a frequency of about 3000 cm^{-1}. What wavelength of electromagnetic radiation is resonant with this vibration? What is the frequency in hertz? What is the force constant of the C—H bond if the alkane is taken to be a stationary mass because of its size and the H atom is assumed to execute simple harmonic motion?
4. Three 10.0-g masses are connected by springs to fixed points as harmonic oscillators shown in Fig. 5-12. The Hooke's law force constants of the springs are $2k$, k, and k as shown, where $k = 2.00\,N\,m^{-1}$. What are the periods and frequencies of oscillation in hertz and radians per second in each of the three cases a, b, and c?

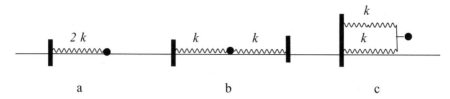

Figure 5-12 Three Harmonic Oscillators.

5. Consider a harmonic oscillator connected to another harmonic oscillator (Fig. 5-13). Write the sum of forces on each mass, m_1 and m_2. This is a classic problem in mechanics, closely related to the double pendulum (one pendulum suspended from another pendulum).
6. Set up the classical equations of motion for the system in Fig. 5-13.

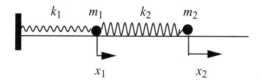

Figure 5-13 A Double Harmonic Oscillator. Displacements x_1 and x_2, shown positive, may also be negative or zero.

7. Write the equations of motion in matrix form. Take m_1 and m_2 to be unit masses (say 1.000 kg each).
8. For simplicity, take the specific case where $k_1 = k_2 = k$. Write the matrix of force constants analogous to matrix (5-29). Diagonalize this matrix. What are the roots? Discuss the motion of the double pendulum in contrast to two coupled, tethered masses (Fig. 5-1).
9. Write an expression for the kinetic energy T in a system of two masses connected by three springs including a central coupling spring. Write an expression for the potential energy V of this system. Write an expression for the total energy E. Note the similarity between this expression and the electronic Hamiltonian for helium, Eq. (8-17).
10. A spring stretches 0.200 m when it supports a mass of 0.250 kg.

    (a) What is the force constant of the spring?

    If the mass is pulled 0.0500 m below its equilibrium point and released:

    (b) What is the frequency of harmonic oscillation of the mass? What is its period of oscillation?

    (c) What is the potential energy of the mass at the extremes of its excursion?

    (d) What is the maximum speed of the mass?

11. Write and run an MM3 input file for methane "from scratch," that is, open an empty file and put in all the necessary information to do the MM3 calculation on CH_4. What is the enthalpy of formation of CH_4? What are the C—H bond lengths and angles?

12. Draw an enthalpy level diagram showing how Kistiakowsky was able to determine the value of $\Delta_{isom}H^{298}$ for the *cis* → *trans* isomerization from enthalpies of hydrogenation.

13. What is the MM3 enthalpy of formation at 298.15 K of styrene? Use the option **Mark** all **pi atoms** to take into account the conjugated double bonds in styrene. Is the minimum-energy structure planar, or does the ethylene group move out of the plane of the benzene ring?

14. Find the MM3 enthalpy of formation of 1- and 2-methyladamantane. Use the **Rings** tool and the **adamant** option to obtain the base structure of adamantane itself. Use the **Build** tool to add the methyl group. 1-Adamantane is the more symmetrical structure of the two isomers.

15. What enthalpy difference would lead to a 25–75% mixture of *syn* and *skew* rotamers of 1-butene? Neglect any entropy change.

CHAPTER

6

Huckel Molecular Orbital Theory I: Eigenvalues

Most problems in chemistry [all, according to Dirac (1929)] could be solved if we had a general method of obtaining exact solutions of the Schroedinger equation

$$\hat{H}\Psi = E\Psi$$

The reason a single equation $\hat{H}\Psi = E\Psi$ can describe all real or hypothetical mechanical systems is that the *Hamiltonian operator* $\hat{H}$ takes a different form for each new system. There is a limitation that accompanies the generality of the Hamiltonian and the Schroedinger equation: We cannot find the exact location of any electron, even in simple systems like the hydrogen atom. We must be satisfied with a probability distribution for the electron's whereabouts, governed by a function Ψ called the *wave function*.

   We cannot solve the Schroedinger equation in closed form for most systems. We have exact solutions for the energy E and the wave function Ψ for only a few of the simplest systems. In the general case, we must accept approximate solutions. The picture is not bleak, however, because approximate solutions are getting systematically better under the impact of contemporary advances in computer hardware and software. We may anticipate an exciting future in this fast-paced field.

Computational Chemistry Using the PC, Third Edition, by Donald W. Rogers
ISBN 0-471-42800-0   Copyright © 2003 John Wiley & Sons, Inc.

Exact Solutions of the Schroedinger Equation

Among the few systems that can be solved exactly are the particle in a one-dimensional "box," the hydrogen atom, and the hydrogen molecule ion H_2^+. Although of limited interest chemically, these systems are part of the foundation of the quantum mechanics we wish to apply to atomic and molecular theory. They also serve as benchmarks for the approximate methods we will use to treat larger systems.

The Hamiltonian $\hat{H}$ in the equation $\hat{H}\Psi = E\Psi$ is an operator, that is, it is an instruction telling you what operation or operations to perform on some function, in this case, the wave function. The Hamiltonian takes a different form for different mechanical systems. One of the simplest forms of $\hat{H}$ is

$$\hat{H} = -\frac{\hbar^2}{2m}\frac{d^2}{dx^2} + V \tag{6-1}$$

for a single particle of mass m constrained to move on the x-axis under a potential energy V. If V is zero for excursions of a particle between two limits in the x direction designated 0 and a but is infinite elsewhere, this form leads to the Schroedinger equation for what is called a particle in a one-dimensional box,

$$-\frac{\hbar^2}{2m}\frac{d^2}{dx^2}\Psi = E\Psi \tag{6-2}$$

The problem is treated in elementary physical chemistry books (e.g., Atkins, 1998) and leads to a set of *eigenvalues* (energies) and *eigenfunctions* (wave functions) as depicted in Fig. 6-1. It is solved by much the same methods as the harmonic oscillator in Chapter 4, and the solutions are sine, cosine, and exponential solutions just as those of the harmonic oscillator are. This gives the wave function in Fig. 6-1 its sinusoidal form.

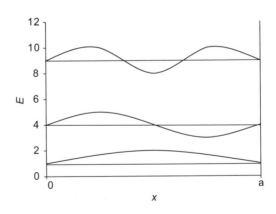

Figure 6-1 Energy Levels E and Wave Functions Ψ for a Particle in a One-Dimensional Box.

The eigenvalues are *discrete* values of E, in this case 1, 4, 9, ... units plotted as horizontal lines on the vertical axis in Fig. 6-1. The unit of energy in Fig. 6-1 is $h^2/8ma^2$, where h is Planck's constant and a is the length of the permitted excursion of the particle along the x-axis. The eigenfunctions appear as sine waves having zero, one, or two *internal nodes* for the three eigenfunctions shown in Fig. 6-1. A *node* is a point where the function crosses the horizontal ($\Psi = 0$). All eigenfunctions for this system have two terminal nodes, one at either end. In general, the higher the energy eigenvalue of a system, the greater the number of nodes in the corresponding eigenfunction. There is an infinite number of possible energy values (solutions) for this system, but this is not the case for many other systems.

Another general observation that comes from consideration of simple systems like this one is that separation of discrete energy levels (quantization) appears when the motion of the particle is restricted in some way, in this case by placing barriers at $x = 0$ and $x = a$. A particle that is not constrained (free particle) is not quantized; its energy can take on any value. These general observations will be useful in understanding and interpreting more complicated systems.

All solutions of the Schroedinger equation lead to a set of integers called *quantum numbers*. In the case of the particle in a box, the quantum numbers are $n = 1, 2, 3, \ldots$. The allowed (quantized) energies are related to the quantum numbers by the equation

$$E = \frac{n^2 h^2}{8ma^2} \tag{6-3}$$

The *hydrogen atom* is a three-dimensional problem in which the attractive force of the nucleus has spherical symmetry. Therefore, it is advantageous to set up and solve the problem in spherical polar coordinates r, θ, and ϕ. The resulting equation can be broken up into three parts, one a function of r only, one a function of θ only, and one a function of ϕ. These can be solved separately and exactly. Each equation leads to a quantum number

$$\begin{aligned} R(r) &\Rightarrow n \\ \Theta(\theta) &\Rightarrow l \\ \Phi(\phi) &\Rightarrow m \end{aligned} \tag{6-4}$$

These are three of the four quantum numbers familiar from general chemistry. The spin quantum number s arises when relativity is included in the problem, introducing a fourth dimension.

The *hydrogen molecule ion* is best set up in confocal elliptical coordinates with the two protons at the foci of the ellipse and one electron moving in their combined potential field. Solution follows in much the same way as it did for the hydrogen atom but with considerably more algebraic detail (Pauling and Wilson, 1935; Grivet, 2002). The solution is exact for this system (Hanna, 1981).

In the few two- and three-dimensional cases that permit exact solution of the Schroedinger equation, the complete equation is separated into one equation in each dimension and the energy of the system is obtained by solving the separated equations and *summing* the eigenvalues. The wave function of the system is the *product* of the wave functions obtained for the separated equations.

Approximate Solutions

The logical order in which to present molecular orbital calculations is *ab initio*, with no approximations, through semiempirical calculations with a restricted number of approximations, to Huckel molecular orbital calculations in which the approximations are numerous and severe. Mathematically, however, the best order of presentation is just the reverse, with the progression from simple to difficult methods being from Huckel methods to *ab initio* calculations. We shall take this order in the following pages so that the mathematical steps can be presented in a graded way.

The simplest molecular orbital method to use, and the one involving the most drastic approximations and assumptions, is the Huckel method. One strength of the Huckel method is that it provides a semiquantitative theoretical treatment of ground-state energies, bond orders, electron densities, and free valences that appeals to the pictorial sense of molecular structure and reactive affinity that most chemists use in their everyday work. Although one rarely sees Huckel calculations in the research literature anymore, they introduce the reader to many of the concepts and much of the nomenclature used in more rigorous molecular orbital calculations.

We have said that the Schroedinger equation for molecules cannot be solved exactly. This is because the exact equation is usually not separable into uncoupled equations involving only one space variable. One strategy for circumventing the problem is to make assumptions that permit us to write approximate forms of the Schroedinger equation for molecules that are separable. There is then a choice as to how to solve the separated equations. The Huckel method is one possibility. The self-consistent field method (Chapter 8) is another.

Three major approximations are made to separate the Schroedinger equation into a set of smaller equations before carrying out Huckel calculations.

1. The Born–Oppenheimer Approximation

We assume that the nuclei are so slow moving relative to electrons that we may regard them as fixed masses. This amounts to separation of the Schroedinger equation into two parts, one for nuclei and one for electrons. We then drop the nuclear kinetic energy operator, but we retain the internuclear repulsion terms, which we know from the nuclear charges and the internuclear distances. We retain all terms that involve electrons, including the potential energy terms due to attractive forces between nuclei and electrons and those due to repulsive forces among electrons. In chemical calculations, we usually write the Schroedinger equation in the same form as the original equation, $\hat{H}\Psi = E\Psi$, with the understanding that it now relates to electronic motion only.

The Born–Oppenheimer approximation is not peculiar to the Huckel molecular orbital method. It is used in virtually all molecular orbital calculations and most atomic energy calculations. It is an excellent approximation in the sense that the approximated energies are very close to the energies we get in test cases on simple systems where the approximation is not made.

Atomic Units

There is a very convenient way of writing the Hamiltonian operator for atomic and molecular systems. One simply writes a kinetic energy part $-\frac{1}{2}\nabla^2$ for each electron and a Coulombic potential $\pm Z/r$ for each interparticle electrostatic interaction. In the Coulombic potential Z is the charge and r is the interparticle distance. The term $\pm Z/r$ is also an operator signifying "multiply by $\pm Z/r$". The sign is $+$ for repulsion and $-$ for attraction.

The symbol ∇^2 is an operator that takes the form

$$\frac{d^2}{dx^2}$$

in Cartesian 1-space,

$$\nabla^2 = \frac{\partial^2}{\partial x^2} + \frac{\partial^2}{\partial y^2} + \frac{\partial^2}{\partial z^2}$$

in Cartesian 3-space, or a more complicated form in other coordinate systems that we might use, such as spherical polar or confocal ellipsoidal coordinates.

Writing $-\frac{1}{2}\nabla^2$ for the kinetic energy of each electron amounts to taking our unit of mass as the mass of the electron instead of the kilogram (which is an arbitrary unit anyway) and defining $\hbar$, the unit of angular momentum, as 1. The same thing can be done with the units of charge in an electrical potential, leaving $V = \pm 1/r$ or $V = \pm Z/r$ for multiply charged species. See Mc Quarrie (1983) for a table of *atomic units* and their SI equivalents. The atomic unit of energy in this system is the hartree, $h = 4.2359 \times 10^{-18}$ joules, and the energy of the ground state of the hydrogen atom is exactly $\frac{1}{2}$ h. Be careful not to confuse the unit h with Planck's constant h.

The sum of two operators is an operator. Thus the Hamiltonian operator for the hydrogen atom has $-\frac{1}{2}\nabla^2$ as the kinetic energy part owing to its single electron plus $-1/r$ as the electrostatic potential energy part, because the charge on the nucleus is $Z = 1$, the force is attractive, and there is one electron at a distance r from the nucleus

$$\hat{H} = -\tfrac{1}{2}\nabla^2 - \frac{1}{r} \tag{6-5}$$

Exercise 6-1

Write the Hamiltonian for the helium atom, which has two electrons, one at a distance r_1 and the other at a distance r_2.

Solution 6-1

$$\hat{H} = -\tfrac{1}{2}\nabla_1^2 - \tfrac{1}{2}\nabla_2^2 - \frac{2}{r_1} - \frac{2}{r_2} + \frac{1}{r_{12}}$$

Notice the interelectronic repulsion term $+1/r_{12}$

These equations are mathematically identical to longer forms such as

$$\hat{H} = -\frac{\hbar^2}{2m}\nabla_1^2 - \frac{\hbar^2}{2m}\nabla_2^2 - \frac{Ze^2}{4\pi\varepsilon_0 r_1} - \frac{Ze^2}{4\pi\varepsilon_0 r_2} + \frac{1}{4\pi\varepsilon_0 r_{12}}$$

for helium.

Using atomic units, the Schroedinger equation for ground-state hydrogen is

$$\left[-\tfrac{1}{2}\nabla^2 - \frac{1}{r}\right]\Psi_{1s} = E\Psi_{1s} \tag{6-6}$$

where ∇^2 is now a kinetic energy operator in 3-space. We can write the Hamiltonian operator for the hydrogen atom in Cartesian 3-space, but the resulting Schroedinger equation is very difficult to solve. Instead, it is converted to spherical polar coordinates by routine but somewhat lengthy manipulations (Barrante, 1998). The Schroedinger equation in the new coordinate system is separated, and the three resulting equations are solved for $R(r)$, $\Theta(\theta)$, and $\Phi(\phi)$. A similar procedure is followed to obtain the exact solution for the hydrogen molecule ion in confocal ellipsoidal coordinates.

By extension of Exercise 6-1, the Hamiltonian for a many-electron molecule has a sum of kinetic energy operators $-\tfrac{1}{2}\nabla^2$, one for each electron. Also, each electron moves in the potential field of the nuclei and all other electrons, each contributing a potential energy V,

$$\hat{H} = \sum -\tfrac{1}{2}\nabla_i^2 + \sum V_i \tag{6-7}$$

where the terms V_i may be $-$ or $+$ according to whether the electron is attracted (to nuclei) or repelled (by other electrons). One can split up the attractive and repulsive terms

$$\hat{H} = \sum [-\tfrac{1}{2}\nabla_i^2 + V_i] + \tfrac{1}{2}\sum \frac{1}{r_{ij}} \tag{6-8}$$

where the V_i terms are now all negative and the repulsive sum is multiplied by $\tfrac{1}{2}$ because we do not want to count repulsions twice, once as i-j repulsions and again as j-i repulsions. Equation (6-8) makes molecular problems look like nothing but a bunch of hydrogen atom problems plus a small interelectronic repulsion term. Unfortunately, this picture is deceptively simple, but it does lead to a useful approximation.

2. The Independent Particle Approximation

By analogy to the atomic orbitals that Schroedinger gave us for the hydrogen atom, we assume that molecules will have orbitals too, that they will define certain electron probability distributions in space, and that they will have specific energies. We call them *molecular orbitals*. Indeed, the exact solution of the Schroedinger equation for the hydrogen molecule ion H_2^+ has exactly these characteristics, a low-energy *bonding* orbital in which the electron probability density is relatively high between the nuclei and a high-energy *antibonding* orbital in which there is a node in electron probability density between the nuclei.

The Hamiltonian operator in Eq. (6-8) is called an *n*-electron Hamiltonian

$$\hat{H} = \hat{H}(1, 2, 3, \ldots, n)$$

The reason the Schroedinger equation for molecules cannot be separated appears in the last term, $\frac{1}{2} \sum \frac{1}{r_{ij}}$, involving a sum of repulsive energies between electrons. To obtain this sum, or even one term of it, say for the *i*th electron, one must know exactly where the *j*th electron is. This is because repulsive force is dependent on distance. The position of the *j*th electron depends on the position of the *i*th electron, however, which is what we are trying to find. Knowing either of these positions exactly (as required by the problem) is a violation of the Heisenberg uncertainty principle for electrons having a kinetic energy within finite limits. The problem is insoluble.

In the *independent particle approximation*, the simplifying assumption is made that $V'(i)$ is an average potential due to a core that consists of the nuclei and all electrons other than electron *i*

$$\hat{H}(i) = -\tfrac{1}{2}\nabla^2(i) + V'(i) \tag{6-9}$$

With this new approximation, the r_{ij} term does not appear [it is hidden in $V'(i)$] and the Schroedinger equation becomes separable into *n* equations, one for each electron

$$\hat{H}(i)\psi(i) = E(i)\psi(i) \tag{6-10}$$

where $\hat{H}(i)$ includes the new potential energy term $V'(i)$. This term is unknown (the r_{ij} problem hasn't gone away); hence, no closed solution exists for Eq. (6-10). We are using the symbol ψ for a wave function for which we do not have an exact solution and ψ for an exact wave function. Throughout, there is an implicit assumption that ψ exists even though it may not be mathematically accessible.

The Basis Set

One way to obtain an approximate solution of Eq. (6-10) is to select any set of *basis functions*,

$$\psi = \sum a_i \phi_i \tag{6-11}$$

as an approximation to the true wave function. If the basis set of the ϕs is judiciously chosen ψ may be a good approximation to the true Ψ. As might be expected, a poorly chosen basis set gives a poor approximation to Ψ. After yet one more simplifying assumption, we will look at one way of choosing an appropriate basis set and we will develop an iterative procedure to obtain the best ψ from any set $\sum a_i \phi_i$.

3. The π-Electron Separation Approximation

It has been known for more than a century that hydrocarbons containing double bonds are more reactive than their counterparts that do not contain double bonds. Alkenes are, in general, more reactive than alkanes. We call electrons in double bonds π electrons and those in the much less reactive C—C or CH bonds σ electrons. In Huckel theory, we assume that the chemistry of unsaturated hydrocarbons is so dominated by the chemistry of their double bonds that we may separate the Schroedinger equation yet again, into an equation for σ electrons and one for π electrons. We assume that we can ignore the σ electrons, as we did the nuclei, except for their contribution to the potential energy. We now have an equation of the same form as Eq. (6-8), but one in which the Hamiltonian for all electrons is replaced by the Hamiltonian for π electrons only

$$\hat{H} = \sum_{i}^{n} [-\tfrac{1}{2}\nabla^2(i) + V_\pi(i)] + \tfrac{1}{2}\sum \frac{1}{r_{ij}} \tag{6-12}$$

with somewhat different meanings for the symbols. Now, n is the number of π electrons of kinetic energy $-\tfrac{1}{2}\nabla^2(i)$, and the potential energy term $V_\pi(i)$ represents the potential energy of a single π electron in the average field of the framework of nuclei and all electrons except electron i. There is one π electron for each C atom participating in a system of C=C double bonds. In Huckel theory, this is now the *core potential*.

In summary, we have made three assumptions 1) the Born–Oppenheimer approximation, 2) the independent particle assumption governing molecular orbitals, and 3) the assumption of π-σ separation. The first two assumptions are characteristic of any molecular orbital theory, but the third is unique to the Huckel molecular orbital method.

The Huckel Method

In the late 1920s, it was shown that the chemical bond existing between two identical hydrogen atoms in H_2 can be described mathematically by taking a linear combination of the $1s$ orbitals ϕ_1 and ϕ_2 of the two H atoms that are partners in the molecule (Heitler and London, 1927). When this is done, the combination

$$\psi = a_1\phi_1 + a_2\phi_2 \tag{6-13}$$

is a new solution of the Schroedinger equation that has the characteristics of a chemical bond. Specifically, the energy calculated from Eq. (6-13) as a function of internuclear distance R goes through a minimum when R approximates the bond distance in H_2. The Heitler–London method is known as the *valence bond approximation* (VB). The right side of Eq. (6-13) is called a *linear combination of atomic orbitals* (LCAO). The orbitals ϕ_1 and ϕ_2 are members of a small basis set. There is also a negative combination of ϕ_1 and ϕ_2 that produces an antibonding solution. Using the VB method, one arrives at a description of chemical bonding from a somewhat different logical premise from that of the molecular orbital (MO) method, but, in their more extended forms, the two methods approach each other, as they must, because they are attempts to describe the same thing. The LCAO method is only one of infinitely many ways that a molecular orbital can be approximated.

A few years later, Huckel (1931, 1932) showed that the LCAO approximation can be applied to the single electrons of the p atomic orbitals of the carbon atoms that are partners in a $C=C$ double bond. The p orbitals are considered to be independent of the σ bonded framework except for the potential energy of charge interaction [Eq. (6-12)]. The linear combination is

$$\psi = \sum a_i p_i \qquad (6\text{-}14)$$

where the p_i are the unhybridized p orbitals of the double-bonded carbon atoms (Fig. 6-2) over two or more conjugated carbon atoms. When this sum is taken for ethylene,

$$\psi_+ = a_1 p_1 + a_2 p_2 \qquad (6\text{-}15a)$$

and

$$\psi_- = a_1 p_1 - a_2 p_2 \qquad (6\text{-}15b)$$

It is a property of linear, homogeneous differential equations, of which the Schroedinger equation is one, that a solution multiplied by a constant is a solution and a solution added to or subtracted from a solution is also a solution. If the solutions p_1 and p_2 in Eq. set (6-15) were exact molecular orbitals, ψ_+ and ψ_- would also be exact. Orbitals p_1 and p_2 are not exact molecular orbitals; they are exact *atomic* orbitals; therefore, ψ is not exact for the ethylene *molecule*.

Figure 6-2 The π orbital of Ethylene. Both interactions ------ between p_z orbitals of carbon ⬡ contribute to a single π orbital of ethylene.

The Expectation Value of the Energy: The Variational Method

Premultiplying each side of the Schroedinger equation by ψ gives

$$\psi E \psi = \psi \hat{H} \psi$$

but E is a scalar; hence,

$$E \psi \psi = \psi \hat{H} \psi$$

for one electronic configuration in ethylene. For all electronic configurations, one must integrate over all space $d\tau$

$$E \int \psi \psi \, d\tau = \int \psi \hat{H} \psi \, d\tau$$

or

$$E = \frac{\int \psi H \psi \, d\tau}{\int \psi^2 d\tau} \tag{6-16}$$

The term *configuration* is used in this context to designate a specific pair of locations for the two electrons in the π bond of ethylene. Even though we may not know exactly where the electrons are, we can know the probability density that they will be in a specific finite region of space by integrating over that region. From Lewis theory (see, e.g., Ebbing and Gammon, 1999) we are accustomed to thinking of an electron pair as constituting a chemical bond. The π bond in ethylene arises from such a pair of electrons, one each from the unhybridized p_z orbitals of carbon, which are not engaged in σ bonding.

It is a fundamental *postulate* of quantum mechanics that E in Eq. (6-16) is the expectation value of the energy for wave function ψ. If the values of Ψ are exact, E is exact. If the ψ values are approximate, as they are in this case, E is an *upper bound* on the true energy. Of two calculated E values, the higher one must be farther from the true value than the lower one, so it is discarded. Minimizing E, which is called *the variational method*, will be used to obtain the best value of ψ (the one that gives the lowest energy) from a given basis set. Criteria other than the energy may be selected, leading to different estimates of how closely ψ approximates its exact value. All properties of the system approach their true values as ψ approaches its exact value.

Exercise 6-2

In 1913 Bohr showed, by an argument that was essentially a combination of classical mechanics and quantum mechanics as it was known at that time, that the energy *spectrum* (ordered set of energy values) of hydrogen is given by

$$E = -\frac{me^4}{8\varepsilon_0^2 n^2 h^2} \tag{6-17}$$

where $n = 1$ for the ground state and $\varepsilon_0 = 8.854 \times 10^{-12}\,C^2N^{-1}m^{-2}$ is a physical constant called the vacuum permittivity. Substitute for the constants in the Bohr equation to obtain a ground state value for E. Give units.

Solution 6-2

The energy of an electron attracted to a proton is negative relative to infinite particle separation,

$$E = -\frac{me^4}{8\varepsilon_0^2 n^2 h^2} = -\frac{9.109 \times 10^{-31}\text{kg}(1.602 \times 10^{-19}\text{C})^4}{8(8.854 \times 10^{-12}C^2N^{-1}m^{-2})^2(6.626 \times 10^{-34}\text{J s})^2}$$
$$= -2.179 \times 10^{-18}\text{J}$$

$$\text{Units:} \quad \frac{\text{kg}\,C^4}{C^4N^{-2}m^{-4}J^2s^2} = \frac{\text{kg m}\,N^2m^3}{s^2}\frac{N^3m^3}{J^2} = \frac{N^3m^3}{J^2} = N\,m = J$$

where the charge is in coulombs (C). The vacuum permittivity ε_0, a constant from Coulomb's law of force f resulting from charge q interaction $f = (1/4\pi\varepsilon_0)(q^2/r^2)$ (Young and Friedman, 2000), is used here essentially as a factor to convert the potential energy between two charges from coulombs^2 per meter to joules. As indicated, the units of ε_0 are ε_0: $C^2m^{-1}/J = C^2N^{-1}m^{-2}$. In atomic units, $E = -2.179 \times 10^{-18}\text{J} = $ exactly $\frac{1}{2}$ hartree.

Exercise 6-3

Schroedinger (1926) showed that the wave function or *orbital* for the hydrogen atom in its ground state can be written

$$\Psi = e^{-\alpha r} \tag{6-18}$$

where r is the radial distance between the proton and the electron and α contains several constants. Find the upper bound of the energy for the hydrogen atom by the variational method.

Solution 6-3

Choosing appropriate units for the charge, the Hamiltonian for radial motion can be written

$$\hat{H} = -\frac{\hbar^2}{2m}\frac{1}{r^2}\frac{d}{dr}r^2\frac{d}{dr} - \frac{e^2}{r} \tag{6-19}$$

where the form of the kinetic energy operator arises from the transformation of ∇^2 from Cartesian coordinates to spherical polar coordinates (Barrante, 1998). This leads to an expression for the upper limit on E [Eq. (6-16)] from the wave function $e^{-\alpha r}$

$$E = \frac{\int_0^\infty e^{-\alpha r}\left(-\frac{\hbar^2}{2m}\frac{1}{r^2}\frac{d}{dr}r^2\frac{d}{dr} - \frac{e^2}{r}\right)e^{-\alpha r}4\pi r^2\,dr}{\int_0^\infty (e^{-\alpha r})^2 4\pi r^2\,dr}$$

where the factor $4\pi r^2 \, dr$ accounts for the radial part of the Schroedinger equation. (The radius vector from the nucleus may be in any direction; hence, its point may be anywhere on the surface of a sphere of area $4\pi r^2$.) After factoring and canceling 4π,

$$E = \frac{\int_0^\infty e^{-\alpha r}\left(-\frac{\hbar^2}{2m}\frac{1}{r^2}\frac{d}{dr}r^2\frac{d}{dr}-\frac{e^2}{r}\right)e^{-\alpha r}r^2\,dr}{\int_0^\infty (e^{-\alpha r})^2 r^2\,dr} \tag{6-20}$$

The kinetic energy operator operating on $e^{-\alpha r}$ gives

$$-\frac{\hbar^2}{2m}\frac{1}{r^2}\frac{d}{dr}r^2\frac{d}{dr}e^{-\alpha r} = -\frac{\hbar^2}{2m}\left(\alpha^2 - \frac{2\alpha}{r}\right)e^{-\alpha r} \tag{6-21}$$

and $(e^{-\alpha r})^2 = e^{-2\alpha r}$, so

$$E = \frac{\int_0^\infty e^{-\alpha r}\left(-\frac{\hbar^2}{2m}\left(\alpha^2 - \frac{2\alpha}{r}\right)e^{-\alpha r} - \frac{e^2}{r}e^{-\alpha r}\right)r^2\,dr}{\int_0^\infty e^{-2\alpha r}r^2\,dr} \tag{6-22}$$

Expanding the numerator we get

$$E = \frac{\int_0^\infty -\frac{\hbar^2\alpha^2}{2m}r^2e^{-2\alpha r}\,dr + \int_0^\infty \frac{\hbar^2\alpha}{m}re^{-2\alpha r}\,dr - \int_0^\infty e^2 re^{-2\alpha r}\,dr}{\int_0^\infty r^2 e^{-2\alpha r}\,dr} \tag{6-23}$$

This looks messy, but we really only need to evaluate two integrals, one with r and the other with r^2. The integrals are of the known form

$$\int_0^\infty x^n e^{-ax}\,dx = \frac{n!}{a^{n+1}}$$

so we get

$$E = \frac{-\frac{\hbar^2\alpha^2}{2m}\left(\frac{1}{4\alpha^3}\right) + \frac{\hbar^2\alpha}{m}\left(\frac{1}{4\alpha^2}\right) - e^2\left(\frac{1}{4\alpha^2}\right)}{\frac{1}{4\alpha^3}} = -\frac{\hbar^2\alpha^2}{2m} + \frac{\hbar^2\alpha}{m}\alpha - e^2\alpha \tag{6-24}$$

or

$$E = \frac{\hbar^2\alpha^2}{2m} - e^2\alpha \tag{6-25}$$

We can minimize the energy of the system with respect to the minimization parameter α by simply taking the first derivative and setting it equal to zero

$$\frac{d}{d\alpha}\left(\frac{\hbar^2\alpha^2}{2m} - e^2\alpha\right) = \frac{\hbar^2\alpha}{m} - e^2 = 0 \tag{6-26}$$

so that

$$\alpha = \frac{me^2}{\hbar^2} \tag{6-27}$$

(In more complicated cases, we shall have to verify that the extremum is a minimum.) Once we know that $\alpha = me^2/\hbar^2$ we can substitute it back into the energy equation and find

$$
\begin{aligned}
E &= \frac{\hbar^2\alpha^2}{2m} - e^2\alpha = \frac{\hbar^2}{2m}\left(\frac{me^2}{\hbar^2}\right)^2 - e^2\left(\frac{me^2}{\hbar^2}\right) \\
&= \frac{1}{2}\left(\frac{me^4}{\hbar^2}\right) - \left(\frac{me^4}{\hbar^2}\right) = -\frac{1}{2}\left(\frac{me^4}{\hbar^2}\right)
\end{aligned}
\tag{6-28}
$$

or, for charge separation in a vacuum and E to be expressed in joules (which brings back $4\pi\varepsilon_0$),

$$
E = -\frac{1}{2}\left(\frac{me^4}{(4\pi\varepsilon_0)^2\hbar^2}\right) = -\frac{me^4}{32\pi^2\varepsilon_0^2\hbar^2}
$$

Recognizing that $\hbar = h/2\pi$, we get

$$
E = -\frac{me^4}{8\varepsilon_0^2h^2}
\tag{6-17}
$$

which is the energy expression we solved for in Exercise 6-2 with $n = 1$ in the ground state. Going back to Eq. (6-27) $m = e = \hbar = 1$ in atomic units, so $E = -\frac{1}{2}$ hartree.

The upper bound in Exercise 6-3 turns out to be exactly the energy of the hydrogen atom in its ground state. This should not come as a surprise, because we started with an exact ground-state orbital. In the general case we will not know Ψ but we will always be able to identify the better of two trial functions because it will give the lower energy. The simple hydrogen orbital $e^{-\alpha r}$ is not normalized. If it had been normalized, we would have had the form $Ne^{-\alpha r}$ where both N and α are collections of constants.

COMPUTER PROJECT 6-1 | *Another Variational Treatment of the Hydrogen Atom*

Part A. In Exercise 6-3 we found that the closed solutions for the integral Eq. (6-23)

$$
E = \frac{\int_0^\infty -\frac{\hbar^2\alpha^2}{2m}r^2e^{-2\alpha r}\,dr + \int_0^\infty \frac{\hbar^2\alpha}{m}re^{-2\alpha r}\,dr - \int_0^\infty e^2 r\,e^{-2\alpha r}\,dr}{\int_0^\infty r^2e^{-2\alpha r}\,dr}
\tag{6-23}
$$

lead to the least upper bound for the energy of the hydrogen atom in the ground state: $E = -\frac{1}{2}$ hartree. Instead of solving three integrals in the numerator and one in the denominator, carry out one numerical integration of the numerator and one integration of the denominator in the expression

$$
E = \frac{\int_0^\infty (-\frac{1}{2}\alpha^2 r^2 + \alpha r - r)\,e^{-2\alpha r}\,dr}{\int_0^\infty r^2e^{-2\alpha r}\,dr}
\tag{6-29}
$$

at several values of α to find out which value of α gives the most negative value of E. Note that Eq. (6-29) is Eq. (6-23) with atomic units $m = e = \hbar = 1$. We shall expect E to be in atomic units of hartrees. In this determination, one can follow a procedure similar to the method of steepest descent used in iterative computer searches for an energy minimum. Pick two values of α and get the direction of descent by going from the more positive to the more negative of the two resulting energies. Now pick another α in the direction of descent and repeat this as many times as needed to find the minimum. Of course, the energy must be a well-behaved function of α for this to work (and it is). The size of the steps to be taken is determined by trial and error; if you overshoot the minimum, go back and take smaller steps.

To make an informed guess for your first value of α, you may wish to reread the section on the Bohr theory of the hydrogen atom and the Schroedinger wave functions for the hydrogen atom in a good physical or general chemistry book (see Bibliography).

A sample determination is

Mathcad

$$\alpha := 1.3$$

$$p := \int_0^{10} \left(-.5 \cdot \alpha^2 \cdot r^2 + \alpha \cdot r - r\right)\exp(-2 \cdot \alpha \cdot r)\, dr$$

$$q := \int_0^{10} r^2 \exp(t - 2 \cdot \alpha \cdot r)\, dr$$

$$E := \frac{p}{q} \qquad E = -0.455$$

This is not E_{min}, of course; you must find the minimum energy by systematic variation of α. Alternatively, a QBASIC or TBASIC program can be written to integrate Eq. (6-28) by Simpson's rule.

Complete Part A of this project by determining about 10 energies at various values of α over a range that is sufficient to prove that E is a well-behaved function of α with a minimum. Report the least upper bound of E and the value of α at which it is found.

Part B. Repeat the entire process of Part A using a *Gaussian* approximation to the wave function for the ground state of the hydrogen atom

$$\psi = e^{-\gamma r^2} \tag{6-30}$$

First decide what the integral equation corresponding to Eq. (6-29) is for the approximate wave function (6-30), then integrate it for various values of γ. Report both γ at the minimum energy and E_{min} for the Gaussian approximation function. This is a least upper bound to the energy of the system. Your report should include a drawing of E as a function of γ (**SigmaPlot** or equivalent) for enough values of γ to make a clear picture of what the minimization function looks like and whether it is well behaved in the vicinity of E_{min}. Comment on the comparison between E_{min} for the ground state approximation function in Part A and E_{min} for the Gaussian approximation function in Part B.

Huckel Theory and the LCAO Approximation

Returning to Huckel theory for ethylene, and substituting the first LCAO [Eq. (6-15a)] for ψ, we have

$$E = \frac{\int (a_1p_1 + a_2p_2)\hat{H}(a_1p_1 + a_2p_2)d\tau}{\int (a_1p_1 + a_2p_2)^2 d\tau} \tag{6-31}$$

which is Eq. (6-16) with the linear combinations of atomic p orbitals substituted for the approximate molecular orbitals ψ. Expanding this equation yields four integrals in the numerator and four in the denominator. This takes a lot of space, so we use the notation

$$\int p_1\hat{H}p_1\,d\tau = \int p_2\hat{H}p_2\,d\tau = \alpha \tag{6-32a}$$

$$\int p_1\hat{H}p_2\,d\tau = \int p_2\hat{H}p_1\,d\tau = \beta \tag{6-32b}$$

$$\int p_1p_1\,d\tau = \int p_2p_2\,d\tau = S_{11} = S_{22} \tag{6-32c}$$

and

$$\int p_1p_2\,d\tau = \int p_2p_1\,d\tau = S_{12} = S_{21} \tag{6-32d}$$

We have assumed that the order of the subscripts on the atomic orbitals p is immaterial in writing α, β, and S. In the general case, these assumptions are not self-evident, especially for β. The interested reader should consult a good quantum mechanics text (e.g., Hanna, 1981; McQuarrie, 1983; Atkins and Friedman, 1997) for their justification or critique.

The expression for the energy, after all assumptions and notational simplifications have been made, is

$$E = \frac{a_1^2\alpha + 2a_1a_2\beta + a_2^2\alpha}{a_1^2S_{11} + 2a_1a_2S_{12} + a_2^2S_{22}} \tag{6-33}$$

If we could evaluate α, β, and S, which are called the *coulomb*, *exchange*, and *overlap integrals* respectively, we could compute E.

We do not know either side of Eq. (6-33), but we do know that E is to be minimized with respect to some minimization parameters. The only arbitrary parameters we have are the a_1 and a_2, which enter into the LCAO. Thus our normal equations are

$$\frac{\partial E}{\partial a_1} = 0 \tag{6-34a}$$

and

$$\frac{\partial E}{\partial a_2} = 0 \tag{6-34b}$$

These minimizations lead to

$$a_1\alpha + a_2\beta = E(a_1 S_{11} + a_2 S_{12}) \tag{6-35a}$$

and

$$a_1\beta + a_2\alpha = E(a_1 S_{12} + a_2 S_{22}) \tag{6-35b}$$

or

$$a_1(\alpha - ES_{11}) + a_2(\beta - ES_{12}) = 0 \tag{6-36a}$$

and

$$a_1(\beta - ES_{12}) + a_2(\alpha - ES_{22}) = 0 \tag{6-36b}$$

A further simplification is made. The wave functions p_1 and p_2, which are *orthogonal* and *normalized* in the hydrogen atom, are assumed to retain their *orthonormality* in the molecule. Orthonormality requires that

$$S_{11} = S_{22} = \int p_1 p_1 \, d\tau = \int p_2 p_2 \, d\tau = 1 \tag{6-37a}$$

and

$$S_{12} = S_{21} = \int p_1 p_2 \, d\tau = \int p_2 p_1 \, d\tau = 0 \tag{6-37b}$$

This yields

$$(\alpha - E)a_1 + \beta a_2 = 0 \tag{6-38a}$$

and

$$\beta a_1 + (\alpha - E)a_2 = 0 \tag{6-38b}$$

as the normal equations having the solution set $\{a_1, a_2\}$. In this context, the normal equations are also called *secular equations*. The exchange integral β is sometimes called the *resonance integral*.

Homogeneous Simultaneous Equations

What we formerly called the *nonhomogeneous vector* (Chapter 2) is zero in the pair of simultaneous normal equations Eq. set (6-38). When this vector vanishes, the pair is *homogeneous*. Let us try to construct a simple set of linearly independent homogeneous simultaneous equations.

$$x + y = 0 \tag{6-39a}$$
$$x + 2y = 0 \tag{6-39b}$$

These equations cannot be true for any solution set other than $\{0, 0\}$. The determinant of the coefficients is not zero

$$\begin{vmatrix} 2 & 1 \\ 1 & 1 \end{vmatrix} = 2 - 1 = 1 \tag{6-39c}$$

Any linearly independent set of simultaneous homogeneous equations we can construct has only the zero vector as its solution set. This is not acceptable, for it means that the wave function vanishes, which is contrary to hypothesis (the electron has to be somewhere). We are driven to the conclusion that the normal equations (6-38) must be *linearly dependent*.

Linearly dependent sets of homogeneous simultaneous equations, for example,

$$x + 2y = 0 \tag{6-40a}$$
$$2x + 4y = 0 \tag{6-40b}$$

are true for any solution set you care to try. They have infinitely many solution sets. The determinant of the coefficients of linearly dependent homogeneous simultaneous equations is zero. For example,

$$\begin{vmatrix} 1 & 2 \\ 2 & 4 \end{vmatrix} = 4 - 4 = 0 \tag{6-40c}$$

which is to say that the matrix of coefficients

$$\begin{pmatrix} 1 & 2 \\ 2 & 4 \end{pmatrix} \tag{6-41}$$

is singular.

To select one from among the infinite number of solution sets, we must have an additional independent nonhomogeneous equation. If the additional equation is

$$x + y = 1 \tag{6-42}$$

the solution set $\{2, -1\}$ satisfies all three equations [one of the pair (6-40) is superfluous] and is the unique solution set for the homogeneous linear simultaneous equation pair plus the additional equation.

In what immediately follows, we will obtain eigenvalues E_1 and E_2 for $\hat{H}\psi = E_i\psi$ from the simultaneous equation set (6-38). Each eigenvalue gives a π-electron energy for the model we used to generate the secular equation set. In the next chapter, we shall apply an additional equation of constraint on the minimization parameters $\{a_1, a_2\}$ so as to obtain their unique solution set.

The Secular Matrix

The coefficient matrix of the normal equations (6-38) for ethylene is

$$\begin{pmatrix} \alpha - E & \beta \\ \beta & \alpha - E \end{pmatrix} \tag{6-43}$$

By the criterion of Exercise 2-9, E is an eigenvalue of the matrix in α and β. There are two secular equations in two unknowns for ethylene. For a system with n conjugated sp^2 carbon atoms, there will be n secular equations leading to n eigenvalues E_i. The family of E_i values is sometimes called the *spectrum* of energies. Each secular equation yields a new eigenvalue and a new eigenvector (see Chapter 7).

If we divide each element of the secular matrix by β and perform the substitution $x = \alpha - E_i/\beta$, we get

$$\begin{pmatrix} x & 1 \\ 1 & x \end{pmatrix} \tag{6-44}$$

as the coefficient matrix of the equation set

$$\begin{pmatrix} x & 1 \\ 1 & x \end{pmatrix} \begin{pmatrix} a_1 \\ a_2 \end{pmatrix} = 0 \tag{6-45}$$

For the equation set to be linearly dependent, the secular determinant must be zero

$$\begin{vmatrix} x & 1 \\ 1 & x \end{vmatrix} = 0 \tag{6-46}$$

Expanding the determinant,

$$x^2 - 1 = 0$$

so that

$$x = \pm 1 \tag{6-47}$$

There are $n = 2$ roots of the polynomial, one for each eigenvalue in the E_i spectrum.

We are free to pick a reference point of energy once, but only once, for each system. Let us choose the reference point α. We have obtained the energy eigenvalues of the π bond in ethylene as one β greater than α (antibonding) and one β lower than α (bonding) (Fig. 6-3).

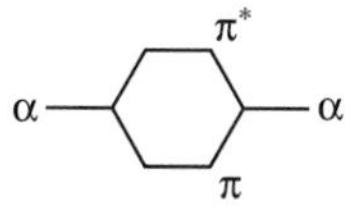

Figure 6-3 The Energy Spectrum of Ethylene. The π orbital is bonding, and the π* orbital is antibonding.

Finding Eigenvalues by Diagonalization

If we drop x from the secular matrix

$$\begin{pmatrix} x & 1 \\ 1 & x \end{pmatrix}$$

we get

$$\begin{pmatrix} 0 & 1 \\ 1 & 0 \end{pmatrix} \tag{6-48}$$

which has the eigenvalues ± 1, as we found by expanding the secular determinant [Eq. (6-47)]. If, using an equivalent method, we diagonalize matrix (6-48), the eigenvalues can be read directly from the principal diagonal of the diagonalized matrix

$$\begin{pmatrix} -1 & 0 \\ 0 & 1 \end{pmatrix} \tag{6-49a}$$

or

$$\begin{pmatrix} 1 & 0 \\ 0 & -1 \end{pmatrix} \tag{6-49b}$$

where the order of the roots on the principal diagonal depends on the method of diagonalization but the eigenvalues do not. If we can diagonalize a matrix comparable to matrix (6-48) by deleting x from any secular matrix, we shall have obtained the eigenvalues for the corresponding π electron system, in units of β, relative to an arbitrary energy α. *Diagonalization does not change the eigenvalues.*

Mathcad

$$A := \begin{pmatrix} 0 & 1 \\ 1 & 0 \end{pmatrix} \qquad\qquad B := \begin{pmatrix} -1 & 0 \\ 0 & 1 \end{pmatrix}$$

$$\text{eigenvals}\,(A) = \begin{pmatrix} -1 \\ 1 \end{pmatrix} \qquad \text{eigenvals}\,(B) = \begin{pmatrix} -1 \\ 1 \end{pmatrix}$$

By substituting back into the definition of x, we get the solution set for the energy spectrum E_i. In ethylene there are two elements on the diagonal, x_{11} and x_{22}, leading to E_1 and E_2. In larger conjugated π systems, there will be more.

If "dropping x," as is usually said, sounds a little arbitrary to you (it does to me, too), what we are really doing is concentrating on one term of a sum

$$\begin{pmatrix} x & 1 \\ 1 & x \end{pmatrix} = \begin{pmatrix} x & 0 \\ 0 & x \end{pmatrix} + \begin{pmatrix} 0 & 1 \\ 1 & 0 \end{pmatrix} \tag{6-50}$$

Diagonalization (the x matrix is already diagonal) yields

$$\begin{pmatrix} x & 0 \\ 0 & x \end{pmatrix} + \begin{pmatrix} -1 & 0 \\ 0 & 1 \end{pmatrix} = \begin{pmatrix} x-1 & 0 \\ 0 & x+1 \end{pmatrix} \tag{6-51}$$

that is, the roots of the secular matrix are

$$\begin{aligned} x \mp 1 &= 0 \\ x &= \pm 1 \end{aligned} \tag{6-52}$$

Polynomial root finding, as in the previous section, has some technical pitfalls that one would like to avoid. It is easier to write reliable software for matrix diagonalization (QMOBAS, TMOBAS) than it is for polynomial root finding; hence, diagonalization is the method of choice for Huckel calculations.

Rotation Matrices

If we premultiply and postmultiply the matrix

$$\begin{pmatrix} 0 & 1 \\ 1 & 0 \end{pmatrix}$$

by the matrix

$$\begin{pmatrix} \cos\theta & \sin\theta \\ \sin\theta & -\cos\theta \end{pmatrix} \tag{6-53}$$

where $\theta = 45°$, the result is

$$\begin{pmatrix} 1 & 0 \\ 0 & -1 \end{pmatrix} \tag{6-54}$$

which is the original matrix rotated one-eighth turn or $45°$, with a sign change in the second row. What we are really doing is rotating the coordinate system that we have arbitrarily imposed on the wave (vector) function ψ in a way that is similar to the coordinate rotation we used to discover the principal axes of an ellipse (see **What's Going on Here?**, Chapter 2).

The premultiplying and postmultiplying matrix is often called a *rotation matrix* **R**. The rotation matrix

$$\mathbf{R} = \begin{pmatrix} \cos\theta & -\sin\theta \\ \sin\theta & \cos\theta \end{pmatrix} \tag{6-55}$$

is widely used to do the same thing as rotation matrix (6-53). It yields

$$\begin{pmatrix} -1 & 0 \\ 0 & 1 \end{pmatrix} \tag{6-56}$$

By rotating it through the proper angle, we have diagonalized matrix (6-48). Diagonalization yields the solution set $x = (\alpha - E)/\beta = \{-1, 1\}$. Multiplying by β,

$$\alpha - E = \mp \beta \tag{6-57a}$$

or

$$E = \alpha \pm \beta \tag{6-57b}$$

which leads to the two-level energy spectrum for ethylene as shown in Fig. 6-3.

Generalization

The advantage of the method just described is that it can be generalized to molecules of any size. Setting up quite complicated secular matrices can be reduced to a simple recipe. A computer scheme can be used to diagonalize the resulting matrices by an iterative series of rotations.

The dimension of the matrix is the number of atoms in the π conjugated system. Let us take the three-carbon system allyl as our next step. Concentrate on the end atom in the system and write an x in the $i = j$ position for that atom. Follow this by writing a 1 in the position corresponding to any atom attached to the x atom. For any atom not attached to x, enter a zero. For allyl, the x goes into the 1, 1 position

C–C–C
↑

This leads to the top row of the secular matrix

x 1   0

Concentrating on the second atom in the allyl chain leads to the row 1 x 1, and concentrating on the third atom gives 0 1 x.
    The full allyl matrix is

$$\begin{pmatrix} x & 1 & 0 \\ 1 & x & 1 \\ 0 & 1 & x \end{pmatrix} \tag{6-58}$$

The zeros in the 1,3 and 3,1 positions correspond physically to the assumption that there is no interaction between π electrons of atoms that are not neighbors, a standard assumption of Huckel theory.
    If we had been interested in the cyclopropenyl system,

$$\underset{\text{C–C}}{\overset{\text{C}}{\diagup\diagdown}}$$

we would have been led to the matrix

$$\begin{pmatrix} x & 1 & 1 \\ 1 & x & 1 \\ 1 & 1 & x \end{pmatrix} \tag{6-59}$$

Butadiene,

$$\text{C}\diagup^\text{C}\diagdown_\text{C}\diagup^\text{C}$$

yields

$$\begin{pmatrix} x & 1 & 0 & 0 \\ 1 & x & 1 & 0 \\ 0 & 1 & x & 1 \\ 0 & 0 & 1 & x \end{pmatrix} \tag{6-60}$$

and so on. We have moved away from notation involving localized double bonds as in ethylene, and we are working from a picture of π bonds delocalized in some way over a carbon atom framework involving all carbon atoms in the conjugated system. We speak of the entire conjugated system as, for example, the *butadienyl system*, meaning butadiene and any ions or free radicals that can be derived from it without moving any carbon atoms. In the next chapter we shall be more specific about *how* the electrons are delocalized over the carbon atom framework.

The rotation matrix **R** must also be given in general form. If the pre- and postmultiplying matrix is contained as a block within a larger matrix containing only ones on the principal diagonal and zeros elsewhere (aside from the rotation block), the corresponding block of the operand matrix is rotated. Elements outside the rotation block are changed, too. For example,

$$
\begin{pmatrix} \cos\theta & \sin\theta & 0 \\ \sin\theta & -\cos\theta & 0 \\ 0 & 0 & 1 \end{pmatrix} \begin{pmatrix} 0 & 1 & 0 \\ 1 & 0 & 1 \\ 0 & 1 & 0 \end{pmatrix} \begin{pmatrix} \cos\theta & \sin\theta & 0 \\ \sin\theta & -\cos\theta & 0 \\ 0 & 0 & 1 \end{pmatrix}
$$

$$
= \begin{pmatrix} 1 & 0 & 0.707 \\ 0 & -1 & -0.707 \\ 0.707 & -0.707 & 0 \end{pmatrix} \tag{6-61}
$$

where θ was once again taken as $45°$. The rotation matrix can be made as large as necessary to conform with any operand matrix. The rotation block may be placed anywhere on the principal diagonal of the rotation matrix **R**. The allyl matrix has not been diagonalized by Eq. (6-61), only part of it has.

The Jacobi Method

The Jacobi method is probably the simplest diagonalization method that is well adapted to computers. It is limited to real symmetric matrices, but that is the only kind we will get by the formula for generating simple Huckel molecular orbital method (HMO) matrices just described. A rotation matrix is defined, for example,

$$
\mathbf{R} = \begin{pmatrix} 1 & 0 & 0 & 0 & \text{etc.} \\ 0 & \cos\theta & \sin\theta & 0 \\ 0 & \sin\theta & -\cos\theta & 0 \\ 0 & 0 & 0 & 1 \\ \text{etc.} & & & & 1 \\ & & & & & \text{etc.} \end{pmatrix} \tag{6-62}
$$

so as to "attack" a block of elements of the operand matrix. In the case of rotation matrix (6-62), the block with a_{22} and a_{33} on the principal diagonal is attacked.

Now,

$$
\tan 2\theta = \frac{2a_{ij}}{a_{ii} - a_{jj}} \tag{6-63}
$$

where a_{ij} denotes the ij element in matrix $\mathbf{A}$ and a_{ii} and a_{jj} are on the principal diagonal. (The double subscript on the elements a_{ij} distinguish them from the solution set $\{a_i\}$.) Initially, element a_{ij} is adjacent to element a_{ii} and above element a_{jj} as it is in the butadienyl system, but this will not be true later in the diagonalization and is not necessary even at the outset. Element a_{ij} can be off the tridiagonal as in the cyclopropenyl matrix, in which case the rotation matrix would be

$$\begin{pmatrix} \cos\theta & 0 & \sin\theta \\ 0 & 1 & 0 \\ \sin\theta & 0 & -\cos\theta \end{pmatrix}$$

The matrix equation

$$\mathbf{RAR} = \mathbf{A}' \tag{6-64}$$

generates a matrix $\mathbf{A}'$ that is *similar* to $\mathbf{A}$ (see section on the transformation matrix in Chapter 2) but has had elements a_{ij} and a_{ji} reduced to zero. The eigenvalues of $\mathbf{A}$ are proportional to the lengths of the corresponding eigenvectors, and orthogonal transformations preserve the lengths of vectors (Chapter 2), so similar matrices have the same eigenvalues.

The good news is that any a_{ij} and a_{ji} elements not on the principal diagonal can be converted to zero by choosing the right $\mathbf{R}$ matrix. The bad news is that each successive $\mathbf{RAR}$ multiplication destroys all zeros previously gained, replacing them with elements that are not zero but *are smaller than their previous value*. Thus the $\mathbf{RAR}$ multiplication must be carried out a number of times that is not just equal to one-half the number of nonzero off-diagonal elements, but is very large, strictly speaking, infinite. The sum of the off-diagonal elements cannot be set equal to zero by the Jacobi method, but it can be made to converge on zero. The Jacobi method is an iterative method.

Let us follow the first few iterations for the *allyl* system by hand calculations. We subtract the matrix $x\mathbf{I}$ from the HMO matrix to obtain the matrix we wish to diagonalize, just as we did with ethylene. With the rotation block in the upper left corner of the $\mathbf{R}$ matrix (we are attacking a_{12} and a_{21}), we wish to find

$$\mathbf{R} \begin{pmatrix} 0 & 1 & 0 \\ 1 & 0 & 1 \\ 0 & 1 & 0 \end{pmatrix} \mathbf{R}$$

First,

$$\tan 2\theta = \frac{2a_{ij}}{a_{ii} - a_{jj}} = \infty$$

$$\theta = 45°$$

$$\sin\theta = \cos\theta = \frac{1}{\sqrt{2}} = 0.7071$$

By the simple HMO procedure, it is always true that $\sin \theta = \cos \theta = 0.7071$ on the first iteration. Now, to eliminate a_{12},

$$\mathbf{RAR} = \mathbf{A'} \tag{6-65}$$

but this is just the multiplication we used as an illustration in the last section. We know that the result is matrix (6-61).

$$\begin{pmatrix} 1 & 0 & 0.707 \\ 0 & -1 & -0.707 \\ 0.707 & -0.707 & 0 \end{pmatrix}$$

Because both matrix $\mathbf{A}$ and the transformation are symmetrical, reducing the a_{12} element to zero also reduces a_{21} to zero. We have gained the zeros we wanted, but we have sacrificed the zeros we had in the 1,3 and 3,1 positions. Other than those eliminated, the off-diagonal elements are no longer zero *but they are less than one*. Attacking the $a_{13} = a_{31} = 0.7071$ element produces

$$\mathbf{A''} = \begin{pmatrix} 1.37 & -0.325 & 0 \\ -0.325 & -1 & 0.628 \\ 0 & 0.628 & -0.37 \end{pmatrix}$$

and so on for further iterations of the method. Again, the zeros previously gained are lost, but they are replaced by nonzero elements that are less than 0.7071. Nine iterations yield

$$\mathbf{A''''''''} = \begin{pmatrix} -1.41 & 0 & 0 \\ 0 & 0 & 0 \\ 0 & 0 & 1.41 \end{pmatrix}$$

where elements that are negligibly small, say 10^{-7}, are recorded as zero. The energy levels or eigenvalues for the three-carbon allyl model are

$$x = -1.41, \ 0, \ 1.41 \tag{6-66}$$

The order of the roots as generated by diagonalization is dependent on the algorithm, as are some of the intermediate matrices generated in the diagonalization procedure. Programs are written to be "opportunistic," that is, to seek a quick means of conversion on the eigenvalues, and the strategy chosen may differ from one program to the next. Many programs, including the one to be described in the next section, have a separate subroutine at the end that takes the eigenvalues in whatever order they are produced by diagonalization and orders them, lowest to highest or vice versa.

In conclusion of this section, it is remarkable that molecular orbitals are never really used in Huckel theory, that is, the integrals α and β are not evaluated. Huckel theory is a scheme for ordering energy levels according to the postulate in Eq. (6-31), the LCAO approximation, and the geometry of the π system. In chemical graph theory, where matrices like the right side of Eq. (6-48) are called adjacency matrices, the relationship between molecular topology and energy is investigated further (Trinajstic, 1983).

Programs QMOBAS and TMOBAS

A simple method of generating the eigenvalues for the general HMO matrix (Dickson, 1968, Rogers et al., 1983) starts by searching the HMO matrix to find the largest off-diagonal element, that is, the one most suitable for attack. Once this element is found, the rotation angle is calculated by Eq. (6-63) and the matrix is partially diagonalized. The search is repeated, the next target element is selected, the rotational angle is calculated, and so on. After each rotation, the off-diagonal elements are tested to see whether they are sufficiently close to zero. Of course, they are not at the beginning, but they get smaller as the rotation is iterated. An arbitrary standard is set up so that when the diagonal elements have been reduced below a certain level, the rotation iterations stop. In QMOBAS and TMOBAS, the criterion for exit from the iterative diagonalization loop is that the root mean square sum of the off-diagonal elements be equal to or less than 10^{-7} times its original value.

Degenerate roots (different roots having the same value) can produce computational difficulties. These problems can usually be circumvented by entering the HMO matrix with elements that are slightly different from 1. For example, 1.0001 might be used.

Heterocyclic and linear heteronuclear π conjugated systems pose a special problem because the heteroatom has an electron density that is greater than or less than the electron density of carbon. The Jacobi procedure suggests an empirical method of compensating for the increased electron density at, for example, nitrogen, in the way in which elements on the principal diagonal are built up by accretion during the iterative diagonalization procedure. If we place a nonzero element on the principal diagonal of the matrix to be diagonalized (after the $x\mathbf{I}$ matrix has been subtracted), when the accretion process is over that position will have an energy lower (or higher depending on the sign of the root) than it otherwise would have had.

Let us take the nitrogen in pyrrole, which is electron rich, as an example. In pyrrole, the value of 1.5 in the 1,1 position causes the lowest energy level to be lowered and the electron density about the nitrogen to be larger than it would be for carbon, which has a zero in the 1,1 position. The value 1.5 is selected by trial and error by comparison to experimental values for spectral transitions, resonance energies, etc. and represents a literature consensus (Strietwieser, 1961). Empirical modifications of off-diagonal entries in the HMO matrix are also used for bonds connecting carbon to atoms other than carbon.

For pyrrole, using QMOBAS with 1.5 in the lead position of the HMO matrix, 31 iterations (system specific) yield a lowest eigenvalue of -2.55β: $E =$

$\{-2.55, 1.20, -1.15, 1.62, -0.62\}$. This will be taken up in more detail in the next chapter.

COMPUTER PROJECT 6-2 | *Energy Levels (Eigenvalues)*

An *energy spectrum* is an ordered set of quantum mechanical energy levels. Each energy level coincides with an eigenvalue of the Schroedinger equation. In Huckel molecular orbital theory, energies are given in units of β relative to α, which is arbitrarily taken to be zero. Energy spectra are often presented as diagrams like Fig. 6-3. The wave function for the higher of the two energies in Fig. 6-3 has one internal node, but the lower energy function has no internal nodes. This is general; the greater the number of nodes, the higher the energy. Energy spectra are usually more complicated than Fig. 6-3 and have several levels for large molecules. The term *spectrum* is more commonly used to describe a pattern of absorption or emission of electromagnetic radiation as bands or lines, but the relation between an electromagnetic radiation spectrum and a molecular or atomic energy spectrum is very close so it is not unreasonable to use the term in this context also.

Procedure. The allyl model is the default in program QMOBAS and TMOBAS. It runs without any modification of the DATA input. Other models require modification. The Huckel matrix for the allyl model has already been given. Its solution yields three eigenvalues and three eigenfunctions, with zero, one, and two internal nodes. Draw the energy level manifold for the allyl model. Any energy below α in energy is bonding; label it π. Any level above α is antibonding; label it π^*. An orbital at the same level as α is nonbonding; label it *n*.

Execute QMOBAS and determine the energy levels (eigenvalues) for the ethylene, allyl, butadienyl, and pentadienyl models.

$$C-C \qquad C-C-C-C \qquad C-C-C$$

$$C-C-C-C-C$$

The upper triangular part of the Huckel matrix for ethylene, Eq. (6-48), exclusive of the diagonal elements consists of only one element. It can be entered into Program QMOBAS by making the dimension of the matrix 2 and changing the data statement to enter 1 in the 1,2 position

DATA 2
DATA 1,2,1,999

The four numbers in the second DATA statement give row, column, entry followed by 999 to show that data input is finished. Only the positive, nonzero, upper triangular part (one element in this case) is entered because the program negates elements and reflects the upper triangular matrix across the diagonal in the statement $A(I,J) = -F$; $A(J,I) = -F$ to give the full Huckel matrix. This is permissible because Huckel matrices are symmetric. The alphabetic input should be changed to

A\$ = "ethylene"

or any other identifying string variable you like. String variables are discussed by Ebert, Ederer ,and Isenhour (1989).

The upper triangular matrix for the butadienyl system

$$
\begin{pmatrix}
0 & 1 & 0 & 0 \\
  & 0 & 1 & 0 \\
  &   & 0 & 1 \\
  &   &   & 0
\end{pmatrix}
$$

is input by means of the DATA statements

    DATA 4
    DATA 1,2,1,0,2,3,1,0,3,4,1,999

along with an appropriate A$ = "...". A zero in the 4,4 position is not necessary.

Each nonzero element in the butadienyl input is 1. The input element 1.0 also works but locations, for example, the 1,2 location, must be specified by integers. In some applications, for example, pyridine (Chapter 7), decimal inputs other than 1.0 are used. Substitute LPRINT for PRINT in QMOBAS to obtain hard copy. Draw diagrams analogous to Fig. 6-3 that show the energy levels in their proper order, lowest to highest. If an energy turns out to be zero (relative to α), label it nonbonding, n. Remember that, because of rounding and a finite number of matrix rotations, that the zero roots may be output as very small values, say 10^{-7} or so.

Using QMOBAS, calculate and order the eigenvalues for the cyclopropenyl, cyclobutadienyl, and cyclopentadienyl models. Draw the energy level manifolds and compare them with the linear models. Two roots with the same energy are said to be degenerate. They are not duplicate solutions to the Schroedinger equation because they have different coefficients (eigenfunctions). See Chapter 7 for a discussion of eigenfunctions. Are there any degenerate roots among these model systems?

Repeat each calculation after having inserted a "counter" into Program QMOBAS to count the number of iterations. The statement ITER = ITER + 1 placed before the GOTO 340 statement increments the contents of memory location ITER, starting from zero, on each iteration. The statement PRINT "ITER", ITER prints out the accumulated number of iterations at the end of the program run. Comment on the number of iterations needed to satisfy the final norm V1 for the different Huckel MO calculations.

Alternative procedure: TMOBAS. The procedure using TMOBAS is the same as for QMOBAS. The TMOBAS program is slightly different in appearance from QMOBAS, but it functions in the same way. See the ***TrueBasic*** documentation for details.

Alternative procedure: Mathcad. Follow the procedure above except that where QMOBAS is indicated, use *Mathcad* instead. Enter the Huckel molecular orbital matrix, modified by subtracting $x\mathbf{I}$, with some letter name. For example, call the modified matrix **A**. Type the command **eigenvals(A)** = with the name of the modified HMO matrix in parentheses. *Mathcad* prints the eigenvalues. The command **eigenvecs(A)** yields the eigenvectors, which are useful in ordering the energy spectrum.

Mathcad

$$A := \begin{pmatrix} 0 & 1 & 0 \\ 1 & 0 & 1 \\ 0 & 1 & 0 \end{pmatrix}$$

$$\text{eigenvals}(A) = \begin{pmatrix} 1.414 \\ 0 \\ -1.414 \end{pmatrix} \quad \text{eigenveces}(A) = \begin{pmatrix} 0.5 & 0.707 & 0.5 \\ 0.707 & 0 & -0.707 \\ 0.5 & -0.707 & 0.5 \end{pmatrix}$$

COMPUTER PROJECT 6-3 | *Huckel MO Calculations of Spectroscopic Transitions*

Linear polyenes (butadiene, hexatriene, etc.) absorb ultraviolet radiation. They have absorption maxima at the approximate wavelengths given in Table 6-1.

Table 6-1 Ultraviolet Absorption Maxima for Polyenes

Ethylene	161 nm
Buta-1,3-diene	217
Hexa-1,3,5-triene	244
Octa-1,3,5,7-tetraene	303

These absorptions are ascribed to π-π^* transitions, that is, transitions of an electron from the highest occupied π molecular orbital (HOMO) to the lowest unoccupied π molecular orbital (LUMO). One can decide which orbitals are the HOMO and LUMO by filling electrons into the molecular energy level diagram from the bottom up, two electrons to each molecular orbital. The number of electrons is the number of sp^2 carbon atoms contributing to the π system of a neutral polyalkene, two for each double bond. In ethylene, there is only one occupied MO and one unoccupied MO. The occupied orbital in ethylene is β below the energy level represented by α, and the unoccupied orbital is β above it. The separation between the only possibilities for the HOMO and LUMO is 2.00β.

Using QMOBAS, TMOBAS, or *Mathcad* and the method from Computer Project 6-2, calculate the energy separation between the HOMO and LUMO in units of β for all compounds in Table 6-1 and enter the results in Table 6-2. Enter the observed energy of ultraviolet radiation absorbed for each compound in units of cm^{-1}. The reciprocal wavelength is often used as a spectroscopic unit of energy.

Table 6-2

Compound	HOMO	LUMO	(LUMO-HOMO)	$\bar{v} = \frac{1}{\lambda}$, cm^{-1}
Ethylene	$\alpha + \beta$	$\alpha - \beta$	-2.00β	...
Buta-1,3-diene...				
...				
etc.				

Radiation of wavelength 161 nm, for example has an energy of $1/161 \times 10^{-9}$m $= 1/161 \times 10^{-7}$cm $= 6.21 \times 10^4$ cm^{-1}.

The quantity β is an energy. The separation in MO levels (2β in the case of ethylene) is a change in energy ΔE and follows Planck's equation $\Delta E = h\nu$. The results in Table 6-2 give four spectroscopic energies of radiation absorbed $\bar{v}$, in units of cm^{-1}, required to promote electrons across four different energy gaps measured in units of β. Plot ΔE in units of β vs. the spectroscopic energy $\bar{v}$. Using Program QLLSQ, TLLSQ, or *TableCurve*, obtain the best slope of the function β vs. $\bar{v}$. This is the amount of energy in cm^{-1} per β, that is, the "size" of the energy unit β. The calculation is approximate because the Huckel approximations are crude, but even an order of magnitude calculation of β is useful. Calculate β in units of joules, kJ mol^{-1}, and electron volts (eV).

PROBLEMS

1. For the atomic orbital $\Psi = e^{-\alpha r}$, where r is the radial distance between the proton and the electron, show that [Eq. (6-21)]

$$-\frac{\hbar^2}{2m}\frac{1}{r^2}\frac{d}{dr}r^2\frac{d}{dr}e^{-\alpha r} = -\frac{\hbar^2}{2m}\left(\alpha^2 - \frac{2\alpha}{r}\right)e^{-\alpha r}$$

2. Evaluate the integrals $\int_0^\infty r^2 e^{-2\alpha r}dr$ and $\int_0^\infty r e^{-2\alpha r}dr$, which are necessary to obtain Eq. (6-25).

3. The expression for the Coulombic potential energy $e/4\pi\varepsilon_0$ can be carried through the entire derivation in Exercise 6-3 to arrive at Eq. (6-17). Show that this is so.

4. For many purposes, it is useful to replace the atomic orbital $\Psi = e^{-\alpha r}$ with a Gaussian function $\psi = e^{-\gamma r^2}$, where γ is a constant. Show that

$$-\frac{\hbar^2}{2m}\frac{1}{r^2}\frac{d}{dr}r^2\frac{d}{dr}e^{-\gamma r^2} = -\frac{\hbar^2}{2m}\left(4\gamma^2 r^2 - 6\gamma\right)e^{-\gamma r^2}$$

5. Show that

$$E = \frac{\int_0^\infty -\frac{\hbar^2\gamma^2}{m}2r^4 e^{-2\gamma r}dr + \int_0^\infty \frac{\hbar^2\gamma}{m}3r^2 e^{-2\gamma r}dr - \int_0^\infty e^2 r e^{-2\gamma r}dr}{\int_0^\infty r^2 e^{-2\gamma r}dr}$$

for the approximate wave function $\psi = e^{-\gamma r^2}$.

6. Using the results from the previous two problems, evaluate the integrals in the answer to Problem 5 and find E as a closed algebraic expression for the Gaussian trial function.

7. Show that, at the minimum energy (least upper bound for the energy arising from the approximate wave function $\psi = e^{-\gamma r^2}$), the minimization parameter γ is $\gamma = \frac{8}{9\pi}$.

8. What is the precise value of the least upper bound of the energy for the approximate wave function $\psi = e^{-\gamma r^2}$?

9. Show that Eqs. (6-35) follow from Eqs. (6-34).

10. Compute the HMO eigenvalues for the cyclobutadienyl system.

11. Draw the energy level diagram for cyclobutadiene.

12. Write the secular matrix for the methylenecyclopropenyl system.

13. Compute the eigenvalues and draw the energy level diagram for methylene-cyclopropene.

14. Write the secular matrix, compute the eigenvalues, and draw the energy level diagram for fulvene.

15. Compute the HMO eigenvalues for benzene and draw its energy level diagram.

16. Draw the energy level diagram for pyrrole.

Place a 2 on the principal diagonal for N. Make no alteration in β for the C—N bond.

17. One convention (Dickson, 1968) for oxygen heterocycles sets the coulomb integral at $\alpha + 2\beta$ and the resonance integral at $\sqrt{2}\beta$. For the oxirane moiety, thought to be important in steroid biosynthesis,

the Huckel matrix is of the same form as the allyl model except that 2 is placed on the principal diagonal in the 1,1 position and $\sqrt{2} = 1.414$ is placed on the off-diagonal for each C—O bond. Run MOBAS or *Mathcad* with the input matrix so modified to find the eigenvalues and coefficients (eigenvectors) for the oxirane model.

18. Long ago, Thiel discovered that cyclopentadiene, heated under N_2 with a dispersion of potassium in benzene, yields potassium cyclopentadienide

but that the analogous reaction with cycloheptatriene does not go

Draw the energy level spectra for the two cyclic models; fill in an appropriate number of electrons for the negative ion for each model. Suggest a reason why one reaction goes and the other does not.

19. In the nineteenth century, Merling treated cycloheptatriene with bromine and obtained a crystalline solid. Reasoning from some information gained in working Problem 15, what might this solid be?

20. Show that $\bar{v} = \Delta E / ch$ has the units of cm^{-1}.

7

Huckel Molecular Orbital Theory II: Eigenvectors

Each eigenvalue has an eigenvector associated with it. The eigenvectors tell us as much about the atomic or molecular system as the eigenvalues do. Like eigenvalues, eigenvectors can be exact for simple systems but in general we know them only approximately. Eigenvectors define a *vector space* that has a number of dimensions equal to the number of basis functions. Unfamiliar and perhaps daunting as you may find spaces with dimensions beyond the three dimensions of Euclidean space, we shall soon be working in many-dimensional spaces so frequently you will find them quite ordinary.

Recapitulation and Generalization

Let us pause to recapitulate and generalize our mathematical position. We assume, in consequence of the single-electron approximation, that there is a unique set of solutions to the Schroedinger equation for each molecule called *orbitals*, each orbital associated with one energy in the energy *spectrum* of the molecule. We assume further (as has been verified by considerable indirect experimental evidence) that this spectrum is analogous to the energy spectrum of the hydrogen atom in that there are many allowed energy levels separated by energy regions that

Computational Chemistry Using the PC, Third Edition, by Donald W. Rogers
ISBN 0-471-42800-0   Copyright © 2003 John Wiley & Sons, Inc.

are not allowed. Conversely, we expect the details of spacing and multiplicity of molecular energy levels to differ from those of atomic energy spectra.

The complexity of molecular systems precludes exact solution for the properties of their orbitals, including their energy levels, except in the very simplest cases. We can, however, approximate the energies of molecular orbitals by the variational method that finds their least upper bounds in the ground state as Eq. (6-16)

$$E_i = \frac{\int \psi H \psi \, d\tau}{\int \psi^2 d\tau}$$

where the subscript i indicates that there are, in general, many levels in the spectrum. Taking normalized orbitals leads to the simplification

$$E_i = \int \psi H \psi \, d\tau$$

Every electron in a molecule has a Coulombic attraction to "its own" nucleus $H_{ii} = \int \psi_i \hat{H} \psi_i d\tau$. In addition, it has an attraction to all other nuclei in the molecule $H_{ij} = \int \psi_i \hat{H} \psi_j d\tau$. Coulombic attraction between nuclei and the electrons normally associated with the nucleus in the pure atomic state is very strong, so the nuclear-electron energy H_{ii} is large. It accounts for most of the energy of the molecule, but it is not what holds the molecule together. The energy holding the molecule together is the bonding energy of attraction H_{ij} between nuclei and electrons that are not in their normal atomic sphere of Coulombic force. Thus we have two distinct kinds of energy integrals in molecules, H_{ii} associated with a large residual energy retained in atoms when they form chemical bonds and a relatively small amount of bonding energy H_{ij}. In addition, orbital overlap integrals are defined $S_{ij} = \int \psi_i \psi_j \, d\tau$ and $S_{ii} = \int \psi_i \psi_i \, d\tau$. The integral $S_{ii} = 1$ for normalized wave functions.

The expansion of any molecular orbital over a *basis set* ϕ_k

$$\psi = \sum_k a_k \phi_k$$

leads to a set of arbitrary expansion coefficients a_k, which we optimize by imposing the conditions generalized from Eq. (6-34a and b)

$$\frac{\partial E}{\partial a_1} = \frac{\partial E}{\partial a_2} =, \ldots \frac{\partial E}{\partial a_k} =, \ldots \frac{\partial E}{\partial a_n} = 0 \quad (k = 1, 2, 3, \ldots, n) \tag{7-1}$$

to find the energy minimum in an n-dimensional vector space.

These expansion coefficients $\{a_k\}$ are the minimization parameters of a set of simultaneous equations. Imposition of all the minimization conditions (7-1) for $k = 1, 2, \ldots, n$ leads to the set of n equations

$$(H_{11} - S_{11}E)a_1 + (H_{12} - S_{12}E)a_2 + (H_{13} - S_{13}E)a_3 + \cdots + (H_{1n} - S_{1n}E)a_n = 0$$
$$(H_{21} - S_{21}E)a_1 + (H_{22} - S_{22}E)a_2 + (H_{23} - S_{23}E)a_3 + \cdots + (H_{2n} - S_{2n}E)a_n = 0$$
$$\vdots \qquad\qquad\qquad\qquad\qquad\qquad\qquad\qquad\qquad \vdots$$
$$(H_{n1} - S_{n1}E)a_1 + (H_{n2} - S_{n2}E)a_2 + (H_{n3} - S_{n3}E)a_3 + \cdots + (H_{nn} - S_{nn}E)a_n = 0$$
$$(7\text{-}2)$$

called the *secular equations*. The secular equations must all be equal to zero because the minimization conditions [Eq. (7-1)] set each derivative equal to zero. The number of equations and unknowns is the number of basis functions. In the simplest case of a linear combination of atomic orbitals (LCAO), the number of basis functions is the same as the number of atoms in the molecule, one basis function to each atom. Under the Huckel approximation that there is no electron exchange or interaction between nonadjacent atoms, some of the coefficients of the secular equations will be zero.

The secular equations can be written in matrix form

$$\begin{pmatrix} H_{11} - S_{11}E & (H_{12} - S_{12}E) & \cdots & (H_{1n} - S_{1n}E) \\ (H_{21} - S_{21}E) & \cdots & & \\ \cdots & & \ddots & \\ (H_{n1} - S_{n1}E) & & & (H_{nn} - S_{nn}E) \end{pmatrix} \begin{pmatrix} a_1 \\ a_2 \\ \vdots \\ a_n \end{pmatrix} = \mathbf{0} \qquad (7\text{-}3)$$

where the ordered set of numbers $\{a_i\}$ is called an *eigenvector*. To Eq. (7-3) there is a corresponding secular determinant

$$\begin{vmatrix} (H_{11} - S_{11}E) & (H_{12} - S_{12}E) & \cdots & (H_{1n} - S_{1n}E) \\ (H_{21} - S_{21}E) & \cdots & & \\ \cdots & & \ddots & \\ (H_{n1} - S_{n1}E) & & & (H_{nn} - S_{nn}E) \end{vmatrix} = 0 \qquad (7\text{-}4)$$

which is set equal to zero to obtain nontrivial solutions to the linearly dependent equation set (see section on the secular matrix in Chapter 6).

The optimization procedure is carried out to find the set of coefficients of the eigenvector that minimizes the energy. These are the best coefficients for the chosen linear combination of basis functions, best in the sense that the linear combination of arbitrarily chosen basis functions with optimized coefficients best approximates the molecular orbital (eigenvector) sought. Usually, some members of the basis set of functions bear a closer resemblance to the "true" molecular orbital than others. If basis function ϕ_k bears a closer resemblance to Ψ than ϕ_{k+1} does, then $a_k > a_{k+1}$.

Matrix Eq. (7-3) applies to only one of the eigenvectors, corresponding to only one eigenvalue among E_n in the energy spectrum. We can arrange the E_n energies in ascending order (ignoring degeneracies for simplicity), to get the diagonal matrix

$$
\begin{pmatrix}
E_1 & 0 & \cdots & 0 \\
0 & E_2 & & 0 \\
\vdots & & \ddots & 0 \\
0 & & 0 & E_n
\end{pmatrix} = \mathbf{E}
\tag{7-5}
$$

corresponding to the matrix of eigenvectors we get by stacking n column vectors next to each other in the order $1, 2, 3, \ldots n$ so that the order of the eigenvalues matches the order of the eigenvectors

$$
\begin{pmatrix}
a_{11} & a_{12} & \cdots & a_{1n} \\
a_{21} & a_{22} & & a_{2n} \\
\vdots & & \ddots & a_{n-1n} \\
a_{n1} & & a_{nn-1} & a_{nn}
\end{pmatrix} = \mathbf{A}
\tag{7-6}
$$

We call this stacked matrix $\mathbf{A}$. Now, Eq. (7-3) has been expanded to include all n eigenvectors.

$$
\begin{pmatrix}
(H_{11}-S_{11}E_j) & (H_{12}-S_{12}E_j) & \cdots & (H_{1n}-S_{1n}E_j) \\
(H_{21}-S_{21}E_j) & \cdots & & \\
\cdots & & \ddots & \\
(H_{n1}-S_{n1}E_j) & & & (H_{nn}-S_{nn}E_j)
\end{pmatrix}
\begin{pmatrix}
a_{11} & a_{21} & \cdots & a_{1n} \\
a_{21} & a_{22} & & a_{2n} \\
\vdots & & \ddots & a_{n-1n} \\
a_{n1} & & a_{nn-1} & a_{nn}
\end{pmatrix} = \mathbf{0}
\tag{7-7}
$$

Notice that matrix element a_{ij} has a subscript i that denotes the *order of coefficients*. The subscript j specifies the vector that corresponds to the energy E_j. The matrix

$$
\begin{pmatrix}
(H_{11} - S_{11}E_j) & (H_{12} - S_{12}E_j) & \cdots & (H_{1n} - S_{1n}E_j) \\
(H_{21} - S_{21}E_j) & \cdots & & \\
\cdots & & \ddots & \\
(H_{n1} - S_{n1}E_j) & & & (H_{nn} - S_{nn}E_j)
\end{pmatrix} = \mathbf{H} - \mathbf{SE}
\tag{7-8}
$$

is the difference between two matrices because each of its elements is the difference between two elements. We might be tempted to write

$$
(\mathbf{H} - \mathbf{SE})\mathbf{A} = \mathbf{0}
\tag{7-9}
$$

$$\begin{pmatrix} H_{11} & H_{12} & \cdots & H_{1n} \\ H_{21} & \cdots & & \\ \cdots & & \ddots & \\ H_{n1} & & & H_{nn} \end{pmatrix} \begin{pmatrix} a_{11} & a_{12} & \cdots & a_{1n} \\ a_{21} & a_{22} & & a_{2n} \\ \vdots & & \ddots & a_{n-1n} \\ a_{n1} & & a_{nn-1} & a_{nn} \end{pmatrix}$$

$$- \begin{pmatrix} S_{11} & S_{12} & \cdots & S_{1n} \\ S_{21} & \cdots & & \\ \cdots & & \ddots & \\ S_{n1} & & & S_{nn} \end{pmatrix} \begin{pmatrix} E_1 & 0 & \cdots & 0 \\ 0 & E_2 & & 0 \\ \vdots & & \ddots & 0 \\ 0 & & 0 & E_n \end{pmatrix} \begin{pmatrix} a_{11} & a_{12} & \cdots & a_{1n} \\ a_{21} & a_{22} & & a_{2n} \\ \vdots & & \ddots & a_{n-1n} \\ a_{n1} & & a_{nn-1} & a_{nn} \end{pmatrix} = \mathbf{0} \quad (7\text{-}10)$$

or

$$\mathbf{HA} - \mathbf{SEA} = \mathbf{0} \tag{7-11}$$

$$\mathbf{HA} = \mathbf{SEA} \tag{7-12}$$

but that would not be quite right because the order of matrices $\mathbf{EA}$ mixes eigenvector components and eigenvalues so that they do not match. To see an example, consider the products $\mathbf{EA}$ and $\mathbf{AE}$, where

$$\mathbf{A} = \begin{pmatrix} a_{11} & a_{12} \\ a_{21} & a_{22} \end{pmatrix} \quad \text{and} \quad \mathbf{E} = \begin{pmatrix} E_1 & 0 \\ 0 & E_2 \end{pmatrix} \tag{7-13}$$

We choose 2×2 matrices for simplicity, but we appreciate that the principle applies in general. The (noncommutative) products are

$$\mathbf{EA} = \begin{pmatrix} E_1 & 0 \\ 0 & E_2 \end{pmatrix} \begin{pmatrix} a_{11} & a_{12} \\ a_{21} & a_{22} \end{pmatrix} = \begin{pmatrix} E_1 a_{11} & E_1 a_{12} \\ E_2 a_{21} & E_2 a_{22} \end{pmatrix} \tag{7-14}$$

and

$$\mathbf{AE} = \begin{pmatrix} a_{11} & a_{12} \\ a_{21} & a_{22} \end{pmatrix} \begin{pmatrix} E_1 & 0 \\ 0 & E_2 \end{pmatrix} = \begin{pmatrix} a_{11} E_1 & a_{12} E_2 \\ a_{21} E_1 & a_{22} E_2 \end{pmatrix} \tag{7-15}$$

Remembering that the *second* subscript, j, on the coefficient a_{ij} identifies the vector, we can see that the second product above (7-15) matches eigenvalue E_1 with the eigenvector having coefficients $\{a_{i1}\}$ and the eigenvalue E_2 matches with the eigenvector having coefficients $\{a_{i2}\}$. Conversely, product (7-14) mixes eigenvectors and eigenvalues. We choose the unmixed order of multiplication. Our vector equations become

$$\mathbf{HA} - \mathbf{SAE} = \mathbf{0} \tag{7-16}$$

$$\mathbf{HA} = \mathbf{SAE} \tag{7-17}$$

written for the proper combination of eigenvectors and eigenvalues.

As of right now, we know none of the matrices $\mathbf{H}$, $\mathbf{A}$, $\mathbf{S}$, or $\mathbf{E}$ in Eq. (7-17) but we do have some critical information about their form, including the integrals defined as

$$\int p_1 p_1 \, d\tau = S_{11}$$

from Eq. (6-32c) and

$$\int p_1 p_2 \, d\tau = \int p_2 p_1 \, d\tau = S_{12} = S_{21}$$

from Eq. (6-32d) in Chapter 6 or more generally as

$$\int \phi_i \phi_i \, d\tau = S_{ii} \tag{7-18}$$

and

$$\int \phi_i \phi_j \, d\tau = \int \phi_j \phi_i \, d\tau = S_{ij} = S_{ji} \tag{7-19}$$

We also know that exact atomic orbitals are orthonormal, that is, $S_{ii} = 1$ and $S_{ij} = 0$ for an LCAO. If we assume that orthonormality is carried from an LCAO into the molecular orbital, then $\mathbf{S} = \mathbf{I}$ and, from Eq. (7-17),

$$\mathbf{HA} = \mathbf{AE} \tag{7-20}$$

If we can find a matrix $\mathbf{A}$ and its inverse $\mathbf{A}^{-1}$ such that, when we premultiply each side of Eq. (7-20) by $\mathbf{A}^{-1}$,

$$\mathbf{A}^{-1}\mathbf{HA} = \mathbf{A}^{-1}\mathbf{AE} = \mathbf{E} \tag{7-21}$$

we shall know that $\mathbf{A}$ is an *orthogonal transform* because we already know that $\mathbf{E}$ is diagonal. We also know that $\mathbf{E}$ is the diagonal matrix of energy eigenvalues of the *similarity transform* (7-21) (see the section on the transformation matrix in Chapter 2); therefore, *the columns of A are ordered eivenvectors of* $\mathbf{E}$, that is, the molecular orbital corresponding to E_j is

$$\psi_j = \sum_i a_{ij} \phi_i \tag{7-22}$$

It also follows that $\mathbf{A}^{-1} = \mathbf{A}^{\mathrm{T}}$, which is a characteristic of similarity transformations, so $\mathbf{A}^{-1}\mathbf{HA} = \mathbf{E}$ implies that

$$\mathbf{A}^{\mathrm{T}}\mathbf{HA} = \mathbf{E} \tag{7-23}$$

Equation (7-23) is a convenience because it is easier to find the transpose of a large matrix than it is to find its inverse. It is also true that in Huckel theory, $\mathbf{A}$ is symmetric, which means that it is equal to its own transpose, leading to the further simplification

$$\mathbf{AHA} = \mathbf{E} \tag{7-24}$$

If we can find $\mathbf{A}$, we shall have found an orthogonal set of eigenvectors. It is interesting and significant to note at this point that $\mathbf{A}$ is only one of many equally valid orthogonal sets of eigenvectors.

The Matrix as Operator

An operator is a mathematical instruction. For example, the operator d/dx is the instruction to differentiate once with respect to x. Matrices in general, and the matrix $\mathbf{R}$ of Chapter 6 in particular, are operators. The matrix $\mathbf{R}$ is an instruction to rotate a part of the operand matrix through a certain angle, θ as in Eq. (6-62).

The product of matrix operators is an operator. For example, rotation through $90°$, followed by another rotation through $90°$ in the same direction and in the same plane, is the same as one rotation through $180°$

$$\mathbf{R}_{90°} \mathbf{R}_{90°} = \mathbf{R}_{180°} \tag{7-25}$$

Thus the Jacobi procedure, by making many rotations of the elements of the operand matrix, ultimately arrives at the operator matrix that diagonalizes it. Mathematically, we can imagine one operator matrix that would have diagonalized the operand matrix $\mathbf{R}_t$ all in one step

$$\mathbf{R}_t = \mathbf{R}_1 \mathbf{R}_2 \mathbf{R}_3 \dots \mathbf{R}_n \tag{7-26}$$

even though in practice, we took n steps to do it. The situation is analogous to tuning a radio. We can imagine one perfect twist of the dial that would land right on the station, but in practice, we make several little twists back and forth across the proper tuning until we find the one we like.

The matrix $\mathbf{A}$ in Eq. (7-21) is comprised of orthogonal vectors. Orthogonal vectors have a dot product of zero. The mutually perpendicular (and independent) Cartesian coordinates of 3-space are orthogonal. An orthogonal $n \times n$ such as matrix $\mathbf{A}$ may be thought of as n columns of n-element vectors that are mutually perpendicular in an n-dimensional vector space.

The Huckel Coefficient Matrix

In Huckel theory, the $\mathbf{H}$ matrix consists of elements $\alpha - E_j$ and β

$$\begin{pmatrix} \alpha - E_j & \beta & \cdots & 0 \\ \beta & \alpha - E_j & \beta & \cdots \\ \cdots & \cdots & \cdots & \cdots \\ 0 & \cdots & \beta & \alpha - E_j \end{pmatrix} \begin{pmatrix} a_{1j} \\ a_{2j} \\ \vdots \\ a_{nj} \end{pmatrix} \tag{7-27}$$

The matrix elements $\alpha - E_j$ and β are not variables in the minimization procedure; they are constants of the secular equations with units of energy. Note that all elements in the matrix and vector are real numbers. The vector $\mathbf{a}_j$ is the set of coefficients for one eigenfunction corresponding to one eigenvalue, E_j. From Eq. (7-24),

$$\mathbf{AHA} = \mathbf{E} \qquad (7\text{-}28)$$

where the set of vectors comprising $\mathbf{A}$ is ordered so that $\mathbf{A}$ is a square, symmetric, orthogonal matrix having the property that it is its own inverse and transpose.

But Eq. (7-28) is the same mathematical operation that we used to obtain the diagonalized matrix of the eigenvalues,

$$\mathbf{RHR} = \mathbf{E} \qquad (7\text{-}29)$$

except that $\mathbf{H}$ was pre- and postmultiplied by the rotation matrix (see, with a slight change in notation, the section on the Jacobi method in Chapter 6). Evidently, the total rotation matrix $\mathbf{R}_t$ is an ordered matrix of eigenvectors. Thus if we keep track of the iterative rotations necessary to arrive at a total rotation matrix $\mathbf{R}_t$, we shall have the matrix of coefficients $\mathbf{A}$. Computationally, this is done by defining a unit matrix $\mathbf{I}$, and each time a partial rotation $\mathbf{R}_i$ is used to move matrix $\mathbf{H}$ toward $\mathbf{E}$, we operate on $\mathbf{I}$ with the same partial rotation matrix. When these iterations are complete, the process that has brought about stepwise transformation of $\mathbf{H}$ into $\mathbf{E}$ has also transformed $\mathbf{I}$ into $\mathbf{A}$

$$\mathbf{R}_1\mathbf{R}_2\mathbf{R}_3\ldots\mathbf{R}_n = \mathbf{R}_t\mathbf{I} = \mathbf{A} \qquad (7\text{-}30)$$

Let us look at the *Mathcad* output for a Huckel matrix in more detail. We select the matrix for ethylene to preserve simplicity.

Mathcad

$$H := \begin{pmatrix} 0 & 1 \\ 1 & 0 \end{pmatrix}$$

$$\text{eigenvals(H)} = \begin{pmatrix} -1 \\ 1 \end{pmatrix} \qquad \text{eigenvecs(H)} = \begin{pmatrix} 0.707 & 0.707 \\ -0.707 & 0.707 \end{pmatrix}$$

$$\text{eigenvec(H, } -1) = \begin{pmatrix} -0.707 \\ 0.707 \end{pmatrix} \qquad \text{eigenvec(H, 1)} = \begin{pmatrix} 0.707 \\ 0.707 \end{pmatrix}$$

In the **Mathcad** calculation of eigenvalues and eigenvectors of the Huckel matrix for ethylene $\begin{pmatrix} 0 & 1 \\ 1 & 0 \end{pmatrix}$, the eigenvalues are given in the order: upper followed by lower. The matrix $\mathbf{E}$ for this order is

$$\mathbf{E} = \begin{pmatrix} -1 & 0 \\ 0 & 1 \end{pmatrix}$$

as given by the first output, eigenvals $(H) = \begin{pmatrix} -1 \\ 1 \end{pmatrix}$. The second output gives the eigenvectors in the correct order, placing the eigenvector with the internal node on the left so that it is associated with the -1 eigenvalue and leaving the 1 eigenvalue to be associated with the eigenvector having no internal nodes. The lower two queries

eigenvec $(H, -1)$

and

eigenvec $(H, 1)$

match the eigenvectors with the eigenvalues in **Mathcad**. They merely confirm this association of eigenvectors with their appropriate eigenvalues, where the first entry in parentheses is the matrix and the second is the eigenvalue.

There are two apparent discrepancies. The order of signs is not the same in the two determinations of the eigenvector for the higher energy, but it doesn't matter which of the two equivalent carbon atoms has the negative coefficient; rotation of the molecule through 180° produces the opposite order of signs. Second, the lower eigenvalue, the one with no internal nodes, is 1 while the higher eigenvalue having one internal node is -1. This is correct because we are measuring a value of bonding energy β, which is negative; hence the eigenvalue 1 leads to a negative bonding energy and -1 leads to a positive antibonding energy. (The roots of the secular determinant have the opposite signs, see also Problem 10 in Chapter 6.)

In a logical sequence, we might expect to try a solution for propene

$$CH_2 = CH - CH_3$$

but under the Huckel approximations, there is no distinction between ethylene and propene because the π electron system covers only the two π carbon atoms in the molecule *as written*. Nothing outside the π system is included by the Huckel method, so the known stabilizing influence of the methyl group is not found at this level of calculation. We do, however, have a solution for the allyl system assumed to have electrons delocalized over all three carbon atoms

$$\overset{\text{-----------}}{C-C-C}$$

which includes the free allyl radical and allyl ions.

Exercise 7-1

A. Find the eigenvalues and eigenvectors for the allyl model.

B. The rationale for replacing x in a simple Huckel matrix with 0 is that the Coulomb integral does not have anything to do with bonding so one can measure β relative to any reference point. Place some small whole numbers, say 2, 3, or 5, on each of the diagonals in the allyl Huckel matrix and determine the eigenvalues. Measure the energies in units of β above and below the selected reference energy. Are they the same as they are for $x = 0$?

Solution 7-1

A. The Huckel matrix for the allyl system was given in Eq. (6-58). Selecting $x = 0$,

$$\begin{pmatrix} 0 & 1 & 0 \\ 1 & 0 & 1 \\ 0 & 1 & 0 \end{pmatrix} \tag{7-31}$$

Arrange the coefficients of the eigenfunctions as the columns of a 3×3 matrix with the low-energy eigenvector on the left and the high-energy eigenvector on the right.

A. *Mathcad*

$$H := \begin{pmatrix} 0 & 1 & 0 \\ 1 & 0 & 1 \\ 0 & 1 & 0 \end{pmatrix}$$

$$\text{eigenvals (H)} = \begin{pmatrix} 1.414 \\ 0 \\ -1.414 \end{pmatrix} \quad \text{eigenvecs (H)} = \begin{pmatrix} 0.5 & 0.707 & 0.5 \\ 0.707 & 0 & -0.707 \\ 0.5 & -0.707 & 0.5 \end{pmatrix}$$

In this case *Mathcad* has already arranged the eigenvectors in their proper order as can be verified by the following queries:

$$\text{eigenvec (H, 1.41)} = \begin{pmatrix} 0.5 \\ 0.707 \\ 0.5 \end{pmatrix} \quad \text{eigenvec (H, 0)} = \begin{pmatrix} -0.707 \\ 0 \\ 0.707 \end{pmatrix}$$

$$\text{eigenvec (H, -1.41)} = \begin{pmatrix} 0.5 \\ -0.707 \\ 0.5 \end{pmatrix}$$

B. *Mathcad*

$$HB := \begin{pmatrix} 2 & 1 & 0 \\ 1 & 2 & 1 \\ 0 & 1 & 2 \end{pmatrix} \quad HB1 := \begin{pmatrix} 3 & 1 & 0 \\ 1 & 3 & 1 \\ 0 & 1 & 3 \end{pmatrix}$$

$$\text{eigenvals (HB)} = \begin{pmatrix} 3.414 \\ 2 \\ 0.586 \end{pmatrix} \quad \text{eigenvals (HB1)} = \begin{pmatrix} 4.414 \\ 3 \\ 1.586 \end{pmatrix}$$

Bond energies relative to energy levels other than $x = 0$ are invariant. The reference point $x = 0$ is an almost universal convention in simple Huckel theory, however, and we shall continue to use it.

Chemical Application: Charge Density

Once the eigenvectors have been found, there is much that can be done to transform them into derived quantities that give us a better intuitive sense of how HMO calculations relate to the physical properties of molecules. One of these quantities is the *charge density*. The magnitude of the coefficient of an orbital a_{ij} at a carbon atom C_i gives the relative amplitude of the wave function at that atom. The square of the wave function is a probability function; hence, the square of the eigenvector coefficient gives a relative probability of finding the electron within orbital j near carbon atom i. This is the relative charge density, too, because a point in the molecule at which there is a high probability of finding electrons is a point of large negative charge density and a portion of the molecule at which electrons are not likely to be found is positively charged relative to the rest of the molecule. Don't forget that, by definition, each *molecular* orbital includes all carbon atoms in the π electron system. There may be one or two electrons in an orbital ($N = 1$, $N = 2$). Unoccupied orbitals make, of course, no contribution to the charge density.

To obtain the total charge density q_i at atom C_i, we must sum over all occupied or partially occupied orbitals and subtract the result from 1.0, the π charge density of the carbon atom alone

$$q_i = 1.0 - \sum Na_i^2 \tag{7-32}$$

The sum $\sum Na_i^2$ is the total electron probability density at C_i and q_i can be positive or negative relative to the neutral situation.

This definition gives distinctly different charge distributions in, for example, the positively charged ion, the free radical, and the negatively charged ion of the allyl system. The low-energy orbital for the allyl model has coefficients given by the leftmost column in the eigenvector matrix in solution 7-1, Part A.

$$
\begin{array}{ccc}
& 0.707 & \\
0.50 & \vdots & 0.50 \\
\vdots & \vdots & \vdots \\
C & \!\!-\!\!C\!\!-\!\! & CH
\end{array}
$$

The positive ion, with two electrons in the lowest (bonding) orbital, has

$$q_1 = q_3 = 1.00 - 2(.50)^2 = .50$$

and

$$q_2 = 1.00 - 2(.707)^2 = 0.00$$

$$
\begin{array}{ll}
\underline{\quad\quad} & \alpha - 1.414\,\beta \\
\text{-- --}\underline{\quad\quad}\text{-- --} & \alpha \\
\underline{\;\;e\quad e\;\;} & \alpha + 1.414\,\beta
\end{array}
$$

where we recall that β is a negative energy. Thus the charge on the ion $CH_3CH{=}CH^+$ is not localized at one end of the molecule but (within the Huckel approximations) is concentrated equally at either end

$$
\begin{array}{ccc}
\delta+ & & \delta+ \\
C & \!\!-\!\!C\!\!-\!\! & C \\
0.50 & 0.00 & 0.50
\end{array}
$$

The allyl free radical with 3 electrons, 2 in the bonding orbital and 1 in the nonbonding orbital, has

$$q_1 = q_3 = 1.00 - 2(.50)^2 - (.707)^2 = 0.00$$
$$q_2 = 1.00 - 2(.707)^2 - (0.00)^2 = 0.00$$

where the sign of an orbital drops out when we square it. Two electrons are in the lowest-energy MO. They give the middle terms in the equations above. The third term arises from the single nonbonding orbital of allyl, middle column matrix **A**.

$$
\begin{array}{ccc}
0.50 & & \\
\vdots & 0.00 & \\
C\!\!-\!\!C & \!\!-\!\! & C \\
& & \vdots \\
& & -0.50
\end{array}
$$

This leads to

$$
\begin{array}{ccc}
C & \!\!-\!\!C\!\!-\!\! & C \\
0.00 & 0.00 & 0.00
\end{array}
$$

as expected from a neutral species.

The negatively charged ion, with 4 electrons, yields

$$
\begin{array}{ccc}
{}^-C & \!\!-\!\!C\!\!-\!\! & C^- \\
0.50 & 0.00 & 0.50
\end{array}
$$

Its charge density distribution is like that of the cation (with sign reversal) because the added electron goes into the nonbonded orbital with a node at the central carbon atom. The probability of finding that electron precisely at the central carbon atom is zero.

The antibonding orbital remains empty for the free radical and both of the ions.

Exercise 7-2

Write out the charge density diagrams for the positive ion, free radical, and negative ion of the cyclopropyl system.

Chemical Application: Dipole Moments

Knowing the charge density q_i at each atom, one can calculate the dipole moment. First, the charge density at each atom is represented as a vector of length q_i from some arbitrary origin in the direction of atom i. If the vector is collinear with a bond and the origin is at an atom, the vector represents a *bond dipole moment*. All vectors need not represent bond dipole moments because they need not all be collinear with bonds. When all charge densities have been represented by vectors, the sum of the vectors is the total dipole moment. By convention, vectors are usually drawn in the positive direction and dipole moments point toward the negative (electron rich) end of the dipole.

As an example, take the triply substituted carbon atom in methylene cyclopropene as the origin for the charge densities in that molecule. The charge densities at each atom are $-0.478, 0.118, 0.180$, and 0.180 according to the numbering in Fig. 7-1. Let carbon atom 2 be taken as the origin. The direction of the bond vector from carbon atom 2 to carbon 1 is reversed in the vector diagram because the charge density at carbon 1 is negative. Taking 140 pm (1 pm $= 10^{-12}$ m) as a reasonable average value for the C–C bond length, the vector diagram in Fig. 7-1 shows that the sum of charge vectors is 112 C pm (coulomb picometers). Multiplying by 4.77×10^{-2} to convert to units of debyes, one obtains 5.3 D with the negative end of the dipole at the methylene carbon. This is certainly too large. The true value is probably between 1 and 2 D. Nevertheless, an order of magnitude has been

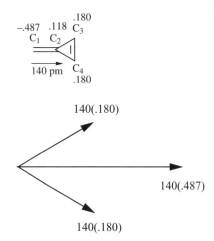

Resultant = 2(140(.180)cos 30) + 140(.487) = 43.6 + 68.2 = 112 C pm

 Dipole moment = 5.3 Debye

Figure 7-1 Dipole Moment Vectors for Methylenecyclopropene. By convention, the dipole moment arrow is drawn in the negative direction.

calculated and the direction of the dipole is correct. We shall treat more refined dipole moment calculations in later chapters.

Chemical Application: Bond Orders

Just as it is possible to calculate the electron probability densities at carbon atoms in a π system, so it is possible to calculate the probability densities *between* atoms. These calculations bear a rough quantitative relationship to the chemical bonds connecting atoms. Bond orders have been correlated with bond lengths and vibrational force constants. Because we are calculating only π electron densities, the results relate only to π bonds. As the reader may anticipate from the discussion to this point, bonds are not localized between atom pairs in MO theory but are delocalized over the entire π system. One can use *Mathcad*, either of the MOBAS programs, or the simple Huckel molecular orbital program SHMO (see section on programs below) to obtain the eigenfunctions and eigenvalues for butadiene shown in Fig. 7-2.

The definition of bond order is

$$P_{jk} = \sum Na_{ij}a_{ik} \tag{7-33}$$

where the summation is over all occupied orbitals connecting atoms C_j and C_k, N is the number of electrons in a single orbital (1 or 2), and a_{ij} and a_{ik} are the coefficients of atoms C_j and C_k. The symbol P_{jk} is given the name *bond order*; it is a measure of the probability of finding a π electron between atoms C_j and C_k. If P_{jk} is large relative to the other bond orders, we anticipate a strong bond. In later

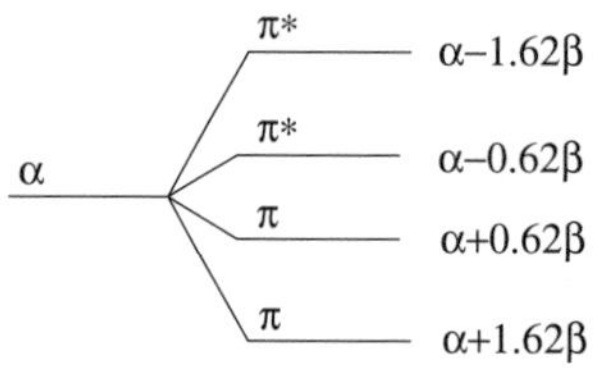

$$H := \begin{pmatrix} 0 & 1 & 0 & 0 \\ 1 & 0 & 1 & 0 \\ 0 & 1 & 0 & 1 \\ 0 & 0 & 1 & 0 \end{pmatrix}$$

$$\text{eigenvals (H)} = \begin{pmatrix} 0.618 \\ 1.618 \\ -0.618 \\ -1.618 \end{pmatrix} \qquad \text{eigenveces (H)} = \begin{pmatrix} 0.602 & 0.372 & 0.602 & -0.372 \\ 0.372 & 0.602 & -0.372 & 0.602 \\ -0.372 & 0.602 & -0.372 & -0.602 \\ -0.602 & 0.372 & 0.602 & 0.372 \end{pmatrix}$$

Figure 7-2   Energy-Level Spectrum for 1,3-Butadiene.

discussions, the term *population* will also be used for this symbol to denote the relative expected electron density.

For buta-1,3-diene, the order of the 1,2 bond is

$$P_{jk} = P_{12} = 2(0.372)(0.602) + 2(0.602)(0.372) = 0.896$$

(Note that the order of internal nodes in the eigenvectors in Fig. 7-2 is 1, 0, 2, and 3 from left to right.)

In the case of the π bond of ethylene, or any isolated π bond, the bond order is 1.0. We may take the value 0.896 for the bond order at the 1,2 position as an indication that the 1,2 π bond in buta-1,3-diene is not exactly the same as the π bond in ethylene $(P_{12} = 1.00)$ but is somewhat diminished by delocalization of electrons over the molecular orbital system. Adding the single σ bond to this result, the "double bond" in 1,3-butadiene is really a "1.896 bond." This delocalization of electrons away from the isolated double bond in the 1,2 position implies an augmentation of the bond order in the 2,3 position; it ought to be more than a single bond by the electron probability density gained from the terminal bonds. The summation of bond orders is not necessarily the same as the number of bonds. For example, in Huckel theory, buta-1,3-diene with 3 σ and 2 π bonds has a total bond order that is 5.23.

Exercise 7-3

Calculate the bond orders for the 2,3 and 3,4 bonds in butadiene. Is the 2,3 bond augmented at the expense of the terminal bonds?

Solution 7-3

$$\underset{1.89 \quad 1.45 \quad 1.89}{C-\!\!-C-\!\!-C-\!\!-C}$$

Yes

Chemical Application: Delocalization Energy

We have already obtained solutions for localized ground-state ethylene leading to the energy $E = 2\alpha + 2\beta$. In looking at allyl, the next more complicated case, we can regard it as an isolated double bond between two sp^2 carbons to which an sp^3 carbon is attached,

C=C–C

or, in valence bond terminology, a resonance hybrid

C=C–C $\leftrightarrow$ C–C=C

In molecular orbital terminology, the hybrid might be represented by one structure with delocalized π electrons spread over the sigma-bonded framework

|||||||||||||||||||
C–C–C

These latter two descriptions are equivalent. Chemical evidence leads us to accept either the valence bond or the molecular orbital representation as preferable to the localized representation. One can calculate the eigenvalue (energy) of the delocalized system and the localized system. Taking $\alpha = 0$ as a reference point, the localized double bond has an energy of 2β but the delocalized model has $E = 2.828\beta$ because it has 2 electrons in the bonding orbital at 1.414β (Exercise 7-1). The difference between the two, 0.828β, is the *delocalization energy* of allyl. This is the Huckel molecular orbital equivalent of the experimentally observed stabilization energy of a double bond by a methyl group.

Exercise 7-4

Write the secular matrix for *localized* buta-1,3-diene

$$CH_2=CHCH=CH_2$$

Solution 7-4

$$\begin{pmatrix} x & 1 & 0 & 0 \\ 1 & x & 0 & 0 \\ 0 & 0 & x & 1 \\ 0 & 0 & 1 & x \end{pmatrix} \tag{7-34}$$

The matrix is the 1,3-butadiene matrix with the ones representing the atoms C_2 and C_3 omitted to reflect the localized nature of the two terminal π bonds. Another way of looking at this matrix is to regard it as representing two ethylene (localized) π bonds in the same linear molecule.

Exercise 7-5

Calculate the delocalization energies of the positive ion, free radical, and negative ion of the allyl model.

Solution 7-5

The energy of the isolated double bond is $2\alpha + 2\beta$. Both ions and the free radical of the allyl system have eigenvalue energies of $2\alpha + 2.828\beta$. The difference is 0.828β in all three cases; hence, the delocalization energies are all 0.828β. The reason the energy is not changed by adding an electron to the allyl positive ion to obtain the free radical is that the electron goes into a nonbonding orbital, which neither augments nor diminishes the energy. The same is true if two electrons are added in to obtain the negative ion. Use of the term *allyl model* or *allyl system* is illustrated by this exercise. The positive ion, the negative ion, the neutral molecule, and the free radical are all represented by the same energy spectrum and the same set of eigenvectors. In the Huckel representation, they differ only in the number of electrons.

Chemical Application: The Free Valency Index

The free valency index F_r is a measure of reactivity, especially of free radicals

$$F_r = 1.732 - \sum P_r \qquad (7-35)$$

where $\sum P_r$ is the sum of bond orders between atom C_r and all atoms to which it is connected. For example, the free valency index for the terminal carbon atoms in 1,3-butadiene is

$$F_1 = F_4 = 1.732 - 0.894 = 0.838$$

Within the predictive capabilities of the models, reactivity is given by F_r. The larger F_r, the more reactive the molecule (or ion or radical). Note that the terminal carbon atoms in buta-1,3-diene are predicted by Huckel theory to be slightly more reactive than the carbon atoms in ethylene. Qualitative correlation with experience is seen for some alkenes and free radicals in Fig. 7-3.

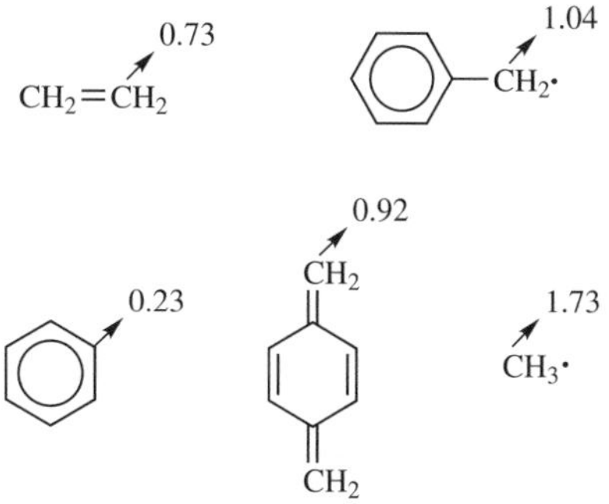

Figure 7-3 Free Valency Indices of Alkenes and Free Radicals.

Chemical Application: Resonance (Stabilization) Energies

The term *resonance energy* has been used in several ways in the literature, but it is generally used to mean the difference between an experimentally determined energy of some relatively complicated molecule and the experimental energy expected by analogy to some relatively simple molecule. For example, the enthalpy of hydrogenation of but-1-ene is -127 kJ mol^{-1}.

$$CH_2=CHCH_2CH_3 + H_2 \rightarrow CH_3CH_2CH_2CH_3$$

from which we can predict the value of $2(-127) = -254$ kJ mol^{-1} for hydrogenation of buta-1,3-diene

$$CH_2=CHCH=CH_2 + H_2 \rightarrow CH_3CH_2CH_2CH_3$$

The actual value of the enthalpy of hydrogenation of 1,3-butadiene is -243 kJ mol^{-1}. Both are hydrogenated to the same product, n-butane; hence the enthalpy diagram (Fig. 7-4) shows that buta-1,3-diene is 11 kJ mol^{-1}. lower in enthalpy than it "ought" to be on the basis of the reference standard, but-1-ene.

The simple molecule (from a π electronic point of view), but-1-ene, is the reference state against which we compare the relatively complicated molecule, buta-1,3-diene. Changing the reference standard to ethylene, however, gives a different value for the resonance energy. Ethylene has an enthalpy of hydrogenation $\Delta_{hyd}H^{298}$ of -136.0 kJ mol^{-1}, which leads to $2(-136) = -272$ kJ mol^{-1} as the anticipated $\Delta_{hyd}H^{298}$ of buta-1,3-diene. Based on this standard, the resonance energy is $272 - 243 = 29$ kJ mol^{-1}. Because different choices of the reference state lead to different values of the resonance energy, historically this has led to different definitions of resonance energy as well (see, e.g., Pauling, 1960, Dewar, 1969). The resonance stabilization of benzene and benzenoid compounds is especially strong. Benzene, which is much less reactive (more stable) than the three double

bonds of its "Kekule structure" would lead us to expect, is said to be *aromatic*.

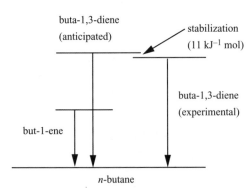

Figure 7-4 Enthalpy Level Diagram for But-1-ene and Buta-1,3-diene. Arrows point down because $\Delta_{hyd}H^{298}$ for hydrogenation is negative (exothermic).

LIBRARY PROJECT 7-1 | *The History of*
Resonance and Aromaticity
Write an essay of approximately 2000 words on the history and the various definitions of the concept of resonance, resonance energy, and aromaticity.

Valence-bond theory explains the stabilization of systems of conjugated double bonds (like those in buta-1,3-diene) with resonance structures that are analogous to the resonating oscillators in the section on normal modes of motion in Chapter 5 (see also Wheland, 1955). This is the origin of the term *resonance* stabilization. The analogy must not be carried too far, however, because energy does not pass from one resonance structure to another as it does from one harmonic oscillator to another. Resonance extremes are hypothetical structures used to describe the unique real molecule. The difference between the enthalpy of hydrogenation anticipated on the basis of a single hypothetical reference structure and the experimental enthalpy of hydrogenation of the real molecule is one definition of the theoretical resonance energy.

We have used the term resonance energy in the heading of this section largely to connect with the older literature. Because of the different ways resonance energy can be defined and the somewhat contentious literature surrounding the term, the more general term *stabilization energy* has come into the molecular orbital literature and is preferable. In this somewhat less restrictive terminology, buta-1,3-diene is said to be stabilized relative to some reference standard. Just as molecules can be *stabilized* by electronic interactions, they can also be destabilized and a *destabilization energy* can be measured relative, once again, to an arbitrary standard state.

Exercise 7-6

Turner (1957) measured $\Delta_{\text{hyd}}H^{298}$ of cyclohexene and found it to be -113 kJ mol^{-1}. Using cyclohexene as the reference standard, calculate the resonance energies of cyclohexa-1,3-diene ($\Delta_{\text{hyd}}H^{298} = -224$ kJ mol^{-1}), cyclohexa-1,4-diene ($\Delta_{\text{hyd}}H^{298} = -225$ kJ mol^{-1}) and benzene ($\Delta_{\text{hyd}}H^{298} = -216$ kJ mol^{-1}). Comment on these results. The value for benzene (Kistiakowsky, 1938) has been corrected to conform with the experimental conditions of Turner's results.

Extended Huckel Theory—Wheland's Method

One restriction imposed by Huckel theory that is rather easy to release is that of zero overlap for nearest-neighbor interactions. One can retain $\alpha - E$ as the diagonal elements in the secular matrix and replace β by $\beta - E_jS$ as nearest-neighbor elements where S is the overlap integral. Now,

$$\begin{pmatrix} \alpha - E_j & \beta - E_jS & \cdots & 0 \\ \beta - E_jS & \alpha - E_j & \beta - E_jS & \cdots \\ \cdots & \cdots & \cdots & \cdots \\ 0 & \cdots & \beta - E_jS & \alpha - E_j \end{pmatrix} \tag{7-36}$$

is the secular matrix.

By analogy to the substitution $x = \alpha - E/\beta$ in the section on the secular matrix in Chapter 6, we can make the substitution

$$x = \frac{\alpha - E_j}{\beta - E_j S} \tag{7-37}$$

which causes the secular matrix to take the same form that it did in the simple Huckel theory, for example,

$$\begin{pmatrix} x & 1 \\ 1 & x \end{pmatrix} \tag{6-44}$$

for ethylene. From the definition of x,

$$E_j = \frac{\alpha - x\beta}{1 - xS} \tag{7-38}$$

Millikan has shown that the overlap integral for hydrogen-like p orbitals in linear hydrocarbons is about 0.27 (Millikan, 1949).

Exercise 7-7

Prove Eq. (7-38) from the definition of x.

Exercise 7-8

What is the energy separation $E_2 - E_1$ of the bonding and antibonding orbitals in ethylene, assuming that the overlap integral S is 0.27?

Solution 7-8

The eigenfunctions are ± 1 as in the simple Huckel calculation for ethylene

$$E_1 = \frac{\alpha + \beta}{1.27}$$

$$E_2 = \frac{\alpha - \beta}{0.73}$$

where, by convention, we have chosen $E_1 < E_2$. If we take

$$\beta = -1, \text{ and } \alpha = 0$$

this leads to

$$E_j = -0.79, 1.37 \qquad (j = 1, 2)$$

The separation is 2.16β, which, considering all the approximations already made, is not greatly different from the separation we found ignoring overlap.

Extended Huckel Theory—Hoffman's EHT Method

Hoffman's extended Huckel theory, EHT (Hoffman, 1963), includes all bonding orbitals in the secular matrix rather than just all π bonding orbitals. This inclusion increases the complexity of the calculations so that they are not practical without a computer. The basis set is a linear combination that includes only valence orbitals

$$\psi_j = \sum a_{ij}\phi_i \tag{7-39}$$

but even for ethylene, this leads to a 12×12 secular matrix because there are 4 valence electrons on each of 2 carbon atoms and 1 on each of the 4 hydrogens.

The orbitals used for methane, for example, are four $1s$ Slater orbitals of hydrogen and one $2s$ and three $2p$ Slater orbitals of carbon, leading to an 8×8 secular matrix. Slater orbitals are systematic approximations to atomic orbitals that are widely used in computer applications. We will investigate Slater orbitals in more detail in later chapters.

We fill the secular matrix $\mathbf{H}$ with elements H_{ij} over the entire set of valence orbitals. The diagonal elements are

$$H_{ii} = \begin{array}{lll} -13.6 & 1s & \text{Hydrogen} \\ -21.4 & 2s & \text{Carbon} \\ -11.4 & 2p & \text{Carbon} \end{array}$$

which are the atomic ionization energies in electron volts (eV). The EHT energies are negative as always for bound states relative to an arbitrary zero of energy defined as the energy of the unbound state. (See Computer Project 3-3 for determination of the ionization energy of hydrogen.)

Off-diagonals are given by

$$H_{ij} = 0.88(H_{ii} + H_{jj})S_{ij}$$

where S_{ij} is, once again, the overlap integral. The off-diagonal H_{ij} is the arithmetic mean of H_{ii} and H_{jj} modified by the overlap integral and multiplied by another empirical factor, $1.75/2 = 0.88$. The S_{ij} are obtained from the Slater orbitals, which closely resemble atomic orbitals in shape. The sum of the occupied EHT orbital energies times 2 (for 2 electrons per orbital in the nondegenerate ground state) is the total energy of the valence electrons in the molecule relative to a reference state of the completely ionized core atoms. The core of C is its nucleus plus its $1s$ electrons.

In the process of diagonalization, the *trace* of the EHT matrix is invariant but the spacing between energy levels changes. Changes in spacing are brought about by the presence of nonzero off-diagonal matrix elements. In methane, for example, the energy of the lowest 4 EHT orbitals gets lower and the energy of the highest 4 orbitals goes up by an equal amount.

Exercise 7-9

Sum H_{ii} for the EHT matrix of methane and so obtain the trace of the EHT matrix.

Solution 7-9

$$\sum H_{ii}(\text{C}, 4\text{H}) = -110\,\text{eV}$$

After diagonalization of the EHT matrix, the lowest 4 orbitals have an energy sum of about -70 eV. The electronic energy for these doubly occupied orbitals is $2(-70) = -140$ eV. The energy gain of the molecule relative to its atoms is $-140 - (-110) = -30\,\text{eV} = -690$ kcal mol^{-1} (1 eV $\cong$ 23 kcal mol^{-1}); therefore, the molecule is stable relative to its atoms. We can envision an energy cycle with three steps (Fig. 7-5):

A. Completely independent core positive ions and electrons come together from infinite separation to form 1 C atom and 4 H atoms. One electron per atomic orbital brings about a total energy change of -110 eV.

B. Completely independent core positive ions and electrons come together to form 1 CH$_4$ molecule. Two electrons per occupied bonding orbital bring about an energy change of about -140 eV.

C. The energy of formation of CH$_4$ in the gaseous state from the gaseous atoms is found from the difference

$$\text{C(g)} + 4\,\text{H(g)} \rightarrow \text{CH}_4(\text{g})$$

By this scheme, the energy of formation of methane from its gaseous independent atoms is about $-30\,\text{eV} = -690$ kcal mol^{-1}. This is not the thermodynamic energy of formation in the standard state as it is usually defined but is closely related to it. The energy of formation defined in this way is the reverse of the energy of atomization of methane. Results of molecular orbital calculations are often reported as atomization energies, particularly in the literature before 1990. The EHT value is quantitatively wrong (the experimental value is -390 kcal mol^{-1}), but the method can be indefinitely improved by self-consistent field iteration (Chapter 8) and reparameterization.

The S_{ij} are geometry dependent; hence, one can try various molecular geometries and select the structure that gives the lowest energy, thereby obtaining the best geometry from among the alternatives tried. We have already seen this procedure used in molecular mechanics. It is a concept that will be used at a sophisticated level in the remaining chapters of this book. Hoffman makes the point, richly confirmed by subsequent studies,

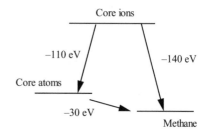

Figure 7-5 EHT Energy Diagram for Methane.

that accurate molecular geometries are easier to determine, and therefore more accurate, than energy or other molecular properties, particularly those dependent upon charge density. As in simple Huckel calculations, but in contrast to the self-consistent field calculations to be described in Chapter 8, there is only one EHT matrix diagonalization.

Applications of EHT include calculation of the rotational barrier in ethane and of the chair-boat conformational energies in cyclohexane. EHT has been largely supplanted by *ab initio* and *semiempirical* calculations, but the method deserves our attention because it introduces Slater orbitals and use of full Huckel matrices as distinct from Huckel matrices that have been simplified by setting most of the elements equal to zero. In his original paper, Hoffman applied his extended Huckel theory to nonplanar molecules and to molecules more diverse and larger than those customarily treated by simple Huckel theory. EHT is the first truly computational molecular orbital method for large molecules, all of the results in the original paper having been obtained on an IBM 7090.

The Programs

MOBAS was written by the author (Rogers, 1983) in BASIC to illustrate matrix inversion in molecular orbital calculations. It is modeled after a program in FORTRAN II given by Dickson (Dickson, 1968).

SHMO is a simple Huckel MO program in FORTRAN that functions much as MOBAS does and is also based on the Dickson program. The SHMO source code must be compiled. Compiled SHMO is in executable code (**.exe**). Run SHMO from the system level (do not go into BASIC) with the single command

>**SHMO**

SHMO responds with a series of prompts. The input format is similar to MOBAS. Matrix elements are entered

  row number, column number, element, 0

except for the last entry, which ends in 99. After receiving the complete input, SHMO prints out the eigenvalues and eigenfunctions for the problem presented to it. Note that data are input to the program from the keyboard and not built into it as in MOBAS.

The output of SHMO is interrupted by a PAUSE statement. This prevents scrolling through the entire output at the end of a calculation. At the PAUSE prompt in the output, hit ENTER to see the first eigenvalue and eigenvector. Hit ENTER again for the second eigenvalue and eigenvector, and so on. To read the program page by page, enter TYPE SHMO.FOR | MORE. To modify SHMO, enter EDIT SHMO.FOR. All of these commands are executed from the system level, at the prompt

  >

If you wish to modify SHMO or run it on a machine that is different from the machine that it was compiled on, you may have to recompile. Compiling FORTRAN programs is outside the scope of this book but is described in detail in the manuals available with any commercial compiler. SHMO was compiled with IBM FORTRAN Compiler 2.0 and also with Microsoft FORTRAN Professional Development System 5.1 (1991).

HMO is a more elaborate Huckel MO program than either MOBAS or SHMO. It calculates charge densities, free valency indices, and bond orders. It is written in FORTRAN and is a modification of a program by Greenwood (Greenwood, 1972). HMO must also be recompiled for different machines. The same two compilers work for HMO as for SHMO. The output of HMO is more compact than that of SHMO. For molecules of eight carbon atoms or fewer, the eigenvalues and eigenvectors fit onto the first screen. After responding to a PAUSE prompt, the total π electron energy, charge densities, free valency indices at each carbon atom, and bond orders are seen on the second screen. As presently formatted, HMO is limited to molecules of six carbon atoms or fewer to make the output compact, but this restriction can be released by changing the FORMAT statements and recompiling. A partial output for buta-1,3-diene is given in Fig. 7-6.

> TOTAL PI-ELECTRON ENERGY= -4.4721
> Charge densities
>   .0000 .0000 .0000 .0000
> Free valency indices
>   .8376 .3904 .3904 .8376
> Bond order matrix
>   1.0000 .8944 1.0000 .0000 .4472 1.0000 -.4472 .0000
>   .8944 1.0000
> Stop - Program terminated.

Figure 7-6   HMO output for Buta-1,3-diene (second screen, condensed).

The total π electron energy is the sum of occupied orbital energies multiplied by two if, as is usually the case, the orbital is doubly occupied. The charge densities and free valency indices were treated in separate sections above. The bond order output should be read as a lower triangular semimatrix. The bond order semimatrix for the butadiene output is shown in Fig. 7-7.

1.0000			
.8944	1.0000		
.0000	.4472	1.0000	
-.4472	.0000	.8944	1.0000

Figure 7-7   The Bond Order Semimatrix for Buta-1,3-diene.

The principal diagonal of the HMO output matrix is the π electron probability density at atom j, $\sum Na_{ij}^2$, where the summation is over all occupied orbitals. This can be thought of as the bond order of atom j with itself $\sum Na_{ij}a_{ij}$. The electronic charge times the electron probability density is the charge density at atom j, *relative to* a charge of 1.000 contributed by the $2p$ electron of the carbon atom, .0000 in the HMO output for buta-1,3-diene. The free valency index follows from Eq. (7-35).

The full bond order matrix is a symmetric tridiagonal matrix (Chapter 2). It is symmetric because the bond order $P_{jk} = \sum Na_{ij}a_{ik}$ is the same as the bond order $P_{kj} = \sum Na_{ik}a_{ij}$. Elements off the tridiagonal ($-.4472$ in the butadiene example) are artifacts of the minimization and should be disregarded. The full bond order matrix for butadiene is

$$
\begin{pmatrix}
1.0000 & 0.8944 & 0 & 0 \\
0.8944 & 1.0000 & 0.4472 & 0 \\
0 & 0.4472 & 1.0000 & 0.8944 \\
0 & 0 & 0.8944 & 1.0000
\end{pmatrix}
$$

There is slight disagreement (about 0.001β) between the bond orders in this output and those from **Mathcad** (see section on bond orders above) because of rounding error, different algorithms being used to determine the eigenvectors, and so on. In an approximate method such as this one, errors of this magnitude are negligible, although we shall soon be carrying out calculations where accuracy to the fifth digit beyond the decimal point is critical.

COMPUTER PROJECT 7-1 | Larger Molecules: Calculations using SHMO

SHMO, being in compiled FORTRAN, is much faster than MOBAS, which is written in standard interpreted BASIC. Solve the Huckel matrix for the hexatriene model using MOBAS and again using SHMO. Record the run time for this problem with each program. If you have a fairly new machine, this part of the project may not work. Both calculations may be so fast that you will not see the difference. From among the numerous problems presented up to this point, find one that is sufficiently demanding of computer iterations that you can tell the difference between the run times for interpreted BASIC and compiled FORTRAN. You will probably have to write your own programs in both BASIC and FORTRAN. Keep them as simple as possible. Coding and compiling FORTRAN programs is one of those things that isn't much fun but should be done at least once or twice for the experience.

Use SHMO to obtain the energy spectrum for the models methylenepentadiene, bicyclohexatriene, and styrene. Draw all three energy level diagrams. Are there degeneracies for these molecules?

Polarographic oxidation entails removing one or more electrons from a molecule M undergoing oxidation at a mercury or similar electrode

$$M \rightarrow M^+ + e^- \quad \text{(oxidation of M)}$$

The more tightly held an electron is, the more difficult it is to remove, hence the higher the electrode potential necessary to remove it. Make the reasonable hypothesis that the electron removed in a one-electron oxidation comes from the highest occupied orbital, HOMO. Using SHMO, determine the HOMO for benzene, biphenyl, and naphthalene.

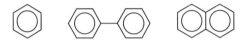

Note that all of the occupied orbitals in these molecules have negative energies, that is, they are below α by an energy measured in units of β. Electrons with the most negative β are most strongly bound and hardest to remove from the molecule. The molecule with the least negative HOMO has electrons that are highest in the energy spectra of these three relatively complicated molecules. These electrons are most easily withdrawn; hence, the molecule with the least negative β is most easily oxidized.

Arrange the three compounds in order of increasing oxidation potential. Plot the experimental value of oxidation potential, 1.3, 1.5, and 2.0 V for naphthalene, biphenyl, and benzene, respectively, vs. β for the HOMO of each compound.

COMPUTER PROJECT 7-2 | *Dipole Moments*

Using Program HMO, calculate the dipole moment of methylenecyclopropene by the HMO method. The program gives total charge densities at each carbon atom, making the calculation of dipole moments essentially a geometric problem. The single charge of the carbon atom must be subtracted from the total charge density to obtain the charge density used in the dipole moment calculation, that is, a computer output at carbon 1 of 1.5 leads to a π charge density at that atom of 0.5. Calculate the dipole moment of fulvene by the HMO method. Assume, for the calculation, that the endocyclic double bonds are parallel as in the diagram below and that the angle at carbon atom 2 is the same as in methylene-cyclopropene. These assumptions are not true, but we will be able to arrive at more accurate geometries and dipole moments by semiempirical and *ab initio* calculations. In which direction is the dipole moment of methylenecyclopropene? In which direction is the dipole moment of fulvene?

methylenecyclopropene   fulvene

Program HMO gives you the option of modifying one or more of the elements input to the semimatrix. Calculate the charge densities of cyclopropeneone by entering the semimatrix

$$
\begin{pmatrix}
2 & & & & \\
1 & 0 & & & \\
0 & 1 & 0 & & \\
0 & 1 & 1 & 0 &
\end{pmatrix}
$$

and selecting the modification option by typing 01 at the prompt: Enter number of elements to be modified I2 format. Respond to the prompt: Enter row (I2), column (I2), and new element (F6.3) by typing

02011.414

This leads to the semimatrix

$$\begin{pmatrix} 2 & & & \\ 1.414 & 0 & & \\ 0 & 1 & 0 & \\ 0 & 1 & 1 & 0 \end{pmatrix}$$

for cyclopropenone. Program HMO automatically loads the full matrix from the semimatrix because Huckel molecular orbital matrices are always symmetrical; hence, the program "knows" what the elements are above the principal diagonal. Calculate the dipole moment of cyclopropenone.

Use the same method to calculate the dipole moment of cyclopentadienone. Assume, for the calculation, that the endocyclic double bonds are parallel and the angle at carbon 2 is the same as in cyclopropenone.

cyclopropenone   cyclopentadienone

There is a substantial difference in dipole moments between methylenecyclopropene and cyclopropenone, but the difference between fulvene and cyclopentadienone is much smaller. Explain.

COMPUTER PROJECT 7-3 | *Conservation of Orbital Symmetry*
Conservation of orbital symmetry is a general principle that requires orbitals of the same phase (sign) to match up in a chemical reaction. For example, if terminal orbitals are to combine with one another in a cyclization reaction as in pattern A, they must rotate in the same direction (conrotatory overlap), but if they combine according to pattern B, they must rotate in opposite directions (disrotatory). In each case, rotation takes place so that overlap is between lobes of the π orbitals that are of the same sign.

Pattern A

Pattern B

For this computer project, obtain the orbitals of butadiene and predict whether the cyclization of butadiene to cyclobutene is conrotatory or disrotatory.

Perform the same calculation for 1,3,5-hexatriene.

Conrotatory and disrotatory concerted reactions can often be distinguished by chemical means. For example, using the results of the previous calculation, predict whether the cyclizations of hexa-2,4-diene will lead to *cis* or *trans* dimethylcyclobutene

Perform the same calculation for cyclization of 2,4,6-octatriene. Which isomer of dimethylcyclohexadiene is formed?

COMPUTER PROJECT 7-4 | *Pyridine*

Heteroatoms have an electron density that is different from carbon. From an HMO point of view, the coefficient of the heteroatom in the secular matrix is larger or smaller according to whether the heteroatom is electronegative or electropositive relative to carbon. Empirical parameters may be used to augment or diminish the wave function through its coefficients, a_i at the heteroatom j. The wave function (hence ψ^2 and the electron probability density) can be augmented at the heteroatom only, as in the problem on pyrrole in Chapter 6, or for both the heteroatom and the carbon atoms immediately adjacent to it. In one approximation (Pilar, 1990), the input matrix for pyridine is

$$
\begin{pmatrix}
x+0.5 & 1.2 & 0 & 0 & 0 & 1.2 \\
1.2 & x & 1 & 0 & 0 & 0 \\
0 & 1 & x & 1 & 0 & 0 \\
0 & 0 & 1 & x & 1 & 0 \\
0 & 0 & 0 & 1 & x & 1 \\
1.2 & 0 & 0 & 0 & 1 & x
\end{pmatrix}
$$

where the modifications $\alpha'_r = \alpha + h_r\beta$ and $\beta'_{rs} = k_{rs}\beta = k_{rs}$ have been made to the secular matrix of benzene to account for the substitution of one nitrogen atom at the 1,1 position for one carbon atom. The parameters h_r and k_{rs} for modifying α and β are based on electronegativity differences from carbon but are not particularly transferable to other problems. Transferability failure is one of the reasons that HMO methods have been largely abandoned in favor of methods described in Chapters 8–10. A thorough understanding of HMO methods is, however, a useful stepping stone to research-level *ab initio* and semiempirical molecular orbital calculations.

Procedure. Subtract $x\mathbf{I}$ from the input matrix above. Load the resulting upper semimatrix into MOBAS. The first element is 1,1,0.5,0. Recall that MOBAS requires entry of only the nonzero elements in the upper semimatrix. Obtain the eigenvalues and eigenvectors.

Repeat the procedure using HMO. HMO requires entry of the entire lower semimatrix, including the diagonal and all zero elements. Because the matrix element format is I1, only one symbol can be entered for each element. The numbers 0.5 and 1.2 cannot be entered in this format; instead enter 1, which will be modified later. The initial unmodified input for pyridine is the same as that for benzene, 010010001000010100010; hence, we can make a trial run on benzene to see if everything is working properly.

Run benzene using HMO. Write out the full bond order matrix, entering zero for any element off the tridiagonal. What is the bond order of benzene? Is there any Kekule-type alternation in this model?

To prepare the input matrix for pyridine, respond to the prompt asking how many elements should be modified with 03. Follow this with 01010.5 to change the 1,1 element to 0.5. Continue for the remaining two element changes. When the last element has been properly modified, the eigenvalues and eigenfunctions are calculated. Rerun to make sure you get the same answer.

Write out the tridiagonal matrix of charge densities (principal diagonal) and bond orders (upper and lower off-diagonals). What is the most active site in pyridine acting as a base? How might Cu^{2+} complex with pyridine? Is there a theory under which this might be regarded as an acid-base reaction? Is there much charge alternation in the pyridine ring? Is there much bond alternation in the pyridine ring?

PROBLEMS

1. Find the determinant of

$$\begin{pmatrix} \cos\theta & \sin\theta \\ \sin\theta & -\cos\theta \end{pmatrix}$$

2. Given the modified Huckel matrix and the orthogonal transform in Exercise 7-1, carry out the multiplication

   AXA

   and identify the product you get.

3. Is the allyl coefficient matrix orthonormal?

4. Write out the coefficients of the butadienyl system, as they are produced by program MOBAS, in matrix form. Is the matrix symmetric? If not, can it be made symmetric by exchanging rows only?

5. Determine the eigenvectors and eigenvalues for methylenecyclobutene.

6. Determine the delocalization energy and dipole moment for methylenecyclobutene.

7. Draw charge density diagrams for the positive ion, free radical, and negative ion of the butadienyl system.

8. Draw bond order and free valency index diagrams for the butadienyl system.

9. Write a "counter" into program MOBAS to determine how many iterations are executed in solving for the allyl system. The number is not the same for all computers or operating systems. Change the convergence criterion (statement 300) to several different values and determine the number of iterations for each.

10. Refer to Computer Project 7-2. Calculate β in units of electron volts using Wheland's extension of Huckel molecular orbital theory.

11. Print Program MOBAS. Identify the statements in the program that generate the eigenvector matrix $\mathbf{A}$ by performing the same rotation operations on $\mathbf{I}$ that it performs on the input matrix to generate the eigenvalue matrix.

12. Determine the dipole moment of cyclobutenone.

13. Assuming a $(2s)^2(2p)^2$ electron distribution for the carbon atoms, calculate the energy of formation of ethylene from the gaseous atoms.

14. Spectroscopically determined values of β vary, but they are usually around -2.4 eV. In the section on resonance stabilization, we saw that thermodynamic measurements of the total resonance stabilization of butadiene yield 11 and 29 kJ mol^{-1} according to the reference standard chosen. Calculate the delocalization energy of buta-1,3-diene in units of β. Determine two values for the "size" of the energy unit β from the thermochemical estimates given. Do these agree well or poorly with the spectroscopic values?

15. The delocalization energy of benzene is 2β (verify this). From information in Exercise 7-6 calculate yet another value for the "size" of the unit β based on the thermodynamic values of the enthalpy of formation of benzene. Does this value agree with the thermodynamic values in Problem 14? Does it agree with the spectroscopic value?

8

Self-Consistent Fields

Because of its severe approximations, in using the Huckel method (1932) one ignores most of the real problems of molecular orbital theory. This is not because Huckel, a first-rate mathematician, did not see them clearly; they were simply beyond the power of primitive mechanical calculators of his day. Huckel theory provided the foundation and stimulus for a generation's research, most notably in organic chemistry. Then, about 1960, digital computers became widely available to the scientific community.

Beyond Huckel Theory

To advance beyond Huckel's method, we look again at the problem of calculating elements of the secular matrix. This is the problem Huckel simply swept away by setting the diagonal elements α equal to zero (as a reference point) and giving the basic energy unit the simplest possible definition: $\beta = 1$ for adjacent carbon atoms and $\beta = 0$ for nonadjacent carbons in alternant hydrocarbons. Huckel molecular orbital energies come out in units of β, but an indication of the inadequacy of the Huckel method is that evaluations of β for different molecular systems do not give the same results (Problems 15 and 16 in Chapter 7). The first systematic evaluation of all elements in the secular matrix for larger molecules by Hoffmann (1963) came

Computational Chemistry Using the PC, Third Edition, by Donald W. Rogers
ISBN 0-471-42800-0   Copyright © 2003 John Wiley & Sons, Inc.

about fully three decades after Huckel's original work. About a decade earlier, Pople (1953) and others introduced the use of self-consistent fields into molecular orbital calculations. Both full-matrix and self-consistent field calculations rely heavily on computers.

Elements of the Secular Matrix

To begin a more general approach to molecular orbital theory, we shall describe a variational solution of the prototypical problem found in most elementary physical chemistry textbooks: the ground-state energy of a particle in a box (McQuarrie, 1983) The particle in a one-dimensional box has an exact solution

$$E = \frac{n^2 h^2}{8ma^2} \tag{8-1}$$

and has an exact wave function

$$\Psi = A \sin(kx) \tag{8-2}$$

Let the dimension of the box be 1 in any units and consider only the ground state, for which $n = 1$. Now

$$E = \frac{h^2}{8m} = 0.125 \frac{h^2}{m} \tag{8-3}$$

but, although we know the answer, we wish to test our approximation method by taking a linear combination of functions

$$\psi = c_1 x(1 - x) + c_2 x^2 (1 - x)^2 \tag{8-4}$$

which we shall write

$$\psi = c_1 f_1 \pm c_2 f_2 \tag{8-5}$$

There are two functions, so we shall obtain two eigenvalues. The ground-state energy will be the lower of the two. The full secular matrix is

$$\begin{pmatrix} H_{11} - E_j S_{11} & H_{12} - E_j S_{12} \\ H_{21} - E_j S_{21} & H_{22} - E_j S_{22} \end{pmatrix} \quad (j = 1, 2) \tag{8-6}$$

If we do not make any simplifying assumptions, we must calculate the matrix elements

$$H_{11} = \int f_1 H f_1 \, d\tau \tag{8-7a}$$

$$H_{12} = \int f_1 H f_2 \, d\tau \tag{8-7b}$$

$$H_{21} = \int f_2 H f_1 \, d\tau \tag{8-7c}$$

and

$$H_{22} = \int f_2 H f_2 \, d\tau \tag{8-7d}$$

where the Hamiltonian operator for a particle in a box with the potential energy $V = 0$ is

$$\frac{-\hbar^2}{2m} \frac{d^2}{dx^2} \tag{8-8}$$

In this problem, the integral over "all space" $d\tau$ is in only one dimension, x. The limits of integration are the dimensions of the box, 0 and 1 in whatever unit was chosen.

Exercise 8-1

Calculate matrix element H_{11}

Solution 8-1

$$f_1 = x(1-x)$$

$$\frac{d^2}{dx^2} x(1-x) = -2$$

$$\frac{-\hbar^2}{2m} \int_0^\infty x(1-x) \frac{d^2}{dx^2} x(1-x) dx = \frac{-\hbar^2}{2m} \int_0^\infty x(1-x)(-2) d\tau \tag{8-9}$$

$$= \frac{\hbar^2}{m} \left(\frac{x^2}{2} - \frac{x^3}{3} \right) \Big|_0^1 = \frac{\hbar^2}{m} \frac{1}{6} = \frac{\hbar^2}{6m}$$

Also, one needs to calculate the matrix elements

$$S_{11} = \int f_1 f_1 \, d\tau \tag{8-10a}$$

$$S_{12} = \int f_1 f_2 \, d\tau \tag{8-10b}$$

$$S_{21} = \int f_2 f_1 \, d\tau \tag{8-10c}$$

$$S_{22} = \int f_2 f_2 \, d\tau \tag{8-10d}$$

Again, the limits of integration are from zero to one.

Exercise 8-2

Calculate S_{11}

Solution 8-2

$$S_{11} = \int_0^1 x(1-x)x(1-x)d\tau$$

$$= \int_0^1 (x^2 - x^3 - x^3 + x^4)d\tau$$

$$= \frac{1}{3} - \frac{1}{4} - \frac{1}{4} + \frac{1}{5} = \frac{1}{30} \qquad (8\text{-}11)$$

Now, calculating all H_{ij} and S_{ij} elements in the same way, and inserting them into the secular matrix, one obtains

$$\begin{pmatrix} \dfrac{\hbar^2}{6m} - \dfrac{E_j}{30} & \dfrac{\hbar^2}{30m} - \dfrac{E}{140} \\[2ex] \dfrac{\hbar^2}{30m} - \dfrac{E}{140} & \dfrac{\hbar^2}{105m} - \dfrac{E_j}{630} \end{pmatrix} \qquad (8\text{-}12)$$

Dividing each element by $\hbar^2/m$ and setting

$$x = \frac{mE_j}{\hbar^2}$$

yields

$$\begin{pmatrix} \dfrac{1}{6} - \dfrac{x}{30} & \dfrac{1}{30} - \dfrac{x}{140} \\[2ex] \dfrac{1}{30} - \dfrac{x}{140} & \dfrac{1}{105} - \dfrac{x}{630} \end{pmatrix} \qquad (8\text{-}13)$$

This can be cleared of fractions by multiplying by 1260 to obtain the secular determinant

$$\begin{vmatrix} 210 - 42x & 42 - 9x \\ 42 - 9x & 12 - 2x \end{vmatrix} = 0 \qquad (8\text{-}14)$$

corresponding to Eq. set (1-9), which was solved iteratively in Computer Project 1-2 to yield the roots

$$x = 4.93487 \text{ and } 51.065 \qquad (8\text{-}15)$$

The lower of the two roots is the one we seek for the ground-state energy of the system.
Thus, $x = mE_j/\hbar^2$; $E_j = x(\hbar^2/m)$, and, recalling that $\hbar = h/2\pi$,

$$E = \frac{4.93487}{(2\pi)^2}\left(\frac{h^2}{m}\right) = 0.125002\left(\frac{h^2}{m}\right) \qquad (8\text{-}16)$$

as contrasted to the exact solution of $0.125(h^2/m)$. Note that the energy obtained from the variational solution is slightly higher than the solution obtained from the exact wave function. [This result makes it clear why we needed to find the roots of Eq. (1-9) to six significant figures.]

One of the things illustrated by this calculation is that a surprisingly good approximation to the eigenvalue can often be obtained from a combination of approximate functions that does not represent the exact eigenfunction very closely. Eigenvalues are not very sensitive to the eigenfunctions. This is one reason why the LCAO approximation and Huckel theory in particular work as well as they do.

Another feature of advanced molecular orbital calculations that we can antici-pate from this simple example is that calculating matrix elements for real molecules can be a formidable task.

The Helium Atom

The helium atom is similar to the hydrogen atom with the critical difference that there are two electrons moving in the potential field of a nucleus with a double positive charge $(Z = 2)$ (Fig. 8-1).

The Hamiltonian for the helium atom,

$$\hat{H} = -\frac{\hbar^2}{2m}\nabla_1^2 - \frac{\hbar^2}{2m}\nabla_2^2 - \frac{Ze^2}{4\pi\varepsilon_0 r_1} - \frac{Ze^2}{4\pi\varepsilon_0 r_2} + \frac{1}{4\pi\varepsilon_0 r_{12}}$$

becomes

$$\hat{H} = -\tfrac{1}{2}\nabla_1^2 - \tfrac{1}{2}\nabla_2^2 - \frac{2}{r_1} - \frac{2}{r_2} + \frac{1}{r_{12}} \tag{8-17}$$

when atomic units are used. Regrouping,

$$\hat{H} = \left[-\tfrac{1}{2}\nabla_1^2 - \frac{2}{r_1}\right] + \left[-\tfrac{1}{2}\nabla_2^2 - \frac{2}{r_2}\right] + \frac{1}{r_{12}}$$

we have two Hamiltonians that are identical to the hydrogen case except for a different nuclear charge, plus an added term $1/r_{12}$ due to electrostatic repulsion of the two electrons acting over the interelectronic distance r_{12}

$$\hat{H}_{\text{He}} = \hat{H}_1 + \hat{H}_2 + \frac{1}{r_{12}} \tag{8-18}$$

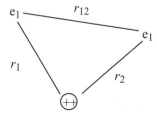

Figure 8-1   Schematic Diagram of a Helium Atom.

If the Hamiltonian were to operate on an exact, normalized wave function for helium, the energy of the system would be obtained

$$E_{He} = \int_0^\infty \Psi(r_1, r_2)\hat{H}_{He}\Psi(r_1, r_2)d\tau \tag{8-19}$$

but the helium atom is a three-particle system for which we cannot obtain an exact orbital. The orbital and the total energy must, of necessity, be approximate.

As a naive or zero-order approximation, we can simply ignore the "r_{12} term" and allow the simplified Hamiltonian to operate on the $1s$ orbital of the H atom. The result is

$$E_{He} = -\frac{2^2}{2} - \frac{2^2}{2} = -4.00 \text{ hartrees} \tag{8-20}$$

which is 8 times the exact energy of the hydrogen atom ($-\frac{1}{2}$ hartree). The 2 in the numerators are the nuclear charge $Z = 2$. In general, the energy of any *hydrogen-like* atom is $-Z^2/2$ hartrees per electron, provided we ignore interelectronic electrostatic repulsion.

We can compare this result with the experimental first and second *ionization potentials* (IPs) for helium

$$\begin{array}{l} He \longrightarrow He^+ + e^- \\ \searrow \\ He^{2+} + e^- \end{array} \tag{8-21}$$

which are energies that must go into the system to bring about ionization and hence are equal in magnitude but opposite in sign to the binding energy of the ionized electron. Helium has two ionization potentials, one for each electron, as shown in reaction (8-21).

If we compare the calculated total ionization potential, IP $= 4.00$ hartrees, with the experimental value, IP $= 2.904$ hartrees, the result is quite poor. The magnitude of the disaster is even more obvious if we subtract the known second ionization potential, $IP_2 = 2.00$, from the total IP to find the *first* ionization potential, IP_1. The calculated value of IP_2, the second step in reaction (8-21) is $IP_2 = Z^2/2 = 2.00$, which is an exact result because the *second* ionization is a one-electron problem. For the first step in reaction (8-21), IP_1 (calculated) $= 2.00$ and IP_1(experimental) $= 2.904 - 2.000 = .904$ hartrees, so the calculation is more than 100% in error. Clearly, we cannot ignore interelectronic repulsion.

A Self-Consistent Field Variational Calculation of IP for the Helium Atom

One approach to the problem of the r_{12} term is a variational self-consistent field approximation. Our treatment here follows that by Rioux (1987), in which he starts from the single electron or *orbital approximation*, assuming that the orbital of helium is separable into two one-electron orbitals $\Phi(1, 2) = \psi(1)\psi(2)$.

The kinetic energy operator for the one-electron system of the H atom is $-(\hbar^2/2m)(1/r^2)(d/dr)r^2(d/dr)$ [Eq. (6-21)] and the potential energy is $-e^2/r$ for attraction of a single electron to the hydrogen nucleus. It is reasonable to use the same operator for a single electron in a separated helium orbital, either $\psi(1)$ or $\psi(2)$. In atomic units we have

$$\hat{H} = -\frac{1}{2r^2}\frac{d}{dr}r^2\frac{d}{dr} \tag{8-22}$$

for each kinetic energy part and $-2/r$ as each potential energy part.

Although we are solving for one-electron orbitals, ψ_1 and ψ_2, we do not want to fall into the trap of the last calculation. We shall include an extra potential energy term V_1 to account for the repulsion between the negative charge on the first electron we consider, electron 1, exerted by the other electron in helium, electron 2. We don't know where electron 2 is, so we must integrate over all possible locations of electron 2

$$V_1 = \int_0^\infty \psi_2 \frac{1}{r_{12}} \psi_2 \, d\tau \tag{8-23}$$

The entire Hamiltonian for electron 1 is

$$\hat{H}_1 = -\frac{1}{2r_1^2}\frac{d}{dr_1}r_1^2\frac{d}{dr_1} - \frac{2}{r_1} + \int_0^\infty \psi_2 \frac{1}{r_{12}} \psi_2 \, d\tau \tag{8-24a}$$

The same treatment produces a similar operator for electron 2.

$$\hat{H}_2 = -\frac{1}{2r_2^2}\frac{d}{dr_2}r_2^2\frac{d}{dr_2} - \frac{2}{r_2} + \int_0^\infty \psi_1 \frac{1}{r_{12}} \psi_1 \, d\tau \tag{8-24b}$$

We do not know the *orbitals* of the electrons either. (An orbital, by the way, is not a ball of fuzz, it is a mathematical function.) We can reasonably assume that the ground-state orbitals of electrons 1 and 2 are similar but not identical to the 1s orbital of hydrogen. The *Slater-type* orbitals

$$\psi_1 = \sqrt{\frac{\alpha^3}{\pi}}e^{-\alpha r_1} \tag{8-25a}$$

and

$$\psi_2 = \sqrt{\frac{\beta^3}{\pi}}e^{-\beta r_2} \tag{8-25b}$$

are chosen to approximate the two electronic orbitals in helium. The integral in Eq. (8-24a), representing the Coulombic interaction between electron 1 at r_1 and electron 2 somewhere in orbital ψ_2, has been evaluated for Slater-type orbitals (Rioux, 1987; McQuarrie, 1983) and is

$$V_1 = \int_0^\infty \psi_2 \frac{1}{r_{12}} \psi_2 \, d\tau = \frac{1}{r_1}[1 - (1 + \beta r_1)e^{-2\beta r_1}] \tag{8-26}$$

Now the approximate Hamiltonian for electron 1 is

$$\hat{H}_1 = -\frac{1}{2r_1^2}\frac{d}{dr_1}r_1^2\frac{d}{dr_1} - \frac{2}{r_1} + \frac{1}{r_1}(1 - (1 + \beta r_1)e^{-2\beta r_1}) \tag{8-27}$$

with a similar expression for $\hat{H}_2$ involving αr_2 in place of βr_1 in the Slater orbital. The orbital is normalized so the energy of electron 1 is

$$E_1 = \int_0^\infty \psi_1 \hat{H}_1 \psi_1 \, d\tau \tag{8-28}$$

with a similar expression for E_2.

Calculating E_1 requires solution of three integrals

$$E_1 = \int_0^\infty \psi_1 (-\tfrac{1}{2}\nabla_1^2)\psi_1 \, d\tau - \int_0^\infty \psi_1 \left(\frac{z}{r_1}\right)\psi_1 \, d\tau + \int_0^\infty \psi_1 (V_1)\psi_1 \, d\tau \tag{8-29}$$

but we already know the first two integrals, $\alpha^2/2$ and $-Z\alpha$ in atomic units, from the solution of Exercise 6-3. We also know the potential energy V_1 from Eq. (8-26). Integration of the third term in Eq. (8-29) (Rioux, 1987) yields the energy of the electron in orbital ψ_1

$$E_1 = \frac{\alpha^2}{2} - Z\alpha + \frac{\alpha\beta(\alpha^2 + 3\alpha\beta + \beta^2)}{(\alpha + \beta)^3} \tag{8-30a}$$

with a similar expression for E_2 except that β replaces α in the first two terms on the right

$$E_2 = \frac{\beta^2}{2} - Z\beta + \frac{\alpha\beta(\alpha^2 + 3\alpha\beta + \beta^2)}{(\alpha + \beta)^3} \tag{8-30b}$$

The parameters α and β in the Slater-type orbitals for electrons 1 and 2 are minimization parameters representing an effective nuclear charge as "experienced" by each electron, partially shielded by the other electron from the full nuclear charge. The SCF strategy is to minimize E_1 using an arbitrary starting β and to find α at the minimum. In general, this α is then put into Eq. (8-30b), which is minimized to give a value for β at the minimum. This value then replaces the starting value of β, and a new minimization cycle produces a new α and so on. This iterative process is repeated until there is no difference in successive values of E_1 and α, that is, until the results of the calculation are self-consistent.

In this particular case, the calculations are completely symmetrical up to Eqs. (8-30). Everything we have said for α we can also say for β. At self-consistency, $\alpha = \beta$ so we can substitute α for β at any point in the iterative process, knowing that as we approach self-consistency for one, we approach the same self-consistent value for the other.

A reasonable step at the end of each iteration would be to calculate the total energy of the atom as the sum of its two electronic energies $E_{He} = E_1 + E_2$, but in so doing, we would be calculating the interelectronic repulsion $\alpha\beta(\alpha^2 + 3\alpha\beta + \beta^2)/(\alpha + \beta)^3$ twice, once as an r_{12} repulsive energy and once as an r_{21} repulsion. The r_{21} repulsion should be dropped to avoid double counting, leaving

$$E_{He} = E_1 + \frac{\beta^2}{2} - Z\beta \qquad (8\text{-}31)$$

as the correct energy of the helium atom.

Exercise 8-3

Use *Mathcad* to calculate the first approximation to the SCF energy of the helium atom

Solution 8-3

$$Z := 2.000 \qquad \alpha := 2.000 \qquad \beta := 2.000$$

$$\varepsilon(\alpha, \beta) := \left[\frac{\alpha^2}{2} - Z \cdot \alpha + \frac{\alpha \cdot \beta \cdot (\alpha^2 + 3 \cdot \alpha \cdot \beta + \beta^2)}{(\alpha + \beta)^3}\right]$$

Find the value of α at which $(d/d\alpha)\varepsilon(\alpha, \beta)$ is zero.

$$\alpha := \text{root}\left(\frac{d}{d\alpha}\varepsilon(\alpha, \beta), \alpha\right)$$

$$\alpha = 1.6 \qquad \varepsilon(\alpha, \beta) = -0.812$$

The first iteration produces an approximation to the first ionization potential of He that is $-(-0.812)$ hartrees, 10.2% too small. This is a great improvement over the $> 100\%$ error we found when the r_{12} term was completely ignored.

Exercise 8-4

Continue the calculation in Exercise 8-3 substituting 1.6 as the initial value of β, minimizing to find a new value of α. How much in error is the calculated value of the first ionization potential of He relative to the experimental value of 0.904 hartrees?

Solution 8-4

$$\alpha = 1.713 \qquad \varepsilon(\alpha, \beta) = -0.925$$

$$E := \varepsilon(\alpha, \beta) + \frac{\beta^2}{2} - Z \cdot \beta \qquad E = -2.845$$

The error, on the second iteration, has been reduced to 6.5%. Notice that the calculated IP_1 on this iteration is too large.

COMPUTER PROJECT 8-1 | *The SCF Energies of First Row Atoms and Ions*

Using as many methods as are available to you for comparison (***Mathcad***, QBASIC, and TRUE BASIC), determine the self-consistent field (SCF) energies of the He atom and of the ions Li^+, Be^{2+}, and B^{3+}. Fill in the SCF column of Table 8-1.

Plot the total SCF energy of the He atom and the three two-electron ions as a function of the nuclear charge Z. Describe the curve (linear, monotonic, etc.) so obtained. Using the GAUSSIAN© package, calculate the energies of the atoms and ions in Table 8-1 at the STO-2G and STO-3G levels. The STO-xG levels of calculation are carried out by using a wave function that is the sum of two or more Gaussian functions with parameters chosen so as to approximate a Slater-type orbital. The STO-xG levels of calculation will be discussed in more detail in the next section, but for now, all you need to know is that the six-line input file

```
# STO-2G

Helium

0 1
He
```

will provide the solutions for this project. GAUSSIAN for WINDOWS has a template that facilitates writing input files, but input files can also be written using a DOS editor. Mind the blank lines 2 and 4.

To fill out Table 8-1, change the element symbols in the last line to Li, Be, or B and designate the charge and spin multiplicities as 1 1, 2 1, 3 1 in that order. In line 5, the first number is the single positive charge and the second number is the spin multiplicity, 1 for paired electronic spins and 2 for an unpaired electron. A

Table 8-1   Energies (in hartrees) of First Row Atoms and Ions

	SCF	STO-2G	STO-3G	Exp.
He	−2.848	−2.702	−2.808	−2.904
Li^+				−7.280
Be^{2+}				−13.657
B^{3+}				−22.035
% error				

space between the charge and spin multiplicity is essential. The spin multiplicities are all 1 here because each problem is a two-electron problem and the two spins are paired. To go from an STO-2G calculation to an STO-3G calculation, change the 2 to a 3 in the *route section* (line 1) of the input file.

COMPUTER PROJECT 8-2 | *A High-Level ab initio Calculation of SCF First IPs of the First Row Atoms*

In contrast to the low-level calculations using the STO-3G basis set, very high level calculations can be carried out on atoms by using the *Complete Basis Set-4* (CBS-4) procedure of Petersson et al. (1991,1994). For atoms more complicated than H or He, the first ionization potential (IP_1) calculation is a many-electron calculation in which we calculate the total energy of an atom and its monopositive ion and determine the IP of the first ionization reaction

$$A \rightarrow A^+ + e^-$$

from the difference $-(E_{A^+} - E_A)$. The CBS-4 "program" is actually a suite of several programs and corrections that are linked to one another so that they are carried out sequentially. The procedure is intended to come very close to the result that would have been obtained by using a complete basis set by extrapolation from a large but (obviously) finite basis set.

Procedure. To go from an STO-3G calculation to a CBS-4 calculation, simply replace STO-3G with CBS-4 in the route section of the program used in Computer Project 8-1. Complete Table 8-2 by filling in the CBS-4 Energies of the atoms and ions listed in columns 1 and 3 of Table 8-2 and put them into columns 2 and 4 of the table. You will notice that some of the simpler atoms (H through Be) do not have a listed CBS-4 Energies, but they do have an SCF energy, which should be used in its place. Calculate the IP and complete column 5. Pay special attention to spin multiplicity and Hund's rule. The spin multiplicity is $n + 1$ where n is the number

Table 8-2 Electronic Energies of Atoms and Single-Positive Ions in the First Row of the Atomic Table

Element	Energy (hartrees)	Ion	Energy (hartrees)	IP	IP(exp)
H	−0.4988	H^+	—	0.4988	
He		He^+			
Li		Li^+			
Be		Be^+			
B		B^+			
C		C^+			
O		O^+			
N		N^+			
F		F^+			
Ne		Ne^+			

of unpaired electrons in the atom. This may require some thought for atoms C through Ne. If necessary, review the subject of ionization potentials in a good general chemistry textbook (e.g., Ebbing, and Gammon, 1999). Experimental values for column 6 can be obtained from most general chemistry textbooks. They may require unit conversion.

Look up the experimental values of the first ionization potential for these atoms and calculate the average difference between experiment and the computed values. Depending on the source of your experimental data, the arithmetic mean difference should be within 0.010 hartrees. Serious departures from this level of agreement may indicate that you have one or more of your spin multiplicities wrong.

Plot the calculated first IPs as a function of the atomic number Z for the elements from H to Ne in the atomic table. The plot has a characteristic shape that should be familiar from earlier courses. These plots are frequently given in the experimental units of electron volts (eV; hartrees $\times$ 27.21 = eV) or kilojoules per mole (kJ mol^{-1}; hartrees $\times$ 2625 = kJ mol^{-1}). Write a paragraph or two in your project report explaining why the graph of IP vs. Z appears as it does.

The STO-xG Basis Set

The true value of Ψ for a many-electron atom or a molecule is unknown. If we could set it equal ("expand" it) to a linear combination of an infinite number of basis functions, each defined in a space of infinite dimensions, we could carry out an exact calculation of Ψ. Such a set of basis functions would be a *complete set*.

The various basis sets used in a calculation of the H and S integrals for a system are attempts to obtain a basis set that is as close as possible to a complete set but to stay within practical limits set by the speed and memory of contemporary computers. One immediately notices that the enterprise is directly dependent on the capabilities of available computers, which have become more powerful over the past several decades. The size and complexity of basis sets in common use have increased accordingly. Whatever basis set we choose, however, we are attempting to strike a balance. If the basis set is too small, it is inaccurate; if it is too large, it exceeds the capabilities of our computer. Whether our basis set is large or small, if we attempt to calculate all the H and S integrals in the secular matrix without any infusion of empirical information, the procedure is described as *ab initio*.

Basis functions are themselves *contractions* of simpler functions called *primitives*. Contractions are used because they are easier for the computer to handle; hence, they economize on computer power, with the obvious advantage that larger problems can be solved with greater accuracy. We shall illustrate this idea by contracting three Gaussian primitives to approximate a Slater-type orbital (STO). The resulting contraction is called an STO-3G basis function. If this basis is used to describe the 1s atomic orbitals of H and He, it is the *minimal basis* because H and He have only 1s electrons in the ground state. A minimal basis set contains the smallest number of basis functions necessary for each atom. The minimal basis set for C is 1s, 2s, and 2p. This is larger than the minimal basis set for H because C has a more complicated electronic structure than H. In some approximations, the 1s electrons of carbon are considered part of the "core" along with the nucleus, and only the $2s$ and $2p$ electrons are included in the minimal basis set. Notice that the number of Gaussians does not determine whether a basis set is minimal; both STO-2G and STO-3G are minimal. In the use of a contracted basis set, the primitives are not manipulated independently; they form a single basis function and are treated by the computer as a unit.

The Hydrogen Atom: An STO-1G "Basis Set"

We shall construct the simplest possible basis function by fitting a single Gaussian to the $1s$ STO for the hydrogen atom, with the intention of building up to the STO-3G basis set later. The task is similar to what was done in Computer Project 6-1. In Part A of that project, we optimized the hydrogen $1s$ orbital $f(\alpha, r) = e^{-\alpha r}$ by the variational method and got the exact value, -0.5000 hartrees, for the ground-state energy. In Part B, we found that the lowest variational energy of the Gaussian function $g(\gamma, r) = e^{-\gamma r^2}$ is obtained when γ has been optimized to about 0.83. The result, -0.424 hartrees, is the lowest variational energy you can get from this function, but it is not as good as the result found when the true hydrogenic orbital was used.

In the following, we shall examine the approximation to the Slater-type $1s$ hydrogen wave function by one Gaussian function using Program GAUSSIAN94W©, a commercial package for Gaussian and related calculations, specifically adapted for a Windows© environment.

The Slater-type orbitals are a family of functions that give us an economical way of approximating various atomic orbitals (which, for atoms other than hydrogen, we don't know anyway) in a single relatively simple form. For the general case, STOs are written

$$f_S(\zeta, n, l, m; r, \theta, \phi) = Nr^{n-1}e^{-\zeta r}Y_l^m(\theta, \phi) \tag{8-32}$$

but for the spherically symmetric $1s$ orbital of hydrogen, variation in the spherical harmonic $Y_l^m(\theta, \phi)$ drops out and $n - 1 = 0$, so

$$f_S(\zeta; r) = Ne^{-\zeta r} \tag{8-33}$$

where N is a constant. This function is really just the exact 1s wave function for hydrogen $\Psi(r) = Ne^{-r}$ because $\zeta = Z = 1$ for this special case, that is, the STO is the same as the wave function Ψ.

The Gaussian function can be written

$$g(\gamma, r) = Cx^m y^n z^l e^{-\gamma r^2} \tag{8-34}$$

but for the spherically symmetric $1s$ case, the Cartesian terms $x^m y^n z^l = 1.0$ so we have

$$g(\gamma, r) = Ce^{-\gamma r^2} \tag{8-35}$$

where C is a constant.

In Computer Projects 8-1 and 8-2, we used the STO and CBS basis sets stored as part of the data base of GAUSSIAN. The *general basis* case (keyword gen) in GAUSSIAN permits us to bypass the stored basis sets (there is no stored STO-1G basis set) and make our own basis functions. To run GAUSSIAN under the general basis input to determine the SCF output for the ground state of the hydrogen atom using a single Gaussian trial function, the input file is

```
# gen

hatom gen

0 2
h

1 0
S       1 1.00
0.282942 1.0
    ****
```

Input File 8-1. The General Basis Input for an STO-1G Calculation of the Ground State Energy of the Hydrogen Atom.

The first line # gen (route section) tells the system that we want to define our own function. The lines 2, 3, and 4 are a blank line, program label (for human readers), and a blank line. The next line that is read by the system is 0 2, specifying that the ground state of H has a 0 charge and is a spin doublet (one unpaired electron). The next line, h, specifies hydrogen, followed by a blank.

The remainder of the input file gives the basis set. The line, 1 0, specifies the atom center 1 (the only atom in this case) and is terminated by 0. The next line contains a shell type, S for the $1s$ orbital, tells the system that there is 1 primitive Gaussian, and gives the scale factor as 1.0 (unscaled). The next line gives $\gamma = 0.282942$ for the Gaussian function and a contraction coefficient. This is the value of γ, the Gaussian exponential parameter that we found in Computer Project 6-1, Part B. [The precise value for γ comes from the closed solution for this problem $8/9\pi$ (McWeeny, 1979).] There is only one function, so the contraction coefficient is 1.0. The line of asterisks tells the system that the input is complete.

When we run this program, we get a good deal more information than we are ready for at this point, but one thing is obvious: the energy, found in the last block of output,

$$HF = -0.4244132$$

This result agrees with Computer Project 6-1, but it is not very good, -0.4244 hartrees, as compared to the exact solution of -0.5000, a 15.1% error. What went wrong?

The Gaussian, with r^2 in the exponent, drops off faster than the true $1s$ orbital, which has r in the exponent. The Gaussian is too "thin" at larger distances r from the nucleus (Fig. 8-2).

$$y(x) := \exp(-x) \qquad\qquad z(x) := \exp(-x^2)$$

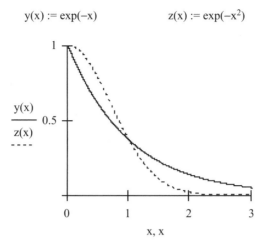

Figure 8-2   The $1s$ Orbital Shape and a Gaussian Approximation (dotted line).

Having obtained a mediocre solution to the problem, we now seek to improve it. The next step is to take two Gaussian functions parameterized so that one fits the STO close to the nucleus and the other contributes to the part of the orbital approximation that was too thin in the STO-1G case, the part away from the nucleus. We now have a function

$$\text{STO-2G} = C_1 e^{-\gamma_1 r^2} + C_2 e^{-\gamma_2 r^2} \tag{8-36}$$

that has two γ parameters, which determine how extended the Gaussian is in the r direction (how "fat" the tail of the function is), and two C parameters, which determine how much of a contribution each Gaussian makes to the final STO approximation (Fig. 8-3).

$$\text{STO-2G} = C_1 e^{-\gamma_1 r^2} + C_2 e^{-\gamma_2 r^2}$$
$$\gamma_1 = 1.31 \qquad\qquad \gamma_2 = 0.233$$
$$C_1 = 0.430 \qquad\qquad C_2 = 0.679$$

It is easy to see that the full shape of the $1s$ orbital is better represented by the sum of these two Gaussians, especially at the tail of the curve where chemical bonding takes place, than it is by one Gaussian. When we run an STO-2G *ab initio* calculation on the hydrogen atom using the GAUSSIAN stored parameters rather than supplying our own, the input file is

```
# sto-2g
hatom
0 2
h
```

Input File 8-2. The Input File for a STO-2G Calculation Using a Stored Basis Set.

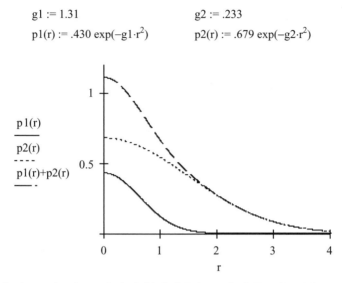

$g1 := 1.31$

$g2 := .233$

$p1(r) := .430 \exp(-g1 \cdot r^2)$

$p2(r) := .679 \exp(-g2 \cdot r^2)$

Figure 8-3   Approximation to the $1s$ Orbital of Hydrogen by 2 Gaussians. The upper curve is the sum of the lower two curves.

We find that there are two Gaussian primitives and one unpaired electron from the output

    1 basis functions   2 primitive gaussians
    1 alpha electrons   0 beta electrons

which agrees with the picture of the STO-2G basis set that we are trying to build. Of course, we want to know what the parameters are for the two Gaussians. The keyword GFinput inserted after # sto-2g in the route section of the input file produces an output file with the added information

    Basis set in the form of general basis input:
    1 0
    S   2 1.00
       0.1309756377D + 01   0.4301284983D + 00
       0.2331359749D + 00   0.6789135305D + 00
    ****

Output File 8-1. Parameters for the STO-2G Basis Set.

The parameterized STO-2G basis function is

$$STO\text{-}2G = 0.4301e^{-1.309\ r^2} + 0.6789e^{-0.233\ r^2} \tag{8-37}$$

which is the function graphed in Fig. 8-2. The smaller exponent contributes to the "tail" of the composite function by causing it to drop off less rapidly with r, and the larger exponent contributes to the function at small values of r. The coefficients `0.4301` and `0.6789` give the intercepts of the two Gaussians at $r = 0$.

We now have two ways of inserting the correct parameters into the STO-2G calculation. We can write them out in a `gen` file like Input File 8-1 or we can use the stored parameters as in Input File 8-2. You may be wondering where all the parameters come from that are stored for use in the STO-xG types of calculation. They were determined a long time ago (Hehre et al, 1969) by curve fitting Gaussian sums to the STO. See Szabo and Ostlund (1989) for more detail. There are parameters for many basis sets in the literature, and many can be simply called up from the GAUSSIAN data base by keywords such as STO-3G, 3-21G, 6-31G*, etc.

But what of the energy?

$$HF = -0.4543974$$

The energy is lower (better) than the STO-1G approximation but not as good as the exact 1s orbital. The error has been reduced from 15.1% to 9.1%.

This process is carried further for the STO-3G approximation (Fig. 8-4).

The energy for the STO-3G approximation is 0.4665819 hartrees, 6.7% in error relative to the exact value. The process has been continued up to STO-6G, the point at which the originators of the procedure stopped because improvement levels out for the larger expansions. One can see this trend in the diminished improvement in our calculations. The first addition brought about a diminution in error by 6%, but addition of a third Gaussian only brought the error down by 2.4%. The calculated energy is approaching a limit, but the limit is not the exact energy. The limit

$$g1 := 3.43 \qquad\qquad g2 := .624 \qquad\qquad g3 := .100$$
$$p1(r) := .154\,\exp(-g1{\cdot}r^2) \qquad p2(r) := .535\,\exp(-g2{\cdot}r^2) \qquad p3(r) := .444\,\exp(-g3{\cdot}r^2)$$

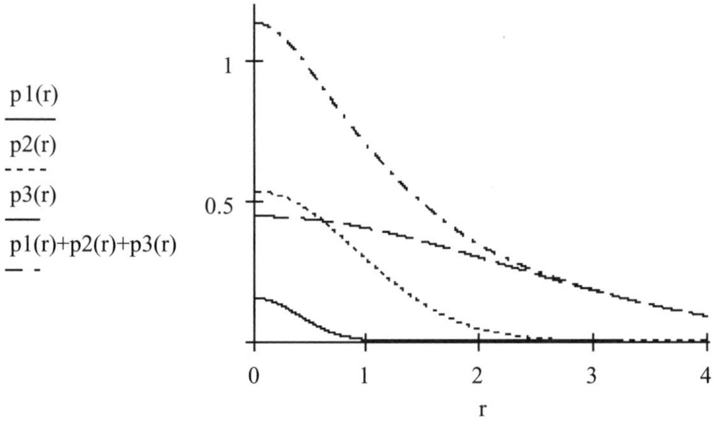

Figure 8-4 A Sum $p1(r) + p2(r) + p3(r)$ (top curve) of Gaussians Used as the STO-3G Approximation the 1s Orbital.

approached is called the *Hartree–Fock limit*, of which more will be said later. The calculation carried out up to this point is called a Hartree–Fock calculation, which is why the energy is labeled HF in the output file.

Semiempirical Methods

If we are willing to use empirical observations in place of the integrals in the secular matrix, we can avoid calculating some or all of the matrix elements. For example, as seen in the EHT method, the negative of the spectroscopic atomic ionization energy makes a good substitute for the calculated Coulomb energy on the logic that the amount of energy necessary to drive a valence electron away from its core is equal and opposite to the amount of energy that held it there in the first place. (This is not strictly true because of rearrangement of the core electrons during the ionization process.) Filling in the matrix elements by fitting them to spectroscopic or other experimental data leads to a *semiempirical* calculation of the eigenvalues and eigenfunctions. Such methods are not fully empirical, even though they use empirical information, because they are rooted in quantum theory as expressed through the variational principle.

Semiempirical methods, of which there are quite a few, differ in the proportion of calculations from first principles and the reliance on empirical substitutions. Different methods of parameterization also lead to different semiempirical methods. Huckel and extended Huckel calculations are among the simplest of the semiempirical methods. In the next two sections, we shall treat a semiempirical method, the self consistent field method, developed by Pariser and Parr (1953) and by Pople (1953), which usually goes under the name of the PPP method.

PPP Self-Consistent Field Calculations

In the Huckel method, we assumed an initial constant π electron density q of one electron per carbon about all carbon atoms in a conjugated π electron system. We also took the electron exchange integrals between atoms to be one arbitrary unit of energy, according to whether the atoms are connected ($\beta = 1$) or not connected ($\beta = 0$). The eigenvectors (coefficients) generated in diagonalizing the secular determinant, however, yield electron densities and bond orders that are in contradiction to the original assumptions. In particular, if a bond order between atoms p and q is large and that between r and s is small, then the resonance integral β is not the same for these atom pairs, but $\beta_{pq} > \beta_{rs}$.

It seems reasonable that, by taking into account the information we have generated in a set of calculated eigenvalues and eigenvectors, we can repeat the calculation and get a new and better set of eigenvalues and eigenvectors. If this works once, it should work many times. There may be convergence to a result that, though not exact, is self-consistent and is a better description of the molecule than the single matrix diagonalization of the Huckel method. This is the essence of the PPP-SCF method.

We have the makings of an iterative computer method. Start by assuming values for the matrix elements and calculate electron densities (charge densities and bond orders). Modify the matrix elements according to the results of the electron density calculations, rediagonalize using the new matrix elements to get new densities, and so on. When the results of one iteration are not different from those of the last by more than some specified small amount, the results are self-consistent.

Both on- and off-diagonal elements are modified, but for simplicity, we shall reset the diagonal elements to zero after each iteration. In this way, orbital energies will be found that are above and below an arbitrary zero energy, stressing the analogy between the PPP-SCF method and the Huckel method. This is an acceptable procedure for hydrocarbons with alternating double and single bonds, called *alternant hydrocarbons*.

The PPP-SCF Method

In PPP-SCF calculations, we make the Born–Oppenheimer, σ-π separation, and single-electron approximations just as we did in Huckel theory (see section on approximate solutions in Chapter 6) but we take into account mutual electrostatic repulsion of π electrons, which was not done in Huckel theory. We write the modified Schroedinger equation in a form similar to Eq. 6.2.6

$$\hat{F}(i)\psi(i) = E(i)\psi(i) \tag{8-38}$$

to emphasize that $\hat{F}$ is an operator similar to the one-electron Hamiltonian operator. The linear combination

$$\psi_i = \sum a_{ij}\phi_j \tag{8-39}$$

is used to generate a secular matrix,

$$\begin{pmatrix} F_{11} - S_{11}E & F_{12} - S_{12}E & \dots \\ F_{21} - S_{21}E & \dots & \\ \dots & & \end{pmatrix} \tag{8-40}$$

analogous to the Huckel matrix. The F matrix can be written succinctly as

$$\begin{pmatrix} F_{11} - E & F_{12} & \dots \\ F_{21} & \dots & \\ \dots & & \end{pmatrix} \tag{8-41}$$

if the S matrix is taken to be **I** (overlap integrals are approximated as zero or one). The corresponding matrix equation

$$\mathbf{FA} = \mathbf{AE} \tag{8-42}$$

leads to $\mathbf{E}$, the diagonal matrix of eigenvalues, and $\mathbf{A}$, the matrix of eigenvector coefficients.

In the Huckel theory of simple hydrocarbons, one assumes that the electron density on a carbon atom and the order of bonds connected to it (which is an electron density between atoms) are uninfluenced by electron densities and bond orders elsewhere in the molecule. In PPP-SCF theory, exchange and electrostatic repulsion among electrons are specifically built into the method by including exchange and electrostatic terms in the elements of the F matrix. A simple example is the 1,3 element of the matrix for the allyl anion, which is zero in the Huckel method but is 1.44 eV due to electron repulsion between the 1 and 3 carbon atoms in one implementation of the PPP-SCF method.

The elements of the F matrix depend on either the charge densities q or the bond orders p, which in turn depend on the elements of the F matrix. This circular dependence means that we must start with some initial F matrix, calculate eigenvectors, use the eigenvectors to calculate q and p, which lead to new elements in the F matrix, calculate new eigenvectors leading to a new F matrix, and so on, until repeated iteration brings about no change in the results. The job now is to fill in the elements of the F matrix.

The *diagonal* matrix element F_{rr} is broken up into three parts

$$F_{rr} = U_{rr} + \tfrac{1}{2}q_r\gamma_{rr} + \sum_{t \neq r} q_t\gamma_{rt} \tag{8-43}$$

where U_{rr} is the localized one-electron Hamiltonian relating to the interaction of electron i with the core at carbon atom r. The term $\tfrac{1}{2}q_r\gamma_{rr}$ is the potential energy of repulsion between electron i and the charge due to all other electrons that can occupy the same orbital. The factor $\tfrac{1}{2}$ appears because, to occupy the same orbital with i, electrons must have opposite spin, that is, they are one-half the total. The sum includes repulsions at all other atoms, $t \neq r$. In Huckel theory, we were free to pick an arbitrary zero of energy α, and in PPP-SCF theory we can do the same thing. The reference point U_{rr} is set equal to zero. This leaves only one term and the sum $\sum_{t \neq r} q_t\gamma_{rt}$ on the right of Eq. (8-43), wherewith to obtain the matrix element F_{rr}.

Let us illustrate the meaning of F_{rr} by the example of carbon atom 1 in the linear, three-carbon allyl anion $C_3H_6^-$. There are two carbon atoms other than C_1, one adjacent and the other nonadjacent. Equation (8-44) has three terms, one for each carbon atom

$$F_{11} = \tfrac{1}{2}q_1\gamma_{11} + q_2\gamma_{12} + q_3\gamma_{13} \tag{8-44}$$

There are similar on-diagonal terms for C_2 and C_3 in the allyl anion. Expect to see these matrix elements again.

The *off-diagonal* elements in the F matrix F_{rt} are defined for neighboring atoms, which are not necessarily adjacent. There are no rr interactions for neighbor atoms.

In the case of adjacent atoms, a bond exists characterized by a bond energy β^{SCF} analogous to the β of Huckel theory but modified by an electron exchange term

$$F_{rt} = \beta^{SCF} - \tfrac{1}{2}p_{rt}\gamma_{rt} \tag{8-45}$$

that is, the value of $\hat{F}$ is made more negative (bonding) by the electron density p_{rt} between atoms r and t times the parameter γ_{rt}. The matrix elements F_{rt} are unlike β in the Huckel treatment in that they change during iteration.

The *parameters* γ_{rr} or γ_{rt} are empirical estimates of how effective repulsion is between an electron in orbital i and the charge clouds on "its own" carbon atom r or the neighboring carbon atoms t in the molecule. For more distant carbon atoms t, γ_{rt} is smaller, as expected for a smaller orbital interaction. Different recipes for obtaining empirical γ_{rt} values have been used (Pilar, 1990). They give similar values. By one scheme, γ_{rr} is taken to be the ionization energy of the carbon atom. More generally, a physical model of interacting negatively-charged spheres is used to calculate repulsive energies $\tfrac{1}{2}p_{rt}\gamma_{rt}$ and the results are fitted to conform with experimental measurements.

Pariser and Parr adjusted the necessary parameters to the empirical singlet and triplet excitation energies in benzene to obtain

$$\begin{aligned}
\gamma_{11} &= 11.35 \text{ eV} \\
\gamma_{12} &= 7.19 \text{ eV} \\
\gamma_{13} &= 5.77 \text{ eV} \\
\gamma_{14} &= 4.79 \text{ eV}
\end{aligned} \tag{8-46}$$

where the subscript 12 indicates nearest neighbors and 13 and 14 are the next most distant carbon atoms, etc. Fitting β^{SCF} to the HOMO-LUMO transitions in benzene in a manner similar to Computer Project 6-2 yields

$$\beta^{SCF} = -2.37 \text{ eV} \tag{8-47}$$

Having filled in all the elements of the F matrix, we use an iterative diagonalization procedure to obtain the eigenvalues by the Jacobi method (Chapter 6) or its equivalent. Initially, the requisite electron densities are not known. They must be given arbitrary values at the start, usually taken from a Huckel calculation. Electron densities are improved as the iterations proceed. Note that the entire diagonalization is carried out many times in a typical problem, and that many iterative matrix multiplications are carried out in each diagonalization. Jensen (1999) refers to an iterative procedure that contains an iterative procedure within it as a "macroiteration." The term is descriptive and we shall use it from time to time.

Like $F_{rt} = \beta^{SCF} - \tfrac{1}{2}p_{rt}\gamma_{rt}$, the zero point, which we may denote α^{SCF}, changes during iteration. Because it is an arbitrary reference point to begin with, we can redefine it as zero after each iteration, ending up with a set of energy levels that qualitatively resembles the set of Huckel energy levels. As in Huckel theory for alternant hydrocarbons (Smith, 1996), orbital energies are symmetrically distributed above and below a (defined) zero, although the calculated values of the energies are not the same. Energy distribution about α^{SCF} is not symmetrical for molecules other than alternant hydrocarbons.

Ethylene

The simplest application is to ethylene. There are only two F_{rr} elements and they are identical, so, completing the analogy with Huckel theory, let us define their energies α^{SCF}. The SCF matrix is

$$\begin{pmatrix} \alpha^{SCF} - E^{SCF} & F_{12} \\ F_{21} & \alpha^{SCF} - E^{SCF} \end{pmatrix} \tag{8-48}$$

We can calculate $F_{12} = F_{21}$ for the first diagonalization as $F_{12} = \beta^{SCF} - \frac{1}{2}p_{12}\gamma_{12}$, where $\beta^{SCF} = -2.37\,\text{eV}$ from Eq. (8-47) and γ_{12} is the repulsion integral for electrons on atoms 1 and 2, adjacent carbon atoms, which are the only kind in ethylene. Equation set (8-46) gives 7.19 eV for γ_{12}. The initial bond order (electron density between atoms) from Huckel theory is 1.00, hence

$$\begin{aligned} F_{12} = F_{21} &= -2.37 - \tfrac{1}{2}1.00(7.19) \\ &= -5.96 \text{ eV} \end{aligned}$$

The form of the SCF matrix is the same as the Huckel matrix; hence, we substitute

$$\begin{pmatrix} 0 & -5.96 \\ -5.96 & 0 \end{pmatrix} \tag{8-49}$$

which is diagonalized as the Huckel matrix was to yield

$$E^{SCF} = \pm F_{12} = \mp 5.96 \text{ eV} \tag{8-50}$$

The solution comes out to be very similar to the Huckel solution for ethylene except that the two energy levels, specified as, are 5.96 eV above and 5.96 eV below the reference level (Fig. 8-5).

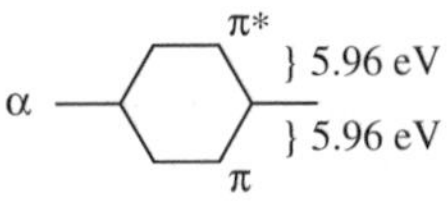

Figure 8-5 The Energy Levels of Ethylene Under the PPP-SCF Parameterization. The π orbital is bonding and the π^* orbital is antibonding.

$$\pi^*$$
$$\} 5.96 \text{ eV}$$
$$\alpha$$
$$\} 5.96 \text{ eV}$$
$$\pi$$

On the basis of this calculation, one would expect to find a $\pi \rightarrow \pi^*$ spectroscopic transition at

$$11.92 \text{ eV}(8065) = 9.61 \times 10^4 \text{cm}^{-1}$$

where 8065 is the conversion factor from eV to cm^{-1}. In fact, in semiquantitative agreement with the calculated $\pi \rightarrow \pi^*$ energy separation, ethylene does have a strong absorption band at $6.21 \times 10^4 \text{cm}^{-1}$ in the vacuum ultraviolet.

In the case of ethylene, we have reached self-consistency in one iteration, that is, the output of the calculation is the same as the input F matrix. In general this will not be true.

Exercise 8-5

Extend the PPP-SCF calculation from ethylene to the allyl anion, $C_3H_6^-$.

$$\overset{-}{C}—C—\overset{-}{C}$$
$$0.50 \ 0.00 \ 0.50$$

Solution 8-5

Find the eigenvectors (eigenfunctions), charge densities q, and bond orders p of $C_3H_6^-$ by the Huckel method. This provides a starting input matrix.

Eigen functions

J=	1	2	3
1	.50000	−.70711	−.50000
2	.70711	.00000	.70711
3	.50000	.70711	−.50000

Charge densities

−.5000	.0000	−.5000

Bond order matrix

1.5000		
0.7071	1.0000	
−.5000	0.7071	1.5000

To form the first SCF input matrix from the HMO calculation, fill the charge densities and bond orders into the matrix

$$\begin{pmatrix} \frac{1}{2}q_1\gamma_{11} + q_2\gamma_{12} + q_3\gamma_{13} & \beta^{SCF} - \frac{1}{2}p_{12}\gamma_{12} & -\frac{1}{2}p_{13}\gamma_{13} \\ \beta^{SCF} - \frac{1}{2}p_{12}\gamma_{12} & q_1\gamma_{12} + \frac{1}{2}q_2\gamma_{11} + q_3\gamma_{23} & \beta^{SCF} - \frac{1}{2}p_{23}\gamma_{23} \\ -\frac{1}{2}p_{13}\gamma_{13} & \beta^{SCF} - \frac{1}{2}p_{23}\gamma_{23} & q_1\gamma_{13} + q_2\gamma_{23} + \frac{1}{2}q_3\gamma_{11} \end{pmatrix}$$

which leads to

$$
\begin{pmatrix}
\frac{1}{4}\gamma_{11} + \frac{1}{2}\gamma_{13} & \beta^{SCF} - \frac{1}{2}0.707\gamma_{12} & \frac{1}{4}\gamma_{13} \\
\beta^{SCF} - \frac{1}{2}0.707\gamma_{12} & \gamma_{12} & \beta^{SCF} - \frac{1}{2}0.707\gamma_{12} \\
\frac{1}{4}\gamma_{13} & \beta^{SCF} - \frac{1}{2}0.707\gamma_{12} & \frac{1}{4}\gamma_{11} + \frac{1}{2}\gamma_{13}
\end{pmatrix}
$$

If we take $\gamma_{ij} = 11.35, 7.19, 5.77$, and $\beta^{SCF} = 2.37$, we get

$$
\begin{pmatrix}
5.72 & -4.91 & 1.44 \\
-4.91 & 7.19 & -4.91 \\
1.44 & -4.91 & 5.72
\end{pmatrix}
$$

which has the roots and eigenvectors

$$
SCF := \begin{pmatrix}
5.72 & -4.91 & 1.44 \\
-4.91 & 7.19 & -4.91 \\
1.44 & -4.91 & 5.72
\end{pmatrix}
$$

$$
\text{eigenvals (SCF)} = \begin{pmatrix} 0.231 \\ 4.28 \\ 14.119 \end{pmatrix} \qquad \text{eigenvecs (SCF)} = \begin{pmatrix} 0.501 & 0.707 & 0.499 \\ 0.706 & 0 & -0.708 \\ 0.501 & -0.707 & 0.499 \end{pmatrix}
$$

as compared to the eigenvectors of the Huckel matrix

$$
HUC := \begin{pmatrix}
0 & 1 & 0 \\
1 & 0 & 1 \\
0 & 1 & 0
\end{pmatrix}
$$

$$
\text{eigenvecs (HUC)} = \begin{pmatrix}
0.5 & 0.707 & 0.5 \\
0.707 & 0 & -0.707 \\
0.5 & -0.707 & 0.5
\end{pmatrix}
$$

Summing over the squares of the coefficients of the lower two orbitals (the upper orbital is unoccupied), we get electron densities of 1.502 at the terminal carbon atoms and 0.997 at the central atom. The charge densities on this iteration are

$$
q_1 = q_3 = -1.502 - (-1.000) = -0.502 \text{ eV}
$$

and

$$
q_2 = -0.997 - (-1.000) = +0.003 \text{ eV}
$$

where the charge density is equal and opposite to the electron density and -1.000 accounts for the single charge brought into the molecule in the p orbital of atomic carbon. Thus the charge densities q are excess charges over all those present within the neutral molecule. There is a slightly higher calculated charge density at the terminal carbon atoms at the expense of the central carbon relative to the electron densities we got by the

HMO method. This is because electron repulsion is taken into account in the SCF calculation whereas it is not taken into account in the Huckel calculation.

Spinorbitals, Slater Determinants, and Configuration Interaction

Because single-electron wave functions are approximate solutions to the Schroedinger equation, one would expect that a linear combination of them would be an approximate solution also. For more than a few basis functions, the number of possible linear combinations can be very large. Fortunately, spin and the Pauli exclusion principle reduce this complexity.

Thus far, we have considered only the space part of one-electron orbitals, but each orbital also has a spin part. An electron is described by four quantum numbers. An orbital can be written as the product of its space part and its spin part

$$\Phi(x,y,z,s) = \phi(x,y,z)\sigma(s) \tag{8-51}$$

where $\Phi(x,y,z,s)$ is called a *spinorbital*. The *Pauli principle* says that no two electrons may be identical in all respects (have all four quantum numbers the same) or, in a more general form, that any electron exchange must be antisymmetric (bring about a change in sign of the spinorbital). Determinants have the property that if two rows or columns are identical, the determinant is zero (vanishes) or when rows or columns are exchanged (Exercise 2-14), the sign of the determinant changes sign. These are just the properties we are looking for.

Slater showed that spinorbitals, arrayed as a determinant, change sign on electron exchange so as to obey the Pauli principle. If we write a linear combination of two spinorbitals as a determinant where we assume the space parts are the same but the spin parts are not the same

$$\Phi(1,2) = \begin{vmatrix} \phi(1,\alpha) & \phi(1,\beta) \\ \phi(2,\alpha) & \phi(2,\beta) \end{vmatrix} \tag{8-52}$$

we get the antisymmetric linear combination

$$\Phi(1,2) = \phi(1,\alpha)\phi(2,\beta) - \phi(2,\alpha)\phi(1,\beta) \tag{8-53}$$

but we do not get the symmetric combination $\Phi(1,2) = \phi(1,\alpha)\phi(2,\beta) + \phi(2,\alpha)\phi(1,\beta)$. Determinants like Eq. (8-52), which can be quite large for large basis sets, are called *Slater determinants*. In the general case, the Slater determinant is antisymmetric. Note also that if we make the spins the same (all four quantum numbers the same)

$$\Phi(1,2) = \begin{vmatrix} \phi(1,\alpha) & \phi(1,\alpha) \\ \phi(2,\alpha) & \phi(2,\alpha) \end{vmatrix}$$

the Slater determinantal wave function vanishes. Linear combinations that do not obey the Pauli principle can be discarded.

Wave functions that do satisfy the Pauli principle are said to be *antisymmetrized*. Just as it is possible to construct linear combinations of simple functions to approximate solutions to the Schroedinger equation, it is also possible to make a linear combination of antisymmetrized determinantal wave functions to give approximate solutions, that is, linear combinations of linear combinations. We shall call these *multiple-determinant wave functions*. By the general principle that larger basis sets more closely approximate the Schroedinger equation than smaller subsets thereof, multiple-determinant wave functions should be better approximations to the solution of the Schroedinger equation than single-determinant wave functions. An infinitely large multiple-determinant wave function converges on the exact wave function Ψ. In practice, multiple-determinant wave functions in PPP-SCF theory affect the excited states; hence, they are important in determining spectra, which involve the transition of an electron from the ground state to an excited state.

As described up to now, the PPP-SCF procedure using a single determinant overestimates electronic repulsion because all electrons are assumed to be locked into the ground state. This error can be diminished by use of multiple-determinant wave functions. Quantum mechanically, occupation of any state has some nonzero probability, although the ground state predominates. Each single-determinant wave function has some unoccupied orbitals called *virtual orbitals*. Replacing a filled orbital with a virtual orbital (promoting electrons to an excited state) gives a new basis function for the linear combination that is to generate the multiple-determinant wave function. More than one substitution of virtual for occupied orbitals can be made (CI doubles, triples, etc.), approaching the full *configuration interaction* solution called a full CI solution. The degree of CI substitution chosen is a trade-off between accuracy required and computer time allowed because CI methods approaching full CI interactions are very time consuming. Antisymmetrization and CI substitution will be treated in more detail in Chapters 9 and 10.

The Programs

The SCF program used here is a modified and cut-down version of the suite of SCOF programs given by Greenwood (1972). It is available in a compiled version SCF.EXE and a source version SCF.FOR. STO-xG and CBS-4 methods are available as part of the GAUSSIAN94W and GAUSSIAN98 suites of programs. Program SCF should be stored in a FORTRAN directory along with SHMO and HMO. Program SCF is run from the system level by the command **scf**.

COMPUTER PROJECT 8-3 | *SCF Calculations of Ultraviolet Spectral Peaks*

This project will familiarize you with the input necessary to carry out calculations using Program SCF. The concept involved is the reverse of that used in Computer

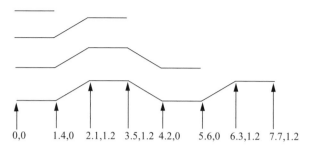

0,0   1.4,0   2.1,1.2 3.5,1.2 4.2,0   5.6,0   6.3,1.2 7.7,1.2

Figure 8-6 Approximate x-y Coordinates for the Alternant Hydrocarbons Ethene Through 1,3,5.7-Octatetraene.

Project 6-2. In the earlier project, you were asked to find a series of LUMO-HOMO energy separations in units of β for four linear conjugated hydrocarbons, ethylene, 1,3-butadiene, 1,3,5-hexatriene, and 1,3,5,7-octatetraene and then to fit a linear function to the four energies in units of β versus the corresponding experimental ultraviolet spectroscopic energies in units of cm^{-1}. The slope of the linear function gave the size of the energy β in units of cm^{-1}. After this it was a simple matter to convert cm^{-1} to other units, eV, kJ mol^{-1}, etc.

In this project, we shall predict the wavelength of the absorption maxima of the same four polyenes using the calculated difference (in units of eV), between the LUMO and HOMO of these four molecules (Fig. 8-6). Bear in mind that this is not an *ab initio* calculation of wavelengths of maximum absorption, because empirically fitted parameters, β^{SCF}, γ_{11}, γ_{12},..., exist within the program or are calculated internally from empirical parameters and the input geometry. In Program SCF, the γ_{ij} parameters are step functions up to the distance 2.81 Å, but are continuous functions beyond that distance. Note that $\gamma_{14} < 4.97$ eV.

Procedure. Carry out SCF calculations for ethylene, 1,3-butadiene, 1,3,5-hexatriene, and 1,3,5,7-octatetraene using Program SCF. The program prints a series of prompts. You will need to designate the number of molecules to be run in any series, for example, by answering NMOLS? with 001. You will need to tell how many molecular orbitals will be calculated and how many are filled, for example, 004002 for 1,3-butadiene. You will need to specify the geometry by giving the x-coordinates of all atoms in the molecule followed by the y-coordinates. The molecule is assumed to be planar; hence, the z-coordinates are all 0. Unformatted inputs are separated by commas, for example, 0,1.4,2.1,3.5 for the x-coordinates of butadiene and 0,0,1.2,1.2 for the y-coordinates. Following this, the atoms are numbered in the same order as the coordinates were entered, obviously, 1,2,3,4 in this case. Use commas to separate unformatted entries. At this point, the coordinates are automatically printed out as an input check. The number of derivatives is asked, NDER?. For now, we are only interested in the parent compound; enter 0001. After more output, you are asked for OPTIONS? For now, choose 0000. Designators like I3 are input formats (Chirlian, 1981). I3 signifies an integer in the rightmost position in a field of 3 digits, for example, 001.

Sketch the molecules on graph paper to help in determining the atomic coordinates. This is the first use of molecular geometry, a property that will become increasingly important as we go on. At this stage, the geometries are approximate; the difference, for example, between *cis* and *trans* isomers is ignored.

Determine the SCF energy difference between the LUMO and HOMO for these molecules. Program output is in eV. Convert to cm^{-1}. Predict the wavelength of the most intense ultraviolet absorption peak from the calculated energy separation. Plot the predicted wavelengths of maximum absorption against the number of double bonds. Plot the experimental values on the same graph. They can be found in Computer Project 6-2. Comment on the agreement between predicted and experimental wavelength maxima, or the lack of it.

COMPUTER PROJECT 8-4 | *SCF Dipole Moments*

Unlike Huckel programs, SCF programs require an input geometry. Because charge densities are calculated, it is a simple matter to combine the atomic coordinates with the SCF charge densities (Chapter 7) to obtain a dipole moment. The charge densities are progressively refined by recalculating the matrix elements during the SCF calculation. In particular, alternation of long single and short double bonds is taken into account (partially) by the calculation. Hence, we expect better agreement between the calculated dipole moment and experiment than Huckel calculations gave.

The implication of self-consistency is that the calculation is iterated until calculated properties (strictly, one selected property) remain constant on renewed iteration. Program SCF, however, stops calculating new matrix elements after 10 iterations. The assumption is that self-consistency will have been reached by then.

We shall also investigate the OPTIONS? prompt that was set to 0 in Computer Project 8-3. At the prompt OPTIONS?, one has the choices 0, 1, 2, or 3. Note the I1 format. Taking the choices in reverse order, 3 permits one to input a value for nuclear charge. This option will not be used here.

Entering 2 in response to the OPTIONS? prompt permits the operator to modify the repulsion integrals γ_{ij} . The matrix of gamma values is called a *G matrix*. On 2 having been entered, the system responds with the prompt NITEM? (I2) requesting (in I2 format) how many matrix elements you want to change. This is followed by a prompt I,J,GAM(I,J)? (2I2,F7.3), requesting the row, column, and value you want to put into the lower triangular form of the G matrix. Modification of the G matrix elements and of the F matrix elements (below) may be tried as an extension to Part B of this project. Having fulfilled this option, one is free to exercise other options until a 0 is entered to signify that there are no more desired changes.

An option input of 1 can be used to alter the F matrix called the *core matrix*. Elements can be entered or altered off the diagonal or, for inclusion of heteroatoms, by designating on-diagonal elements. The prompts and response formats for the F matrix are similar to those of the G matrix. This option is similar to heteroatom inclusion in HMO. We shall use option 1 in Part C below. After all options have been entered, the final option is 0, which causes the calculation to be carried out. In

Computer Project 8-3, this was the only option entry; hence, the parent molecule itself was run.

Procedure

A. Calculate the dipole moment of methylenecyclopropene using the geometry given in Chapter 7.

We found it convenient to take the central carbon atom as the origin of an x-y coordinate system measured in angstroms. Take all the bonds in the σ framework to be about 1.4 Å long. A short exercise in high-school trigonometry gives the coordinates of all the carbon atoms for entry into Program SCF. Is the SCF dipole moment closer to the (interpolated) experimental value of 1-2 D than the Huckel calculation?

B. Investigate the influence of geometry on the dipole moment by "stretching" and "compressing" the methylene bond, that is, calculate μ at a methylene bond length of 0.7 to 1.4 Å at intervals of 0.1 Å. Plot dipole moment vs. methylene bond length from these results. Leave the rest of the geometry the same as it is in Part A. What is the change in μ in debyes per angstrom (D Å^{-1})? There is a discontinuity between 1.4 and 1.5 Å. Why?

Investigate the F and G matrices using a methylene bond length of 1.5 Å. Is something wrong with them? What is it? Why does this change appear? As an extension to this part of the project, correct the F and G matrices using options 1 and 2. When altering off-diagonals of the G matrix, your entries will be in units of β, that is, enter -1.0 to get an element of -2.37 eV. Is the dipole moment improved? Is the problem completely solved? Why not?

C. Calculate the dipole moment of cyclopropenone using the OPTIONS input to change the 1,1 matrix element to -2.0 for electronegative oxygen. Use the results to infer the direction of the dipole moment toward or away from the oxygen atom.

PROBLEMS

1. Calculate the integrals H_{12}, H_{21}, and H_{22} that go with the integral calculated in Exercise 8-1.
2. Calculate the integrals S_{12}, S_{21}, and S_{22} that go with the integral calculated in Exercise 8-2.
3. Express the determinant $\begin{vmatrix} 2 & 3 \\ 4 & 5 \end{vmatrix}$ as a single scalar. Exchange the columns and express the determinant as a single scalar again. What happens to the sign?
4. Given the linear combination

$$\psi = c_1 x(1-x)^2 + c_2 x^2(1-x)$$

analogous to Eq. (9-18), find H_{11} by the method used in Exercise 8-1.

5. Given the linear combination in Problem 4, compute H_{12}, H_{21}, and H_{22}.
6. Fill out the H matrix for the linear combination in Problem 4.
7. Determine S_{11}, S_{12}, S_{21}, and S_{22} for the linear combination in Problem 4.
8. Fill out the secular determinant for the linear combination in Problem 4.
9. Find both roots of the secular equation

$$(28 - 4x)(28 - 4x) - (7 - 3x)^2 = 0$$

which arises from the expansion of the determinant in Problem 8.

10. From the lower root of the equation solved in Problem 8, determine the eigenvalue (energy) for the linear combination in Problem 4. This eigenvalue is analogous to the eigenvalue in Eq. (8-16).

11. Calculate $E_1 = \frac{\alpha^2}{2} - Z\alpha + \alpha\beta(\alpha^2 + 3\alpha\beta + \beta^2)/(\alpha + \beta)^3$ for $\alpha = 2.00$ and $\beta = 2.00$. Calculate $E(\alpha,\beta)$ for $\alpha = 2.00$ and $\beta = 1.60$. Which values of α and β give better Slater orbitals?

12. Show that, under the constraint $\alpha = \beta$, $\alpha\beta(\alpha^2 + 3\alpha\beta + \beta^2)/(\alpha + \beta)^3 = \frac{5}{8}\alpha$ and $E_1 = \frac{\alpha^2}{2} - Z\alpha + \frac{5}{8}\alpha$. This amounts to reducing the number of minimization (shielding) parameters from two to one in the Slater orbitals.

13. Using Program SCF for ethylene and 1,3,5-hexatriene, list the electron repulsion integrals in the form γ_{11}, γ_{12}, and so on. Take the coordinates from Figure 8-6. Try small variations in the atomic coordinates to see what their influence is on γ_{ij}.

14. How many iterations does it take to achieve self-consistency for the helium problem treated (partially) in Exercises 8-3 and 8-4? What is the % discrepancy between the calculated value of the first ionization potential and the experimental value of 0.904 hartrees when the solution has been brought to self-consistency?

15. In treating the energy of helium as in Exercises 8-3 and 8-4, make the "mistake" of entering $Z = 3.0$ instead of $Z = 2.0$. What do you get? Suppose you make the "mistakes" of entering $Z = 4.0$ and $Z = 5.0$. What do you get?

16. Run an STO-2G determination of the energy of the hydrogen atom using the coefficients

$$STO\text{-}2G = C_1 e^{-\gamma_1 r^2} + C_2 e^{-\gamma_2 r^2}$$

$$\gamma_1 = 1.5 \qquad \gamma_2 = 0.2$$

$$C_1 = 0.4 \qquad C_2 = 0.7$$

What is the % difference between your result and the result using the GAUSIAN stored parameters? What is the % difference between your result and the exact result for the hydrogen atom, 0.500 hartrees? Make small changes in C_i and γ_i and recalculate the energy. Is the energy a sensitive function of the STO parameters C_i and γ_i?

17. Verify the result in Eq. (8-50) by finding the eigenvalues of matrix (8-49). What are the eigenvector coefficients?

18. There is considerable variation in the values assigned to the electron repulsion integrals in Exercise 8.9.1. Salem (1966) points out that calculation using Slater orbitals leads to

$$\gamma_{11} = 17 \text{ eV}$$
$$\gamma_{12} = 9.0 \text{ eV}$$
$$\gamma_{13} = 5.6 \text{ eV}$$

and

$$\beta = 2.3 \text{ eV}$$

(Note that agreement with Pariser and Parr's empirical value is better for γ_{13} than for γ_{11}.) Use Salem's values to calculate electron densities on the three carbon atoms of the allyl anion for one iteration beyond the initial Huckel values, as was done in Exercise 8.9.1. Comment on the results you get, as to the qualitative picture of the anion, the influence of electron repulsion on the charge densities, and agreement or lack of agreement with the results already obtained with the Pariser and Parr parameters.

19. 2,3-Dimethyl-2-butene and 2,5-dimethylhexadiene have absorption peaks at 192 and 243 nm in the ultraviolet. Which peak corresponds to which compound? What are the approximate HOMO-LUMO separations in electron volts?

20. Write the STO-4G basis function as a sum of exponentials similar to Eq. (8-37) but with 4 terms on the right.

21. What is the PPP-SCF dipole moment of methylenecyclobutene?

22. What is the PPP-SCF dipole moment of cyclobutenone?

9

Semiempirical Calculations on Larger Molecules

Semiempirical molecular orbital calculations have gone through many stages of refinement and elaboration since Pople's 1965 papers on CNDO. Programs like PM3, which is widely used in contemporary research, are the cumulative achievement of numerous authors including Michael Dewar (1977), Walter Thiel (1998), James Stewart (1990), and their coworkers.

The Hartree Equation

The cornerstone of semiempirical and *ab initio* molecular orbital methods is the Hartree equation and its extensions and variants, the Hartree–Fock and Roothaan–Hall equations. We have seen that the Hamiltonian for the hydrogen atom,

$$\hat{H} = -\tfrac{1}{2}\nabla^2 - \frac{1}{r} \tag{9-1}$$

can be expanded to forms like

$$\hat{H} = -\tfrac{1}{2}\nabla_1^2 - \tfrac{1}{2}\nabla_2^2 - \frac{2}{r_1} - \frac{2}{r_2} + \frac{1}{r_{12}} \tag{9-2}$$

Figure 9-1 Schematic Diagram
of a Helium Atom.

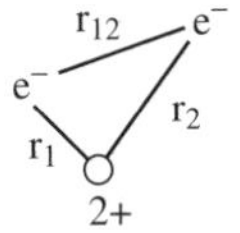

for the helium atom. In this form, there are two kinetic energy operators, one for each of the two electrons, and three potential energy operators, two for electron-nucleus attraction and one for interelectronic repulsion (Fig. 9-1). Strictly speaking, the helium Hamiltonian should include the kinetic energy of the helium nucleus

$$\hat{H} = -\tfrac{1}{2}\nabla_{He}^2 - \tfrac{1}{2}\nabla_1^2 - \tfrac{1}{2}\nabla_2^2 - \frac{2}{r_1} - \frac{2}{r_2} + \frac{1}{r_{12}} \tag{9-3}$$

although one rarely sees it written this way.

For a larger molecule with N nuclei and n electrons, we write

$$\hat{H} = - \sum_{I=1}^{N} \tfrac{1}{2}\nabla_I^2 - \sum_{i=1}^{n} \tfrac{1}{2}\nabla_i^2 - \sum_{I=1}^{N}\sum_{i=1}^{n} \frac{Z_I}{r_i} + \sum_{i=1}^{n}\sum_{j<i}^{n} \frac{1}{r_{ij}} + \sum_{I=1}^{N}\sum_{J<I}^{N} \frac{Z_I Z_J}{R_{IJ}} \tag{9-4}$$

where Eq. (9-4) is simply an extended form of Eq. (9-3). The indexes I and J are for counting nuclei, i and j are for electrons, and Z is the nuclear charge. The double sums are set up in the way that they are to avoid double counting. The term $(Z_I Z_J)/R_{IJ}$ for the potential energy of internuclear repulsion did not appear in Eq. (9-3) because there is only one nucleus in the helium atom.

Under the *Born–Oppenheimer approximation*, the nuclei are assumed to be so much more massive and slow moving than the electrons that their motions are independent and can be treated separately. This permits the Hamiltonian in Eq. (9-4) to be separated into two parts, one that refers to nuclei only

$$\hat{H}(R_I) = - \sum_{I=1}^{N} \tfrac{1}{2}\nabla_I^2 + \sum_{I=1}^{N}\sum_{J<I}^{N} \frac{Z_I Z_J}{R_{IJ}} \tag{9-5}$$

and one that refers to electrons only

$$\hat{H}_i(r_i) = -\tfrac{1}{2} \sum_{i=1}^{n} \nabla_i^2 - \sum_{I=1}^{N}\sum_{i=1}^{n} \frac{Z_I}{r_1} + \sum_{i=1}^{n}\sum_{J<I}^{n} \frac{1}{r_{ij}} \tag{9-6}$$

Later, when we use the nuclear Hamiltonian (9-5) to treat molecular vibrational spectra, zero point energies, and heat capacities, it will include a term E_{el} to account for the electronic binding energy holding the molecule together. We shall ignore this part of the Hamiltonian for the time being.

Even when we are working solely with the electronic Hamiltonian (9-6), we must remember that the nuclei do move, albeit relatively slowly, with respect to each other and that each new value of R_{IJ} provides a new solution to the Schroedinger equation. In a simple diatomic molecule, for example, we expect that the locus of energies at different values of R will pass through a minimum at the equilibrium bond length R_{eq}.

Returning to the electronic equation, we make the standard *orbital assumption* that the molecular orbital is a product of single electron orbitals

$$\psi(r_1, r_2, r_3, \ldots) = \phi(r_1)\phi(r_2)\phi(r_3)\ldots \tag{9-7}$$

Each orbital is calculated for one electron moving in an average field of the nuclei and all other electrons. The field that influences electron 1 in helium, for example, is

$$U_1(r_1) = \int \phi(r_2)\frac{1}{r_{12}}\phi(r_2)dr_2 \tag{9-8}$$

Under the single-electron approximation, Hamiltonian (9-6) becomes

$$\hat{H}(r_i) = -\frac{1}{2}\sum_{i=1}^{n}\nabla_i^2 - \sum_{I=1}^{N}\sum_{i=1}^{n}\frac{Z_I}{r_1} + U(r_i) \tag{9-9}$$

where $U(r_i)$ is the effective or average potential that electron i experiences from all nuclei I and all electrons $j \neq i$. When this Hamiltonian is used in an eigenvalue equation where ε_i designates an orbital energy, the result is the *Hartree equation* for electron i

$$\hat{H}_i(r_i)\phi(r_i) = \varepsilon_i\phi(r_i) \tag{9-10}$$

For the helium atom ground state, which we shall later generalize to many electron atoms and molecules,

$$E_{\text{He}} = \int \psi(r_1)\psi(r_2)\,\hat{H}\psi(r_1)\,\psi(r_2)\,d\tau \tag{9-11}$$

for normalized wave functions. The integral over all space $d\tau$ in Eq. (9-11) is a double integral over dr_1 and dr_2, and the Hamiltonian is given by Eq. (9-2), which assumes the Born–Oppenheimer approximation with $Z = 2$. The five terms in Eq. (9-2) yield three integrals

$$\int \psi(r_1)\left[-\tfrac{1}{2}\nabla_1^2 - \frac{2}{r_1}\right]\psi(r_1)\int\psi(r_1)\psi(r_1)\,dr_1 \tag{9-12a}$$

$$\int \psi(r_2)\left[-\tfrac{1}{2}\nabla_2^2 - \frac{2}{r_2}\right]\psi(r_2)\int\psi(r_2)\psi(r_2)dr_2 \tag{9-12b}$$

and

$$\int \int \psi(r_1)\psi(r_2)\frac{1}{r_{12}}\psi(r_1)\psi(r_2)\,dr_1\,dr_2 \qquad (9\text{-}12\text{c})$$

The first two integrals are simplified by the fact that orthonormal functions yield

$$\int \psi(r_1)\psi(r_1)\,dr_1 = 1 \qquad (9\text{-}13\text{a})$$

and

$$\int \psi(r_2)\psi(r_2)\,dr_2 = 1 \qquad (9\text{-}13\text{b})$$

Under this simplification, the three integrals (9-12a, b, and c) are called I_1, I_2, and J_{12}. Now,

$$E_{He} = I_1 + I_2 + J_{12} \qquad (9\text{-}14)$$

Using more flexible trial functions (polynomials in r_1, r_2, and r_{12} perhaps) for $\psi(r_i)$, one can calculate very accurate energies for He by the SCF method. This gives us confidence that we can make a valid generalization to larger systems.

Exchange Symmetry

Strangely enough, the universe appears to be comprised of only two kinds of particles, *bosons* and *fermions*. Bosons are symmetrical under exchange, and fermions are antisymmetrical under exchange. This bit of abstract physics relates to our quantum molecular problems because electrons are fermions.

By Max Born's postulate, the product of $\psi(x)$ and its *complex conjugate* $\psi^*(x)$ times an infinitesimal volume element d^3x is proportional to the probability that a particle will be in the volume element d^3x

$$\psi(x)\psi^*(x)d^3x \propto \text{prob} \qquad (9\text{-}15)$$

A two-particle wave function $\psi(x_1, x_2)$

$$\psi(x_1,x_2)\psi^*(x_1,x_2)d^3x_1 d^3x_2 = \psi(x_2,x_1)\psi^*(x_2,x_1)d^3x_1 d^3x_2 \qquad (9\text{-}16)$$

gives the probability that two particles will occupy volume elements d^3x_1 and d^3x_2 simultaneously.

For the probable location of two particles to be identical under the *exchange operation*, $x_1 x_2 \rightarrow x_2 x_1$, the wave function before exchange must be exactly equal to the wave function after exchange times a phase factor $e^{i\theta}$

$$\psi(x_1,x_2) = |\psi(x_2,x_1)|e^{i\theta} \qquad (9\text{-}17)$$

Equality between the 1, 2 wave function and the modulus of the 2, 1 wave function, $|\psi(x_2, x_1)|$, shows that they have the same curve shape in space after exchange as they did before, which is necessary if their probable locations are to be the same. The phase factor orients one wave function relative to the other in the complex plane, but Eq. (9-17) is simplified by one more condition that is always true for particle exchange. When exchange is carried out twice on the same particle pair, the operation must produce the original configuration of particles

$$\psi(x_1, x_2) \rightarrow \psi(x_2, x_1) \rightarrow \psi(x_1, x_2) \tag{9-18}$$

There are only two ways this can be true. Either the phase factor is 1 or it is -1, that is,

$$\psi(x_1, x_2) = \psi(x_2, x_1) \tag{9-19a}$$

or

$$\psi(x_1, x_2) = -\psi(x_2, x_1) \tag{9-19b}$$

The first allowable equation holds for bosons, and the second holds for fermions. If we form a linear combination of wave functions for bosons,

$$\Psi_{Bose}(x_1, x_2) = \phi_1(x_1)\phi_2(x_2) + \phi_2(x_1)\phi_1(x_2) \tag{9-20}$$

everything is OK because, taking into account particle indistinguishability, this is the same as

$$\Psi_{Bose}(x_1, x_2) = \phi_1(x_1)\phi_2(x_2) + \phi_2(x_1)\phi_1(x_2) = 2[\phi_1(x_1)\phi_2(x_2)] \tag{9-21}$$

but if we do the same thing with fermions,

$$\Psi_{Fermi}(x_1, x_2) = \phi_1(x_1)\phi_2(x_2) - \phi_2(x_1)\phi_1(x_2) = 0 \tag{9-22}$$

and we have lost the orbital. Wave functions can be identical for bosons but not for fermions. This is the origin of the statement that no two fermions can occupy the same orbital. It is a generalization of Pauli's original exclusion principle for electrons, which are fermions.

Electron Spin

All of our orbitals have disappeared. How do we escape this terrible dilemma? We insist that no two electrons may have the same wave function. In the case of electrons in spatially different orbitals, say, 1s and 2s orbitals, there is no problem, but for the two electrons in the 1s orbital of the helium atom, the *space* orbital is the same for both. Here we must recognize an extra dimension of relativistic space-time that adds a new quantum number $s = \pm\frac{1}{2}$ for what is commonly called the *electron spin*.

If two electrons occupy the same space orbital but have different spins, we can write

$$\psi(1, 2) = 1\,s\alpha(1)1\,s\beta(2) \tag{9-23a}$$

indicating that the $1s$ parts of the orbital $\psi(1, 2)$ are the same but the spins are different, α for one-electron and β for the other. We can also write

$$\psi(2, 1) = 1\,s\alpha(2)1\,s\beta(1) \tag{9-23b}$$

for the same configuration with the spins reversed.

These equations are legitimate *spinorbitals*, but neither is acceptable because they both imply that we can somehow "label" electrons, α for one and β for the other. This violates the principle of indistinguishability, but there is an easy way out of the problem; we simply write the orbitals as linear combinations

$$\psi_1 = 1\,s\alpha(1)1\,s\beta(2) + 1\,s\alpha(2)1\,s\beta(1) = \psi(1, 2) + \psi(2, 1) \tag{9-24a}$$

and

$$\psi_2 = 1\,s\alpha(1)1\,s\beta(2) - 1\,s\alpha(2)1\,s\beta(1) = \psi(1, 2) - \psi(2, 1) \tag{9-24b}$$

Neither of these equations tells us which spin is on which electron. They merely say that there are two spins and the probability that the 1, 2 spin combination is α, β is equal to the probability that the 2, 1 spin combination is α, β. The two linear combinations $\psi(1, 2) \pm \psi(2, 1)$ are perfectly legitimate wave functions (sums and differences of solutions of linear differential equations with constant coefficients are also solutions), but neither implies that we know which electron has the "label" α or β.

Now that we have selected two wave functions that do not violate the principle of indistinguishability, let us look at their exchange properties. The linear combinations are

$$\psi_1(1, 2) = \psi(1, 2) + \psi(2, 1) \tag{9-25a}$$

and

$$\psi_2(1, 2) = \psi(1, 2) - \psi(2, 1) \tag{9-25b}$$

On exchange,

$$\psi_1(2, 1) = \psi(2, 1) + \psi(1, 2) = \psi_1(1, 2) \tag{9-26a}$$

which is acceptable for bosons but not for fermions. Similarly,

$$\psi_2(2, 1) = \psi(2, 1) - \psi(1, 2) = -[\psi(1, 2) - \psi(2, 1)] = -\psi_2(1, 2) \tag{9-26b}$$

which is acceptable for fermions but not for bosons. Because electrons are fermions, we are driven to the conclusion that the linear combination (9-24b) is the only combination that properly describes the ground-state $1s$ orbital of helium. This is true of higher atomic and molecular orbitals as well.

Slater Determinants

While idly dreaming over these equations (theoreticians call it "working") we might happen to notice that the linear combination (9-24b) we have selected for ground-state helium is the same as the expansion of a 2×2 determinant

$$\begin{vmatrix} 1\,s\alpha(1) & 1\,s\beta(1) \\ 1\,s\alpha(2) & 1\,s\beta(2) \end{vmatrix} = 1\,s\alpha(1)\,1\,s\beta(2) - 1\,s\beta(1)\,1\,s\alpha(2) \qquad (9\text{-}27)$$

which, with slight notational changes, is Eq. (8-52). Might it be a general principle that legitimate wave functions which obey the Pauli principle for electrons are expanded determinants?

Consider lithium. We know from a century of empirical chemistry that one electron in the active metal Li is very different from those in inert helium He. Reasoning from the pattern of hydrogen energy levels developed by Bohr, we do the reasonable thing and put one electron in the high-energy $2s$ orbital of Li and leave the other two electrons in the $1s$ orbital. This agrees with the empirical evidence that ground-state Li, like all alkali metals, loses one electron easily, but only one, to go to the single positive ion Li^{+}.

Manipulating linear combinations for Li, one soon discovers that the only one that satisfies the indistinguishability principle for electrons is the expansion of a *Slater determinant*

$$\begin{vmatrix} 1\,s\alpha(1) & 1\,s\beta(1) & 2\,s\alpha(1) \\ 1\,s\alpha(2) & 1\,s\beta(2) & 2\,s\alpha(2) \\ 1\,s\alpha(3) & 1\,s\beta(3) & 2\,s\alpha(3) \end{vmatrix} \qquad (9\text{-}28)$$

Moreover, there are 2 terms in the expansion of the Slater determinant for He but there are 6 terms for Li. Looking at beryllium, we find 24 terms. This is the beginning of the factorial series

$$1!, 2!, 3!, 4!, \ldots = 1, 2, 6, 24, \ldots$$

When we *square* the wave function, we expect to find a probability $P = 1$ over all space, so the $n!$ terms in the expanded determinant must be multiplied by the factor $1/\sqrt{n!}$ to obtain the determinantal wave function normalized to 1.

In summary:

1. The top row of the Slater determinant shows no preference for any spinorbital ϕ_i over any other; the electron may be in any one of them with equal

probability. The same is true for electron 2 as shown in the second row, and so on. The Slater determinant

$$\psi(1,2,\ldots,n) = \frac{1}{\sqrt{n!}} \begin{vmatrix} \phi_1(\mathbf{r}_1)\alpha(1) & \phi_1(\mathbf{r}_1)\beta(1) & \phi_2(\mathbf{r}_1)\alpha(1) & \alpha_2(\mathbf{r}_1)\beta(1) & \cdots & \phi_n(\mathbf{r}_1)\alpha(1) & \phi_n(\mathbf{r}_1)\beta(1) \\ \phi_1(\mathbf{r}_2)\alpha(2) & \phi_1(\mathbf{r}_2)\beta(2) & \cdots & & & & \\ & \cdots & & & & & \\ & \cdots & & & & & \\ \phi_1(\mathbf{r}_n)\alpha(n) & \phi_1(\mathbf{r}_n)\beta(n) & \cdots & & & & \end{vmatrix}$$

$$(9\text{-}29)$$

"mixes" probabilities for all electrons in all orbitals equally to find the molecular orbital $\psi(1, 2, \ldots, n)$. The radius vector from the nucleus to the electron $\mathbf{r}_i$ in Eq. (9-29) will usually be represented by its scalar magnitude r_i.

2. The Slater determinant changes sign on exchange of any two rows (electrons), so it satisfies the principle of antisymmetrical fermion exchange.

3. In short, the Slater determinantal molecular orbital and only the Slater determinantal molecular orbital satisfies the two great generalizations of quantum chemistry, uncertainty (indistinguishability) and fermion exchange antisymmetry.

4. We shall assume antisymmetrized orbitals from this point on when we write $\psi(1, 2, \ldots, n)$.

Exercise 9-1

Show that the atomic determinantal wave function

$$\psi(1,2) = \frac{1}{\sqrt{2}} \begin{vmatrix} 1\,s\alpha(1) & 1\,s\beta(1) \\ 1\,s\alpha(2) & 1\,s\beta(2) \end{vmatrix} = \frac{1}{\sqrt{2}}[1\,s\alpha(1)1\,s\beta(2) - 1\,s\beta(1)1\,s\alpha(2)]$$

is normalized if the $1s$ orbitals are normalized.

Solution 9-1

In the notation of Eq. (9-29), $\phi_1(\mathbf{r}_1) = 1\,s$. If the $1s$ orbitals are normalized, then the spinorbitals $1\,s\alpha(1)$, etc. are normalized because α and β are normalized. If we take just the expanded determinant for two electrons without $1/\sqrt{2}$, the normalization constant, and (omitting complex conjugate notation for the moment) integrate over all space

$$\int\int \psi(1,2)\psi(1,2)dx_1\,dx_2$$

$$\int\int [1\,s\alpha(1)1\,s\beta(2) - 1\,s\beta(1)1\,s\alpha(2)][1\,s\alpha(1)1\,s\beta(2) - 1\,s\beta(1)1\,s\alpha(2)]dx_1\,dx_2$$

we get a sum of four integrals

$$\int\int 1\,s\alpha(1)1\,s\beta(2)1\,s\alpha(1)1\,s\beta(2)dx_1\,dx_2 - \int\int 1\,s\beta(1)1\,s\alpha(2)1\,s\alpha(1)1\,s\beta(2)dx_1\,dx_2$$

$$- \int\int 1\,s\alpha(1)1\,s\beta(2)1\,s\beta(1)1\,s\alpha(2)dx_1\,dx_2 + \int\int 1\,s\beta(1)1\,s\alpha(2)1\,s\beta(1)1\,s\alpha(2)dx_1\,dx_2$$

The spin eigenfunctions are orthogonal

$$\int \alpha\alpha \, d\sigma = \int \beta\beta \, d\sigma = 1$$

and

$$\int \alpha\beta \, d\sigma = \int \beta\alpha \, d\sigma = 0$$

over all spin space σ.

Regrouping terms in the first integral,

$$\int\int \underbrace{1\,s\alpha(1)1\,s\alpha(1)}_{1} \underbrace{1\,s\beta(2)1\,s\beta(2)}_{1} \, dx_1 \, dx_2 = 1 \times 1 = 1$$

because the spinorbitals are normalized. The same thing happens in integral 4. Regrouping terms in the second integral,

$$\int\int \underbrace{1\,s\alpha(1)1\,s\beta(1)}_{0} \underbrace{1\,s\alpha(2)1\,s\beta(2)}_{0} \, dx_1 \, dx_2 = 0 \times 0 = 0$$

because the spinorbitals are orthogonal, and the same thing happens to integral 3.

The sum of the four integrals is

$$1 - 0 - 0 + 1 = 2$$

showing that the determinant by itself is normalized but it is not normalized to 1.

If we premultiply the determinant by $1/\sqrt{n!} = 1/\sqrt{2}$ in this case, it carries into each term in the integral, giving

$$\frac{1}{\sqrt{2}}\frac{1}{\sqrt{2}}[1 - 0 - 0 + 1] = \tfrac{1}{2} - 0 - 0 + \tfrac{1}{2} = 1$$

so that the full wave function $\frac{1}{\sqrt{2}}\begin{vmatrix} 1\,s\alpha(1) & 1\,s\beta(1) \\ 1\,s\alpha(2) & 1\,s\beta(2) \end{vmatrix}$ is normalized to 1. (It is possible to normalize to numbers other than 1, but this is rarely done and is not useful for our purposes because we seek a probability of certainty ($P = 1$) that a particle is somewhere in all of space-time.)

Exercise 9-2

(This exercise is just a reminder to those who may need one.) How many permutations (arrangements) can you make of three books, one red, one green, and one yellow, on your bookshelf?

Solution 9-2

There are $3! = 6$ permutations,
RGB   RBG   GRB   GBR   BRG   BGR

Exercise 9-3

Generalize the solution of Exercise 9-1 to the case of a many-electron wave function [Eq. (9-29)] yielding P_m permutations.

Solution 9-3

Each of the products $1\,s\alpha(1)1\,s\beta(2)$ and $1\,s\beta(1)1\,s\alpha(2)$ is a different permutation of electrons over orbitals, call them P_1 and P_2, respectively. Resuming conventional complex conjugate notation, call the permutations over the complex conjugate $\psi^*(1,2)$ Q_1 and Q_2, respectively. The integration $\int\int \psi^*(1,2)\psi(1,2)dx_1 dx_2$ can be written

$$\int\int [Q_1 - Q_2][P_1 - P_2]d\tau$$

where $d\tau$ designates integration over all space available to the system. This integral produces a sum

$$\int\int Q_1 P_1\,d\tau - \int\int Q_2 P_1\,d\tau - \int\int Q_1 P_2\,d\tau + \int\int Q_2 P_2\,d\tau$$

After regrouping, we found that the first and fourth integrals with P and Q the same $(Q_1 P_1, Q_2 P_2)$ integrated to 1 but the second and third integrals $(Q_2 P_1, Q_1 P_2)$ yielded 0. Following that, the sum of the double integrals was multiplied by $1/\sqrt{n!} = 1/2$ to see to it that $\psi(1,2)$ was properly normalized to 1, not some other number.

In the general case, expanding determinant (9-29) gives

$$\psi(1,2,\ldots,n) = \sum c_P \int \cdots \int \phi_1(P_1)\phi_2(P_2)\ldots\phi_M(P_M)d\tau$$

where $M = n!$ and c_P is the number of exchanges necessary to go from P_1 to some other permutation. For example, for permuting books, RGB $\rightarrow$ GRB $\rightarrow$ GBR involves two exchanges. The number of exchanges determines the sign of each term in the sum: the term is positive if c_P is even and negative if it is odd.

The complex conjugate of $\psi(1,2,\ldots,n)$ can be written

$$\psi^*(1,2,\ldots,n) = \sum c_Q \int \cdots \int \phi_1^*(Q1)\phi_2^*(Q2)\ldots\phi_M^*(QM)d\tau$$

whereupon the multiple integral

$$\int \cdots \int \psi^*(1,2,\ldots,n)\psi(1,2,\ldots,n)d\tau$$

produces a sum of possibly very many terms

$$\sum c_P \sum c_Q \int \cdots \int [\phi_1^*(Q1)\phi_2^*(Q2)\ldots\phi_M^*(QM)][\phi_1(P1)\phi_2(P2)\ldots\phi_M(PM)]d\tau$$

Those terms with identical P and Q integrate to 1 and those with nonidentical P and Q integrate to 0. There are $n!$ nonzero integrals yielding a sum of $n!$ terms. The normalization factor is $1/\sqrt{n!}$ for permutation P and $1/\sqrt{n!}$ for permutation Q. This gives $(1/\sqrt{n!})(1/\sqrt{n!})n! = 1$.

The Hartree–Fock Equation

In later work, both Hartree and, independently, Fock (1930) used antisymmetrized orbitals in what we now know as the *Hartree–Fock equation*, an extension of the Hartree equation. When we treat many-electron atoms or molecules by the variational method using *antisymmetrized* orbitals, a new term, K_{ij}, appears in the energy equation, Eq. (9-14)

$$K_{ij} = \int\int \phi_i^*(r_1)\phi_j^*(r_2)\frac{1}{r_{12}}\phi_i(r_2)\phi_j(r_1)dr_1 dr_2 \qquad (9\text{-}30)$$

due to the possibility of exchanging electrons between different spinorbitals ϕ_i and ϕ_j.

Picking the helium atom again as our prototypical system (Atkins and Friedman, 1997),

$$E = I_1 + I_2 + J_{12} \pm K_{12} \qquad (9\text{-}31)$$

in the excited state where I is the one-electron orbital energy and J is the Coulomb energy. The new integral K_{12} is similar to J_{12} in Eqs. (9-12c) and (9-14) except that the electrons have been *exchanged* $(r_1 \rightarrow r_2$ and $r_2 \rightarrow r_1)$ on the right of integral (9-30). The integral K_{ij} is the *exchange integral* corresponding to the linear combination

$$\psi_\pm(r_1, r_2) = \frac{1}{\sqrt{2}}\phi_1^*(r_1)\phi_2^*(r_2) \pm \phi_1(r_2)\phi_2(r_1) \qquad (9\text{-}32)$$

for He in the excited state, He$_{\text{exc}}$. It appears on antisymmetrization of the atomic orbital; therefore, it is a quantum mechanical term, as distinct from I and J, which are classical. The $\pm$ sign shows that the exchange integral can enhance or diminish stability. In helium, the magnitude of K is about $\frac{1}{10}$ that of the excitation energy from the $1s$ orbital to the $2s$ orbital (Levine, 1991) (Fig. 9-2).

To see how and under what conditions stability is enhanced or diminished, we need to consider the symmetry of the orbital (9-32). Electrons in the antisymmetric orbital ψ_- have a zero probability of occurring at the node in ψ_- where $r_1 = r_2$. Electron mutual avoidance of the node due to *spin correlation* reduces the total energy of the system because it reduces electron repulsion energy due to *charge*

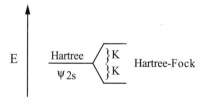

Figure 9-2   The Influence of Particle Exchange on the Energy of He$_{\text{exc}}$.

correlation. Do not confuse the two correlations; spin correlation is a fundamental characteristic of bosons and fermions. Chemists call it the Pauli exclusion principle. It operates *independently* of charge correlation, but it influences the potential energy of charge correlation according to whether electrons are close together or far apart due to their spin. When the electrons in He_{exc} avoid each other due to spin correlation, the energy is decreased

$$E = I_1 + I_2 + J_{12} - K_{12} \tag{9-33a}$$

When electrons attract each other, as in the symmetric spin case, the total energy is increased due to increased charge repulsion

$$E = I_1 + I_2 + J_{12} + K_{12} \tag{9-33b}$$

The secular determinant in the Hartree–Fock procedure is

$$\begin{vmatrix} H_{11} - E_1 & H_{12} \\ H_{21} & H_{22} - E_2 \end{vmatrix} = 0 \tag{9-34}$$

assuming $\mathbf{S} = \mathbf{I}$. Therefore, by the method given in the section on the secular matrix in Chapter 6, the elements of the determinant must be

$$H_{11} = H_{22} = I_1 + I_2 - J_{12} \tag{9-35}$$

and

$$H_{12} = H_{21} = K_{12} \tag{9-36}$$

to obtain the required roots

$$E_i = I_1 + I_2 + J_{12} \pm K_{12}$$

When an antisymmetrized orbital is used in place of a single orbital for many-electron systems, the energy of the ground state is "better," that is, lower, than the Hartree energy by the exchange energy. This is consistent with (but does not prove) the qualitative idea that replacing an orbital with a new orbital that better represents the physics of the system lowers its calculated energy. In this case, the antisymmetrized orbital better represents the true orbital (even though we do not know exactly what that is) because the Pauli principle is a valid part of the physics of electron interaction.

For the record, we should point out that the equations developed in this chapter are extensions of the nonrelativistic, time-independent Schroedinger equation. The Pauli principle arises from a relativistic treatment of the problem, but we shall follow the custom of most chemists and accept it as a postulate, "proven" because it gives the right answers.

In the general case of an electronic Hamiltonian for atoms or molecules under the Born–Oppenheimer approximation,

$$\hat{H}_i(r_i) = -\frac{1}{2}\sum_{i=1}^{n}\nabla_i^2 - \sum_{i=1}^{n}\frac{Z_I}{r_1} + \sum_{i=1}^{n}\sum_{J<I}^{n}\frac{1}{r_{ij}} \tag{9-37}$$

use of the variational method with antisymmetrized orbitals

$$E = \int\cdots\int\psi^*(1,2,\ldots,n)\sum\hat{H}_i(r_i)\psi(1,2,\ldots,n)d\tau \tag{9-38}$$

produces very many integrals, but most of them drop out as they did in Exercises 9-1 and 9-3. Two classes of integrals arise from two groupings of terms in the Hamiltonian for a many-particle system, one from a sum of one-electron terms

$$\sum\hat{H}_i(r_i) = \sum\left[-\frac{1}{2}\nabla_i^2 - \frac{Z}{r_i}\right]$$

and the other from a sum of two-electron terms

$$\sum\hat{H}_i(r_i) = \sum\frac{1}{r_{ij}}$$

In the first case, permutations P and Q must be identical for nonzero terms in the antisymmetrized sum just as they were in Exercise 9-3, leaving only

$$\int\phi_i(r_i)\left(-\frac{1}{2}\nabla_i^2 - \frac{Z}{r_i}\right)\phi_i(r_i)d\tau \tag{9-39}$$

as the integral that contributes to the energy of the system. This integral is given the symbol I_i.

In the second case

$$E = \int\cdots\int\psi^*(1,2,\ldots,n)\sum\frac{1}{r_{ij}}\psi(1,2,\ldots,n)d\tau \tag{9-40}$$

the operator is only a premultiplicative factor $1/r_{ij}$, so evaluation of the multiple integral is again similar to the normalization in Exercise 9-3. Terms remain under the condition P = Q, leaving double integrals in place of multiple integrals

$$\int\int\phi_i^*(r_i)\phi_j^*(r_j)\frac{1}{r_{ij}}\phi_i(r_i)\phi_j(r_j)d\tau \tag{9-41}$$

where integration is over all space r_i and r_j. We give this integral the name J_{ij}.

The arbitrary labels i and j that we use in our equations have no influence on the physics of the real system, so the labels ϕ_j and ϕ_i are just as valid as the labels ϕ_i and ϕ_j for a two-electron interaction. Thus, in addition to the set of J_{ij} integrals for which P = Q, we get a nonzero set of double integrals contributing to the basis set in which P and Q are related by a one electron exchange

$$\int \int \phi_i(r_i)\phi_j(r_j)\frac{1}{r_{ij}}\phi_i(r_j)\phi_j(r_i)d\tau \tag{9-42}$$

This integral is called K_{ij}. If we sum over *doubly occupied orbitals*, the I and J integrals contribute twice to the energy of each orbital, once for each electron but the K_{ij} integral contributes only once because there can be only one exchange of two electrons; hence the ratio of 2:1 for integrals I and J versus the integral K. Also the K_{ij} integral is negative because it arises from a single exchange.

In summary, the Hartree–Fock equation for antisymmetrized orbitals is written

$$E = 2\sum_{i=1}^{N} I_i + \sum_{i=1}^{N}\sum_{j=1}^{N}(2J_{ij} - K_{ij}) \tag{9-43}$$

where

$$I_i = \phi_i^*(r_i)\left(-\tfrac{1}{2}\nabla_i^2 - \frac{Z}{r_i}\right)\phi_i(r_i)d\tau \tag{9-44}$$

$$J_{ij} = \int \int \phi_i^*(r_i)\phi_j^*(r_j)\frac{1}{r_{ij}}\phi_i(r_i)\phi_j(r_j)d\tau \tag{9-45}$$

and

$$K_{ij} = \int \int \phi_i^*(r_i)\phi_j^*(r_j)\frac{1}{r_{ij}}\phi_i(r_j)\phi_j(r_i)\,d\tau \tag{9-46}$$

Note that integrals can be referred to as energies in this context because of Eq (9-38).

The Fock Equation

By this time, we have introduced so many approximations and restrictions on our wave function and energy spectrum that is no longer quite legitimate to call it a "Schroedinger equation" (Schroedinger's initial paper treated the hydrogen atom only.) We now write

$$\mathbf{F}\psi = \varepsilon\psi \tag{9-47}$$

as the Hartree–Fock equation. One-electron orbitals obey the equation

$$\hat{F}_i \phi_i = \varepsilon_i \phi_i \tag{9-48}$$

where $\hat{F}_i$ is called the *Fock operator.*

The Fock operator

$$\hat{F} = \hat{f}_i + \sum_j (2\hat{J}_j - \hat{K}_j) \tag{9-49}$$

is made up of a one-electron part

$$\hat{f}_i = -\tfrac{1}{2}\nabla_i^2 - \frac{Z}{r_i} \tag{9-50}$$

and two two-electron parts

$$\hat{J}_j(r_1) = \int \phi_j^*(r_2) \frac{1}{r_{12}} \phi_j(r_2) d\tau \tag{9-51}$$

$$\hat{K}_j(r_1) = \int \phi_j^*(r_2) \frac{1}{r_{12}} \phi_i(r_2) d\tau \tag{9-52}$$

Note that $\hat{J}_j(r_1)$ and $\hat{K}_j(r_1)$ are operators that go to make up the Fock operator. They operate on functions. One often sees the notation

$$\int \phi_j^*(r_2) \frac{1}{r_{12}} \phi_j(r_2) d\tau \phi_i(r_1) \tag{9-53}$$

and

$$\int \phi_j^*(r_2) \frac{1}{r_{12}} \phi_i(r_2) d\tau \phi_j(r_1) \tag{9-54}$$

or something similar, used to show their operator nature. The notation $\hat{J}_j(r_1)$ and $\hat{K}_j(r_1)$ emphasizes that the Fock operators are functions of the (probable) locations of the electrons, which are known only through their orbitals. The orbitals, in turn, are obtained through the secular matrix of Fock operators. Once again, we arrive at a circular calculation starting with a set of assumed orbitals, calculating the elements of the F matrix leading to a new set of orbitals, and so on to self-consistency.

Having the Slater atomic orbitals, the linear combination approximation to molecular orbitals, and the SCF method as applied to the Fock matrix, we are in a position to calculate properties of atoms and molecules *ab initio*, at the Hartree–Fock level of accuracy. Before doing that, however, we shall continue in the spirit of semiempirical calculations by postponing the *ab initio* method to Chapter 10 and invoking a rather sophisticated set of approximations and empirical substitutions that permit calculation of molecular properties with great computational efficiency. The semiempirical methods so arrived at are probably the most widespread research-level molecular orbital calculations carried out in both academic and industrial laboratories.

The Roothaan–Hall Equations

Application of the variational self-consistent field method to the Hartree–Fock equations with a linear combination of atomic orbitals leads to the Roothaan–Hall equation set published contemporaneously and independently by Roothaan and Hall in 1951. For a minimal basis set, there are as many matrix elements as there are atoms, but there may be many more elements if the basis set is not minimal.

The LCAO approximation for the wave functions in the Hartree–Fock equations

$$\phi_i = \sum a_{ij}\chi_j \tag{9-55}$$

which is essentially Eq. (7-22), gives the Roothaan equations

$$F_i \sum a_{ij}\chi_j = \varepsilon_i \sum a_{ij}\chi_j \tag{9-56}$$

where the χ_j are LCAO basis functions. In the more general case the basis functions need not be atomic orbitals. The Roothaan equations are simultaneous equations in the minimization parameters a_{ij}. The normal equations are

$$(F_{11} - S_{11}\varepsilon_1)a_{11} + (F_{12} - S_{12}\varepsilon_1)a_{12} + (F_{13} - S_{13}\varepsilon_1)a_{13} + \cdots + (F_{1n} - S_{1n}\varepsilon_1)a_{1n} = 0$$
$$(F_{21} - S_{21}\varepsilon_2)a_{21} + (F_{22} - S_{22}\varepsilon_2)a_{22} + (F_{23} - S_{23}\varepsilon_2)a_{23} + \cdots + (F_{2n} - S_{2n}\varepsilon_2)a_{2n} = 0$$
$$\vdots \qquad\qquad\qquad\qquad\qquad\qquad\qquad\qquad \vdots$$
$$(F_{n1} - S_{n1}\varepsilon_n)a_{n1} + (F_{n2} - S_{n2}\varepsilon_n)a_{n2} + (F_{n3} - S_{n3}\varepsilon_n)a_{n3} + \cdots + (F_{nn} - S_{nn}\varepsilon_n)a_{nn} = 0$$
$$\tag{9-57}$$

These are just the secular equations shown in equation set (7-2) with F in place of H and the "stacked matrix" Eq. (7-6) of eigenvectors in place of a single eigenvector. In matrix notation

$$\begin{pmatrix} (F_{11}-S_{11}\varepsilon_j) & (F_{12}-S_{12}\varepsilon_j) & \cdots & (F_{1n}-S_{1n}\varepsilon_j) \\ (F_{21}-S_{21}\varepsilon_j) & & \cdots & \\ \cdots & & \ddots & \\ (F_{n1}-S_{n1}\varepsilon_j) & & & (F_{nn}-S_{nn}\varepsilon_j) \end{pmatrix} \begin{pmatrix} a_{11} & a_{12} & \cdots & a_{1n} \\ a_{21} & a_{22} & & a_{2n} \\ \vdots & & \ddots & a_{n-1n} \\ a_{n1} & & a_{nn-1} & a_{nn} \end{pmatrix} = 0 \tag{9-58}$$

that is,

$$(\mathbf{F} - \mathbf{S}\varepsilon)\mathbf{A} = 0 \tag{9-59}$$

or

$$\mathbf{FA} = \mathbf{SA\epsilon} \tag{9-60}$$

which is the same as Eq. (7-17) except that the Fock matrix replaces the Huckel matrix. Given the Fock operator [(Eq. (9-49)] and a basis set, we can calculate the Fock matrix elements $F_{ij} = \int \phi_i \hat{F} \phi_j d\tau$ and the overlap elements $S_{ij} = \int \phi_i \phi_j d\tau$. An initial approximate eigenvalue spectrum ϵ, which might come from a PPP-SCF calculation, gives us everything we need to calculate $(\mathbf{F} - \mathbf{S\epsilon})$; hence, we can solve Eq. (9-59) for the matrix of eigenvectors $\mathbf{A}$. Matrix $\mathbf{A}$ gives us the coefficients for new linear combinations ϕ_i, ϕ_j, new Fock operators, a Fock matrix for the next iteration, which leads to an improved orbital energy spectrum, and so on.

The Roothaan–Hall equation set (9-57) is often written in the notation

$$\sum_{\nu=1}^{N} (F_{\mu\nu} - \epsilon_i S_{\mu\nu}) c_{\nu i} = 0 \tag{9-61}$$

leading to the one-orbital energies ϵ_i where $F_{\mu\nu}$ is a Fock matrix element and $S_{\mu\nu}$ is an element in the overlap matrix. These equations become zero when the determinant $|F_{\mu\nu} - \epsilon_i S_{\mu\nu}|$ becomes zero because, in general, $c_{\nu i} \neq 0$. (For a brief, readable account of the development of this equation, see Zerner, 2000.)

The Semiempirical Model and Its Approximations: MNDO, AM1, and PM3

If we assume that $\mathbf{S} = \mathbf{I}$ (which is not true in general), the matrix form of the Fock equation can be written

$$\mathbf{FA} = \mathbf{AE} \tag{9-62}$$

where $\mathbf{A}$, is the matrix of molecular orbital coefficients, having elements a_{ij} in the basis set expansion. The assumption that $\mathbf{S} = \mathbf{I}$ is not necessary, but it saves on computer resources (time and memory). This is the *neglect of diatomic differential overlap* (NDDO) simplification. If the NDDO approximation is made for two-center terms in the Fock matrix elements, it must hold for 3- and 4-center terms. This approximation is made in all three MOPAC methods, MNDO, AM1, and PM3. (The names are trivial: Modified Neglect of Differential Overlap, Austin (TX) Method 1, and Parameterized Method 3.)

Once the format of the Fock matrix is known, the semiempirical molecular problem (and it is a considerable one) is finding a way to make valid approximations to the elements in the Fock matrix so as to avoid the many integrations necessary in *ab initio* evaluation of equations like $F_{ij} = \int \phi_i \hat{F} \phi_j d\tau$. After this has been done, the matrix equation (9-62) is solved by self-consistent methods not unlike the PPP-SCF methods we have already used. Results from a semiempirical calculation include or may include the optimized molecular geometry, the energy values of all the quantum levels in the system, charge densities and bond strengths, electronic and vibrational spectral transitions, and derived information like ionization potentials and dipole moments.

For the purpose of approximating and parameterizing the numerous integrals necessary to obtain F_{ij}, the energies representing electronic and nuclear interactions are broken up into categories. Valence electrons are separated from nonvalence electrons. Nonvalence electrons are taken to be part of a *core* of nuclei and electrons influencing valence electrons through their classical Coulombic force field. Further semiempirical approximations to the elements F_{ij} may be made in many ways.

A widely used protocol (Thiel, 1998) is as follows:

1. A specific carbon atom attracts "its own" electron by what is called a one-center, one-electron interaction.
2. Valence electrons are repelled by other electrons in valence orbitals of the same carbon atom, a one-center, two-electron repulsion. These interactions are often parameterized with spectroscopic transition energies.
3. Two-center, one-electron resonance (bonding) is treated essentially as in PPP theory; two-center electronic repulsion is treated classically (as a Coulombic repulsion). Two-center repulsion integrals represent the energy of interaction between the charge distribution on different atoms. Based on a classical model of charge interaction, Dewar has obtained a semiempirical function of distance $f(R_{ij})$ between point charges i and j where the distance R_{ij} is determined from the internuclear distance between atoms and the function is fitted to give correct values in the limits of $R = 0$ and $R = \infty$.
4. One-electron-core integrals are parameterized classically.
5. Core-core interactions are also parameterized classically.

The result is an essentially classical model with the exception of the two-center, one-electron resonance integral, which is of quantum mechanical origin. Although the parameterization is dominated by classical interactions, it is used within the quantum mechanical framework of the Fock matrix. Thus MNDO, AM1, and PM3 are legitimate quantum mechanical molecular orbital methods with a strong influx of classical empirical parameterization. Typically, a molecular orbital package (MOPAC) contains the parameters necessary for each class of calculation, MNDO, AM1, or PM3. Such a package is said to contain the MNDO, AM1, and PM3 *Hamiltonians* (strictly, Fock operators), which are called up with the appropriate *keyword* as the first line of the input program. The point here is that the three methods are really the same except for different parameterizations. They produce the same information, but the computed numerical values they arrive at are different.

Beyond these approximations as to what the actual integrals in the F matrix are, in some more recent semiempirical methods further adjustment of any or all of the elements of the F matrix is used to bring the calculated results into the best possible agreement with standard thermochemical results, largely $\Delta_f H^{298}$. The number of parameters per element is 5–7 in MNDO and 18 in PM3 (Thiel, 1998).

Methods parameterized with thermochemical $\Delta_f H^{298}$ produce, of course, $\Delta_f H$ results at 298 K, as distinct from *ab initio* results, which are total energies at 0 K, E^0. Results at 298 K are the data most useful in practical applications, for example, in the determination of standard free energy changes of reaction and equilibrium constants. Having the data in immediately useful form can be merely a convenience or, in some cases, a real advantage, especially if one does not believe that calculated heat capacity data necessary for correction from $\Delta_f E^0$ to $\Delta_f H^{298}$ are reliable.

Exercise 9-4

Use MNDO, AM1, and PM3 (MOPAC, *ccl.net*) to determine the ionization potential of the hydrogen atom

$$H \rightarrow H^+ + e^-$$

Note that this exercise refers to the standard MOPAC implementation, which does not have a graphical user interface (gui).

Solution 9-4

The input file for this calculation consists of only four lines, two of which may be blank

```
MNDO doublet


h
```

The first line specifies the method and gives the spin multiplicity for one electron $(n + 1) = 2$. The second line may be blank or it can be used for an identifying message like

```
Semiempirical treatment of the hydrogen atom
```

This identifier will be echoed in the output file. The third line is a spacer or a second comment line, and the fourth line identifies the atom (in either upper or lower case).

The MOPAC executable can be run from DOS by using the command **mopac**. Respond to the prompt asking for an input file with the full input filename, including the file extension if any. For example, this might be **h.txt** if you used a text editor to create the input file. Some systems show only the filename without the extension but you still need the extension for MOPAC. The **dir** command in DOS will give you the full filename. Alternatively, you can use **rename h.txt h** to obtain the input file **h** with no **.txt** extension. After editing an input file, for example, changing the keyword from MNDO to AM1, you will need a new filename, say, **h1**, to avoid redundancy with the output files created during the MOPAC run. Redundant files are not normally erased by a new run.

The MNDO output from this four-line input file contains the ionization energy along with other information (Fig. 9-3). The results for the three methods are MNDO: 11.91, AM1: 11.40, and PM3: 13.07 eV. The experimental value is 13.61 eV.

```
HEAT OF FORMATION        =    52.102000 KCAL
ELECTRONIC ENERGY        =   -11.906076 EV
CORE-CORE REPULSION      =     .000000 EV
DIPOLE                   =     .000000 DEBYE
NO. OF FILLED LEVELS     =    0
AND NO. OF OPEN LEVELS   =    1
IONIZATION POTENTIAL     =    11.906276 EV
MOLECULAR WEIGHT         =    1.008
SCF CALCULATIONS         =    2
COMPUTATION TIME         =     .880 SECONDS
```

Figure 9-3 Partial Output from the MNDO Calculation of the Ionization Potential of Hydrogen.

Exercise 9-5

Find the ground-state energy and the equilibrium bond distance (length) for the hydrogen molecule H_2 with the *Arguslab* implementation of MOPAC (**arguslab.com**) and the AM1 Hamiltonian. The *Arguslab* implementation of MOPAC has a gui.

Solution 9-5

Build C_2 by clicking on **File→new** and following the steps in the **help** tutorial (right click, drag, control right click). Getting used to a new molecular structure package takes some time. Don't be discouraged if you have to return to the tutorial frequently at first.

When your C_2 molecule appears correctly in the window, go to selection mode (arrow). Change the default sp^3 C atoms to H by right clicking on one atom followed by a left click on **change atom→H[s]→H_hydrogen**. Repeat for the other atom. Left click **calculation→energy→AM1→OK**, followed by **calculation→run**. If a `Save a Molecule` screen opens up, save under a unique filename like your initials. You will probably get something like E = ~ −0.7 au (hartrees), but this depends on the bond length, which hasn't been optimized yet.

Go through the same run routine, clicking **calculation→optimize geometry** etc. Your initial geometry may not converge. If it doesn't, use the geometry cleaning tool at the upper right of the *arguslab* window. Optimize the geometry again. (The geometry cleaning tool isn't exact.) After any run, you can get bond information by right clicking on the bond in the molecular diagram, which gives a series of options including **Bond Info**. Left clicking on **Bond Info** shows you the bond length = 0.7081 Å. The complete information file on the run is obtained by left clicking **Edit→latest output file**.

Repeat the energy calculation (**calculation→energy→AM1→OK**, followed by **calculation→run**) at the optimum geometry. You should get −1.011 hartrees (Fig. 9-4).

Exercise 9-6

Repeat Exercise 9-5 using the *Arguslab* package and the MNDO and PM3 Hamiltonians. Determine the total energy of atomization to the separated stationary atoms and electrons in hartrees.

```
************* Final Geometry *************

H   1.18325673   0.00000000   0.00000000   1
H   1.89132627   0.00000000   0.00000000   1
```

Figure 9-4 Cartesian (x, y, z) Geometric Output from Arguslab AM1 Calculation. The difference between the coordinates on the x-axis is 0.708 angstroms.

Solution 9-6

	MNDO	AM1	PM3	experiment
r, Å	0.663	0.708	0.699	0.741
E, hartrees	−1.040	−1.011	−1.148	−1.174

The Programs

MOPAC is available from many commercial and freeware sources. Among the commercial sources are Serena Software and Arguslab (See Appendix Sources). A freeware source is **ccl.net** (**http://ccl.net/cca/software/MS-DOS/mopac_for_dos/index.shtml**). Quantum Chemistry Program Exchange (QCPE) is between commercial and freeware sources because it charges a small fee for handling and software storage. MOPAC underwent a long evolution up to MOPAC 6.0 under government financial support; hence, it is public domain software. We shall distinguish between standard MOPAC, which does not have a gui, and commercial MOPAC, which usually does. (What you pay for is the gui. Writing your own gui is beyond the capabilities of most nonspecialists.)

The standard MOPAC 6.0 implementation has an identifier similar to the following:

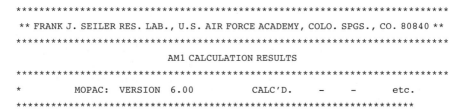

```
***********************************************************************
** FRANK J. SEILER RES. LAB., U.S. AIR FORCE ACADEMY, COLO. SPGS., CO. 80840 **
***********************************************************************
                       AM1 CALCULATION RESULTS
***********************************************************************
*        MOPAC: VERSION 6.00        CALC'D.    -    -      etc.
***********************************************************************
```

Commercial PCMODEL (Serena) does not include MOPAC as part of the package, but Serena makes it available as a collateral program. PCMODEL has a gui that permits MM optimization of a molecular geometry followed by a **save** option in **mopac** format that saves the file as **filename.mop** in the correct format for input to standard MOPAC 6.0. This option is virtually essential for molecules larger than four or five heavy (nonhydrogen) atoms, which would be daunting to input by hand.

COMPUTER PROJECT 9-1 | *Semiempirical Calculations on Small Molecules: HF to HI*

A. Repeat Exercises 9-5 and 9-6 for the hydrogen halides HF to HI using the **Arguslab** package. To identify an atom, right click on it and left click on **Atom Info**. To change an atom, do the same thing except that you left click on **change atom**. You will find that several calculations give you the error message `Unsupported Element`, indicating that the parameters have not been included for that particular calculation. Merely put a * in your table when this happens. You will get at least one calculation (PM3) for each hydride. You should have 4 tables with 3 columns each.

B. Make tables similar to the one in Part A for the *dihydrides* of the group 6 elements, starting with oxygen, H_2O and going to selenium. Include the simple bond angle of the dihydride, for example, H—O—H = 107.6 (PM3), with the geometric result. To find a simple angle, hold down shift, left click on **atoms H, O**, and **H**, left click on **Monitor→Angle**.

C. Find the MNDO, AM1, and PM3 estimates of the dipole moment and $\Delta_f H^{298}$ of formaldehyde H_2O=O. The calculation is run just like a geometric optimization, but be sure the **Dipole moment** box is checked before the program run.

COMPUTER PROJECT 9-2 | *Vibration of the Nitrogen Molecule*

Using MOPAC and the MNDO Hamiltonian, calculate the energy and equilibrium bond length of N_2 to 4 significant figures. The input file is

```
mndo
nitrogen

N
N     1.1     1   1     0     0
```

where the value 1.1 in the fifth line of the file is a starting, approximate, value for the equilibrium bond length. The third and fourth entries in that line are a designator that the second N atom is connected to atom 1 (also N) and a switch saying *do* optimize.

Increase the bond length of the molecule by 0.001 Å and recalculate the energy at this fixed value. Fixing a bond length at the value given in the second entry of the fifth line of the input file (preventing optimization to the equilibrium bond length) is brought about by changing the switch from 1 to 0 in the fourth entry of that line. Leaving a blank in place of 1 would accomplish the same purpose, but it is good practice to enter 0 as a placeholder. Increase the bond length from 1.100 to 1.110 in steps of 0.001 Å. Plot the energy E, calculated by MNDO, as a function of bond length r. What is the mathematical function you observe? Is this consistent with what you know about the harmonic oscillator? According to MNDO, is it reasonable to regard the N—N bond as a harmonic (Hooke's law) spring? Print out the input data to your E vs. r plot and use the file to answer the questions below.

Questions

Suppose that we agree to regard the N—N molecule as a classical (nonquantum) harmonic oscillator and we stipulate that each N atom makes a maximum excursion away from its equilibrium bond length of 0.006 Å during each vibration.

1. What is the minimum atomic speed? Where in their excursion do the atoms reach their minimum speed?
2. Where in their excursion is the maximum atomic speed attained?
3. What is the interatomic separation at the minimum potential energy? For convenience, define the minimum potential energy of the system as zero at the minimum of the potential well.
4. Can the two-mass N—N system be regarded as a one-mass system having a reduced mass μ? If so, what is μ?
5. What is the maximum potential energy of the N—N system as defined?
6. What is the maximum kinetic energy of the N—N system as defined?
7. Calculate
   (a) The maximum excursion $r_{max} - r_0$ in meters.
   (b) The reduced mass μ in kilograms.
   (c) The maximum potential energy per N_2 molecule ε in joules.
   (d) The force constant k in newtons per meter.
   (e) The force on the virtual oscillator of mass μ at excursion r_{max}.
   (f) The frequency of oscillation in hertz.
   (g) The period of oscillation in seconds.
   (h) The frequency of oscillation in cm^{-1}.
8. How long does it take for each N atom to get from its minimum speed to its maximum speed?
9. What is the maximum speed of the nitrogen atoms as they move toward and away from one another?
10. What is the speed of a nitrogen atom as it passes through the point $r = 1.106$?

Normal Coordinates

The "neglected" part of the molecular Schroedinger equation, after making the Born–Oppenheimer separation in the first section of this chapter, is

$$\hat{H}(R_I) = -\sum_{I=1}^{N} \tfrac{1}{2}\nabla_I^2 + \sum_{I=1}^{N} \sum_{J<I}^{N} \frac{Z_I Z_J}{R_{IJ}} + E_{el}$$

where E_{el} is negative (binding). It governs motions of the nuclei, in particular, vibrational motion. Like any other Hamiltonian, the nuclear Hamiltonian can, in principle, be separated into a sum of partial Hamiltonians leading to a sum of energies and a product of wave functions. Expressed in arbitrary coordinates, separation of the Hamiltonian is not feasible because of the cross terms it contains, but expressed in terms of its normal coordinates, the Hamiltonian can be separated.

We shall concentrate on the potential energy term of the nuclear Hamiltonian and adopt a strategy similar to the one used in simplifying the equation of an ellipse in Chapter 2. There we found that an arbitrary elliptical orbit can be described with an arbitrarily oriented pair of coordinates (for two degrees of freedom) but that we must expect cross terms like $8xy$ in Eq. (2-40)

$$5x^2 + 8xy + 5y^2 = 9$$

If, instead of making an arbitrary choice of the coordinate system, we choose more wisely, the ellipse can be expressed more simply, without cross terms [Eq. 2-43)]

$$x'^2 + 9y'^2 = 9$$

The new coordinates are found by rotation of the old ones in the x-y plane such that they lie along the principal axes of the ellipse.

Here we shall consider a homonuclear diatomic molecule restricted to a one-dimensional x-space (Starzak, 1989) (Fig. 9-5). Although there is only one space coordinate, there are two degrees of freedom. The whole molecule can undergo motion (translation), and it can vibrate.

The force on one nucleus due to stretching or compressing the bond is equal to the force constant of the bond k times the distance between the nuclei $(x_2 - x_1)$. It is equal and opposite to the force acting on the other nucleus, and it is also equal to the mass times the acceleration $\ddot{x}$ by Newton's second law (see section on the harmonic oscillator in Chapter 4). The equations of motion are

$$m\ddot{x}_1 = -k(x_1 - x_2) = -kx_1 + kx_2$$
$$m\ddot{x}_2 = k(x_1 - x_2) = kx_1 - kx_2$$

or, in matrix form,

$$\begin{pmatrix} m & 0 \\ 0 & m \end{pmatrix} \begin{pmatrix} \ddot{x}_1 \\ \ddot{x}_2 \end{pmatrix} = \begin{pmatrix} -k & k \\ k & -k \end{pmatrix} \begin{pmatrix} x_1 \\ x_2 \end{pmatrix}$$

There is no force constant for translation because it encounters no opposing force.

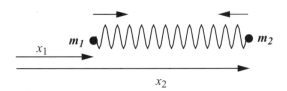

Figure 9-5 A Diatomic Molecule. The molecule can undergo translation without changing the distance $(x_2 - x_1)$, or it can undergo vibration, in which $(x_2 - x_1)$ changes, or it can undergo translation while vibrating.

The m matrix is already diagonalized. Take the masses and the force constant to be 1 arbitrary unit for simplicity and concentrate on the force constant matrix. We can diagonalize the k matrix

$$K := \begin{pmatrix} -1 & 1 \\ 1 & -1 \end{pmatrix}$$

$$\text{eigenvals } (K) = \begin{pmatrix} -2 \\ 0 \end{pmatrix} \quad \text{eigenvecs } (K) = \begin{pmatrix} 0.707 & 0.707 \\ -0.707 & 0.707 \end{pmatrix}$$

We have found the principal axes from the equation of motion in an arbitrary coordinate system by means of a *similarity transformation* $\mathbf{S}^{-1}\mathbf{KS}$ (Chapter 2) on the coefficient matrix for the quadratic containing the mixed terms

$$K := \begin{pmatrix} -1 & 1 \\ 1 & -1 \end{pmatrix} \quad S := \begin{pmatrix} .707 & .707 \\ -.707 & .707 \end{pmatrix}$$

$$S^{-1} = \begin{pmatrix} 0.707 & -0.707 \\ 0.707 & 0.707 \end{pmatrix}$$

$$S^{-1} \cdot K \cdot S = \begin{pmatrix} -2 & 0 \\ 0 & 0 \end{pmatrix}$$

The equations of motion in the transformed coordinates are

$$m\ddot{x}_1 = -2kx_1$$
$$m\ddot{x}_2 = 0$$

The first equation, for vibration, is

$$\ddot{x} = -\frac{2k}{m}x = -\frac{k}{\mu}x$$

where we have used the reduced mass $\mu = (m_1 m_2)/(m_1 + m_2) = 1/2$ in place of unit mass m, as we did in the section on the two-mass problem in Chapter 4 to convert the two-mass vibrational problem into a pseudo one-mass vibrational problem. This sound mathematical technique is often presented essentially as a trick in elementary physical chemistry books (as, indeed, it was in Chapter 4). The second equation says that the acceleration along the translational axis is zero, which is Newton's first law: The system will continue in whatever translational state of motion it is until acted upon by an external force.

Diagonalizing the K matrix converts arbitrary systems in generalized coordinate systems q

$$m\ddot{q}_1 = -kq_1 + kq_2$$
$$m\ddot{q}_2 = kq_1 - kq_2$$

into systems expressed in terms of their normal coordinates Q

$$m\ddot{Q}_1 = \kappa_1 Q_1$$
$$m\ddot{Q}_2 = \kappa_2 Q_2$$

If this can be done in a two-dimensional space, it can (in principle) be done in an n-space.

Polyatomic molecules vibrate in a very complicated way, but, expressed in terms of their normal coordinates, atoms or groups of atoms vibrate sinusoidally *in phase, with the same frequency*. Each mode of motion functions as an independent harmonic oscillator and, provided certain selection rules are satisfied, contributes a band to the vibrational spectrum. There will be at least as many bands as there are degrees of freedom, but the frequencies of the normal coordinates will dominate the vibrational spectrum for simple molecules. An example is water, which has a pair of infrared absorption maxima centered at about 3780 cm^{-1} and a single peak at about 1580 cm^{-1} (**nist webbook**).

Exercise 9-7

Run a MOPAC calculation using the PM3 Hamiltonian to determine the *normal vibrational* modes of the H_2O molecule.

Solution 9-7

One valid form of the input file is the z-matrix form usually associated with GAUSSIAN calculations

```
pm3 force
water

o
h 1 r
h 1 r 2 a

r 1.00
a 105.
```

The keywords call up the PM3 Hamiltonian and a force constant calculation necessary for the vibrational analysis. Line 5, the second line of the z-matrix, stipulates that one hydrogen atom is connected to atom 1 (oxygen) at a distance of r angstroms. Line 6 stipulates the same distance for the second hydrogen atom and that it makes an angle of a with atom 2 (the first hydrogen). This is enough information to completely specify the geometry of the system provided that r and a are specified as they are in lines 8 and 9. Skipping a line (here line 7) is essential. It is a good idea to skip a line after the entire file has been written, here line 10. Some systems require it as a signal to the computer system that the file is complete.

```
        NORMAL COORDINATE ANALYSIS
   ROOT NO.    1           2           3
        1743.06437    3868.95047    3991.10929    etc.
```

where "etc." indicates that there are other roots but that they are of negligible size. The three large numbers are the normal mode vibrational frequencies. The experimental values (corrected for anharmonicity) are 1648, 3832, and 3943. All frequencies are in units of cm^{-1}.

Dipole Moments

MOPAC calculations yield dipole moments on molecules that are far more accurate than those found by simpler methods because the geometries are more accurate. These calculations are complicated enough to discourage use of the methods already shown. In this context, a file-building program is necessary. Note that calculating an input file from a gui drawing is a job consisting of very many very simple calculations—just the thing a computer is good at. The same can be said for adding up all the vectors to find a dipole moment once the geometry is known. It is a very simple task, but you wouldn't want to do it yourself.

COMPUTER PROJECT 9-3 | *Dipole Moments (Again)*
Using PCMODEL or a similar file-constructing gui, construct the input files for cyclohexanone, 1,2-diketocyclohexane, 1,3-diketocyclohexane, and 1,4-diketocy-clohexane. Given the experimental dipole moment of cyclohexanone of 2.87 debyes and the rules of vector addition, estimate the dipole moments for the four target molecules on the simplifying assumption that the cyclohexane ring is planar (it isn't).

Run MOPAC using the MNDO, AM1, and PM3 Hamiltonians to calculate the dipole moments for these four molecules.

Energies of Larger Molecules

At present, pharmaceutical and biochemical applications are probably the most important practical applications of molecular orbital theory. Pharmaceutical and biochemical applications usually involve rather large molecules. (The meaning of the word "large" in reference to molecules depends to a large extent on your point of view; what we call a large molecule is small to a protein chemist.) The principal advantage of the MOPAC suite of programs is that they can be used to obtain structures, energies, and electronic properties of larger molecules than can be treated *ab initio*. The drawback is that the accuracy of calculated properties is subject to the accuracy of parameterization, which may be open to debate. Any change in parameterization necessitates revision of all prior results.

Until recently, naphthalene was something of an outpost for the *ab initio* method. Most calculations were carried out on molecules far smaller than naphthalene, which lends itself to more extensive calculation only because of its simple planar structure and its symmetry (hydrogens not shown)

When naphthalene is completely hydrogenated, its structure becomes much more complicated. The rings take on a three-dimensional configuration and the product molecule, *decalin*, exists as *cis-* and *trans*-isomers as determined by whether the hydrogens add across the central bond on the same side or on opposite sides.

As a first step in molecular orbital calculations on larger molecules we shall examine the energies and structures of *cis-* and *trans*-decalin.

Exercise 9-8

Use a gui to produce the *cis* and *trans* forms of decalin. Run a PM3 calculation of the energies of these two forms. What is the *cis-trans* isomerization energy as calculated by PM3?

Solution 9-8

A partial (edited) output from this program is

```
PM3
cis-decalin
        HEAT OF FORMATION    =    -42.688632 KCAL
        SCF CALCULATIONS     =    20
        COMPUTATION TIME     =    7.790 SECONDS
PM3
trans-decalin
        HEAT OF FORMATION    =    -44.442856 KCAL
        SCF CALCULATIONS     =    27
        COMPUTATION TIME     =    9.230 SECONDS
```

which leads to an isomerization enthalpy of -1.7 kcal mol^{-1}. ("Heat of formation" should be taken to mean *enthalpy* of formation in this context.) Entropy effects being

Figure 9-6 Decalin (PCMODEL, Serena Software). The *cis* decalin molecule is two "chair" forms of cyclohexane fused at a common bond.

negligible, the *trans* form is more stable than the *cis* form. Experimental values are -40.4 ± 1.0 and -43.5 ± 1.0 kcal mol^{-1}, respectively, leading to an isomerization enthalpy of -3.1 kcal mol^{-1}, with the same conclusion as to which isomer is the more stable of the two. This is a good example of the meaning of the term "qualitative agreement." The calculation is not identical with experimental results, nor should we expect it to be. Calculated results for the enthalpy of isomerization are within combined experimental error, and they give the correct order of isomer stability.

A structural diagram of the *cis* form shows the increase in complexity brought about by hydrogenation (Fig. 9-6).

COMPUTER PROJECT 9-4 | *Large Molecules: Carcinogenesis*

It has been known for a long time that polycyclic hydrocarbons are potent carcinogens and that their carcinogenic activity is related to their electronic structure (Pullman, 1955). By one hypothesis, DNA is attacked by electron-rich portions of the carcinogen to form a complex that is so strong as to interfere with the process of transcription of genetic information from one generation of cells to the next. As a result of this faulty transcription process, cells grow in an uncontrolled way.

The purpose of this computer project is to examine several polynuclear aromatic hydrocarbons and to relate their electron density patterns to their carcinogenic activity. If nucleophilic binding to DNA is a significant step in blocking the normal transcription process of DNA, electron density in the hydrocarbon should be positively correlated to its carcinogenic potency. To begin with, we shall rely on clinical evidence that benzene, naphthalene, and phenanthrene

benzene   naphthalene   phenanthrene

are significantly lower in their carcinogenic activity (if any) than 10-methylbenzanthracene and benzo[*a*]pyrene (a known carcinogen found in tobacco smoke)

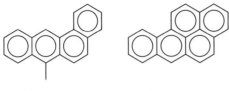

10-Methylbenzanthracene   Benzopyrene

Procedure

Using the **Rings** tool to generate input files, carry out PM3 calculations on these five molecules. Use the keywords PM3 bonds to generate PM3 bond order matrices, the elements of which are directly proportional to electron probability densities. Scan the bond order matrices (near the end of the output file) and locate the bond in each molecule with the highest electron probability density. Arrange the five molecules in order of the highest bond order observed in each. Does this order coincide with the clinical evidence for increasing carcinogenicity, highest bond order most carcinogenic? On the basis of the bond with the highest electron probability density in the two most carcinogenic molecules, identify the region of suspected carcinogenic activity. This region is called the *K-region*. Take care to keep the numbering system of your molecule straight. These files may not follow the conventional numbering system. Atom numbering can be found in PCMODEL by going to the **View** menu and activating **labels→atom nos→OK**.

Exercise 9-9: MOPAC Molecular Energies Using GAUSSIAN94-W

Using GAUSSIAN for Windows, we can carry out a MNDO, AM1, or PM3 optimization, of, for example, the HF molecule, starting from any reasonable H–F bond distance. The input file is similar to the standard MOPAC input file

```
# pm3 opt

pm3 optimization of hf

0 1
h
f,1,1.1
```

File 9-1. PM3 Input File for Optimizing the Energy of HF Starting From a Bond Length of 1.1 Å. Note that the file is not case sensitive; lower case and upper case letters are equally valid.

The first line in File 9-1 is the *route section* calling for a PM3 *optimization*. The next three lines are: a blank line, a program label (human input not read by the system), and a blank line. The input 0 1 indicates that the charge on the molecule is 0 and the spin multiplicity is 1 (paired electrons). The starting geometry is given in the next two lines; h establishes hydrogen as a fixed point and f,1,1.1 says that the fluorine atom is attached to atom 1 (hydrogen) at a distance of 1.1 Å, an initial guess at the bond length. File fragment 9-2 shows that the PM3 optimized bond length is 0.934 Å.

The initial program run produces an energy HF = −1.0000 hartrees (by sheer coincidence) in the penultimate line of the energy output. The computed HF = −1.0000 hartrees = −62.8 kcal mol^{-1}. The experimental value is −65.1 kcal mol^{-1}. The final energy output is just above the "cookie," by Voltaire, in this case.

```
   -- Stationary point found.
                  ---------------------------
                  ! Optimized Parameters !
                  ! (Angstroms and Degrees) !
------------------                  -----------------
! Name  Definition        Value      Derivative Info.           !
--------------------------------------------------------------
! R1    R(2,1)           0.9378      -DE/DX =    0.               !
--------------------------------------------------------------
          Population analysis using the SCF density.
     **************************************************
     Total atomic charges:
              1
  1 H    0.179562
  2 F   -0.179562


  1|1|GINC-UNK|FOpt|RPM3|ZDO|F1H1|PCUSER|01-Feb-1903|0||#PM3
  OPT||pm3 optimization of hf||0,1|H,0.,0.,-0.8440032928|
  F,0.,0., 0.0937781436||Version=486-Windows-G94RevB.2|
  State=1-SG|HF=-0.1000037|RMSD=0.000e+000|RMSF=
  6.008e-006|Dipole=0.,0.,-0.5524105|PG=C*V [C*(H1F1)]||@
  Normal termination of Gaussian 94
  COMMON SENSE IS NOT SO COMMON. -- VOLTAIRE
```

File Fragment 9-2. Partial Energy Output File (Edited) for an Optimized PM3 Calculation on HF Using GAUSSIAN94-W.

Cookies are different for each run. They make it easier for us to distinguish between different runs on the same or similar input files. Some are apt and humorous. They can lighten a long day's work.

PROBLEMS

1. Draw a two-dimensional graph with the horizontal axis representing all real numbers and the vertical axis representing all imaginary numbers (an Argand

diagram). The plane established by these two axes is called the *complex plane*. Locate the point

$$z = 3 + 4i$$

on the complex plane. The distance from the origin to z is called the modulus of z, $|z|$. What is the modulus $|z|$ for the point $z = 3 + 4i$?

2. The point z can also be located by establishing polar coordinates in the complex plane where r is the radius vector and θ is the *phase angle*. Draw suitable polar coordinates for the Argand plane. What is r for the point $z = 3 + 4i$? What is θ in degrees and radians?

3. Show that $z^*z = |z|^2$ where z^* is the *complex conjugate* of z

$$z = x + iy \qquad z^* = x - iy$$

and show that $|z| = r$.

4. Show that

$$e^{i\theta} = \cos\theta + i\sin\theta$$

and that

$$e^{-i\theta} = \cos\theta - i\sin\theta$$

5. Show that

$$z = re^{i\theta}$$

and that

$$z^* = re^{-i\theta}$$

6. Show that

$$\begin{vmatrix} 1\ s\alpha(1) & 1\ s\alpha(2) \\ 1\ s\beta(1) & 1\ s\beta(2) \end{vmatrix} = \begin{vmatrix} 1\ s\alpha(1) & 1\ s\beta(1) \\ 1\ s\alpha(2) & 1\ s\beta(2) \end{vmatrix}$$

7. Recall the definition of a matrix transpose (section on special matrices in Chapter 2). Transpose the matrices

$$\mathbf{A} = \begin{pmatrix} 2 & 3 \\ 4 & 5 \end{pmatrix}$$

and

$$\mathbf{B} = \begin{pmatrix} a & b & c \\ d & e & f \\ g & h & i \end{pmatrix}$$

Are the determinants corresponding to the matrix and its transpose equal in these two cases?

8. Calculate the determinant of the symmetric matrix

$$D := \begin{pmatrix} 1 & 2 & 3 \\ 4 & 5 & 5.5 \\ 7 & 5.5 & 9 \end{pmatrix}$$

and the determinant of its transpose.

9. The wave function for a particle in a one-dimensional infinite potential well (particle in a box) is

$$\Psi = A \sin \frac{n\pi x}{a}$$

where n is a quantum number $n = 0, 1, 2, \ldots$ and a is the dimension of the box in the x direction. Normalize Ψ to 1.

10. One spin combination allowable in excited state helium is $\alpha(1)\alpha(2)$, which is symmetric. There are three others. What are they? Indicate which are symmetric (s) and which are antisymmetric (a).

11. To satisfy the Pauli exclusion principle, the electronic wave function must be antisymmetric. This condition can be met in the excited state of the helium atom by taking the product of an antisymmetric space part such as

$$\frac{1}{\sqrt{2}}[1\,s(1)1\,s(2) - 2\,s(1)1\,s(2)]$$

times a symmetric spin part or by taking the product of a symmetric space part times an antisymmetric spin part. For example,

$$\frac{1}{\sqrt{2}}[1\,s(1)1\,s(2) - 2\,s(1)1\,s(2)]\alpha(1)\alpha(2) \qquad \Psi_{a,s}(1,2)$$

is antisymmetric $\times$ symmetric $=$ antisymmetric, as denoted by the subscripts on $\Psi_{a,s}(1,2)$. It is a legitimate wave function. How many other such legitimate products are there? Write them out.

12. The Hamiltonian for the helium atom is given in Eq. (9-2). All spin parts of the wave function in the answer to Problem 9 are normalized so they all integrate to 1, leaving only the space parts, of which there are four. These four orbitals can be abbreviated as two $\pm$ combinations, $1\,s^*(1)2\,s^*(2) \pm 2\,s^*(1)1\,s^*(2)$ and $1\,s(1)2\,s(2) \pm 2\,s(1)1\,s(2)$. Write out the variational expression $E = \int \Psi^* \hat{H} \Psi d\tau$ for the energy using the abbreviated space orbitals and the full Hamiltonian.

13. "All space" $\int d\tau$ can be (artificially) subdivided into a space for electron 1 $dv(1)$ and a space for electron 2 $dv(2)$, whereupon the answer to Problem 12 becomes

$$E = \frac{1}{2}\int\int [1\,s^*(1)2\,s^*(2) \pm 2\,s^*(1)1\,s^*(2)]\left[-\tfrac{1}{2}\nabla_1^2 - \tfrac{1}{2}\nabla_2^2 - \frac{2}{r_1} - \frac{2}{r_2} + \frac{1}{r_{12}}\right]$$
$$[1\,s(1)2\,s(2) \pm 2\,s(1)1\,s(2)]dv(1)dv(2)$$

This enables us to separate the double integral into a sum of products of two integrals, the first collecting terms to be integrated over the space containing electron 1 and the second consisting of terms that are integrated over the space containing electron 2. The first such product is

$$\int 1\,s^*(1)\left[-\tfrac{1}{2}\nabla_1^2\right]1\,s(1)dv(1)\int 2\,s^*(2)2\,s(2)dv(2)$$

How many more integral products are there like this one for the kinetic energy operator $-\tfrac{1}{2}\nabla_1^2$? How many of them are there for the kinetic energy operator $-\tfrac{1}{2}\nabla_2^2$? Write them out as a sum of integral products equal to the total kinetic energy, E_{kin}.

14. The linear combination of atomic orbitals is orthonormal, hence $\int 1\,s^*(1)2\,s(1)dv(1) = 0$, etc. and $\int 1\,s^*(1)1\,s(1)dv(1) = 1$, etc. Show that the eight integral products in Problem 9.9.7 can be reduced to

$$E_{\text{kin}} = \int 1\,s^*(1)\left[-\tfrac{1}{2}\nabla_1^2\right]1\,s(1)dv(1) + \int 2\,s^*(1)\left[-\tfrac{1}{2}\nabla_1^2\right]2\,s(1)dv(1)$$

15. Proceeding by analogy to the expansion over the two kinetic energy operators in Problems 13 and 14, obtain a sum of eight single integral products that is the energy contribution from the single-electron coulombic operators $-2/r_1$ and $-2/r_2$. Drop those products that contain a zero integral due to orthogonality and retain those for which $\int 1\,s^*(1)1\,s(1)dv(1) = 1$, etc. due to normalization. Show that

$$E_{\text{coul}} = \int 1\,s^*(1)\left(-\frac{2}{r_1}\right)1\,s(1)dv(1) + \int 2\,s^*(1)\left(-\frac{2}{r_2}\right)2\,s(1)dv(1)$$

16. The remaining task in expanding the variational expression

$$E = \frac{1}{2}\int\int[1\,s^*(1)2\,s^*(2)\pm 2\,s^*(1)1\,s^*(2)]\left[-\tfrac{1}{2}\nabla_1^2 -\tfrac{1}{2}\nabla_2^2 -\frac{2}{r_1}-\frac{2}{r_2}+\frac{1}{r_{12}}\right]$$
$$[1\,s(1)2\,s(2)\pm 2\,s(1)1\,s(2)]dv(1)dv(2)$$

involves only the multiplication

$$E_{r_{12}} = \frac{1}{2}\int\int[1\,s^*(1)2\,s^*(2)\pm 2\,s^*(1)1\,s^*(2)]\left(\frac{1}{r_{12}}\right)[1\,s(1)2\,s(2)$$
$$\pm 2\,s(1)1\,s(2)]dv(1)dv(2)$$

which is simpler than what we have done and follows the pattern $(a \pm b)(c)(d \pm e) = ac \pm bc(d \pm e) = acd \pm bcd \pm ace + bce$. Write out the result.

17. Now collect all terms from Problems 9.9.8 through 9.9.10 and show that they add up to the Hartree–Fock equation

$$E_i = I_1 + I_2 + J_{12} \pm K_{12}$$

for the excited state of the helium atom.

18. If the heat of formation of H is calculated by MNDO as 52.10 kcal mol^{-1} (Fig. 9.8.7) what is the bond dissociation energy of H_2?

19. Based on the output file in Fig. 9-3, what is the energy of formation of H^+ as determined by MNDO?

20. Calculate the bond length, ionization potential, and dipole moment of carbon monoxide by MNDO, AM1, and PM3.

21. What are the bond lengths in HCN according to a MNDO calculation? Repeat the calculation using AM1 and PM3 in both the MOPAC and GAUSSIAN for Windows implementations.

22. Run the following rather curious-looking 6-line input file in the MOPAC implementation.

```
1
1 1.
```

(Note that lines 1, 2, 3, and 6 are blank.) Does it run? What do the results mean? What two generalizations can you make about MOPAC input files from this program run?

23. If this absurdly diminished input file runs, can we carry the absurdity a step further and run the file

```
1
```

(Note that lines 1, 2, 3, and 5 are blank.) If so, what does the resulting output describe?

How would we obtain the MNDO approximations to the properties of the nitrogen atom?

24. What is the enthalpy of isomerization of propene (C_3H_6) to cyclopropane at 298 K by a semiempirical calculation?

25. Obtain the dipole moment of methylenecyclopropene by a MNDO calculation and compare your answer with the result obtained from Huckel molecular orbital calculations.

10

Ab Initio Molecular Orbital Calculations

Once having the Hartree–Fock equation and the Slater determinantal method of producing correctly antisymmetrized orbitals, it would seem that we should be able to approach the correct wave function by finding better and better basis sets, however laborious that process might be. In fact, basis set improvement leads to a limiting value for ψ and an energy that is above the experimental energy for the molecule. This limit is called the Hartree–Fock limit. We shall find ways of approaching the Hartree–Fock limit and then examine two ways of getting past it, one from Moeller and Plesset and the other a *density functional* method from Becke.

The GAUSSIAN Implementation

To go from a semiempirical calculation in the GAUSSIAN implementation (File 9-1) to an *ab initio* calculation, one need only change PM3 in the route section of the input file to sto-3g for a single point calculation or sto-3g opt for an optimization. We have made this change in File 10-1 along with the substitution of h for f in the second line of the geometry section to calculate the molecular

Computational Chemistry Using the PC, Third Edition, by Donald W. Rogers
ISBN 0-471-42800-0   Copyright © 2003 John Wiley & Sons, Inc.

properties of H_2. This avoids potential confusion of the Hartree-Fock energy HF with the hydrogen fluoride molecule HF.

```
# sto-3g opt

sto-3g optimization of h2

0 1
h
h,1,1.1
```

File 10-1. Input file for an STO-3G *ab initio* optimization of H_2.

Exercise 10-1

Calculate the H—H bond length in ground-state H_2 using the STO-3G basis set in the GAUSSIAN for Windows implementation.

Solution 10-1

The input file is File 10-1. The bond length is given in File Segment 10-2.

```
Optimization completed.
  – Stationary point found.
                    – – – – – – – – – – – – – – – – – –
                    !     Optimized Parameters   !
                    !     (Angstroms and Degrees) !
 – – – – – – – – – – – – –                    – – – – – – – – – – –
 ! Name  Definition      Value      Derivative Info.        !
 – – – – – – – – – – – – – – – – – – – – – – – – – – – – – – – – – –
 ! R1   R(2,1)          0.712       –DE/DX  =  0.0003       !
 – – – – – – – – – – – – – – – – – – – – – – – – – – – – – – – – – –
```

File Segment 10-2. The STO-3G Estimate of the H—H Bond Length. The experimental value is 0.742 Å.

Exercise 10-2

A class of 20 students has access to only one copy of GAUSSIAN. Need they wait in line to write their input files on this one machine?

Solution 10-2

Certainly not. Input File 10-1 was written using the DOS editor and saved on a 3.5″ floppy disk. You can write your input files at home on a laptop if you like, and then run them when your GAUSSIAN is not otherwise in use. Use .gjf (gaussian job file) as your file extension. If your editor gives the .txt extension or some such, use the **rename** command in DOS. If you run your file directly from the **a:** drive, the output will be stored on the **a:** drive as well. That can be an advantage because you now have a permanent record of the output file for writing papers and reports. Use a fresh floppy so you have enough room for the output file. Later, your output files may exceed the capacity of a floppy. Then you need to go to a ZIP drive or a CD.

Exercise 10-3

Create File 10-1 using an editor independent of the GAUSSIAN system and use it to solve Exercise 10-1.

How Do We Determine Molecular Energies?

We shall examine the simplest possible molecular orbital problem, calculation of the bond energy and bond length of the hydrogen molecule ion H_2^+. Although of no practical significance, H_2^+ is of theoretical importance because the complete quantum mechanical calculation of its bond energy can be carried out by both exact and approximate methods. This permits comparison of the exact quantum mechanical solution with the solution obtained by various approximate techniques so that a judgment can be made as to the efficacy of the approximate methods. Exact quantum mechanical calculations cannot be carried out on more complicated molecular systems, hence the importance of the one exact molecular solution we do have. We wish to have a three-way comparison i) exact theoretical, ii) experimental, and iii) approximate theoretical.

Exact Theoretical. The exact solution is found by solving the problem in ellipsoidal polar coordinates with the nuclei at the foci of the ellipses, in a way similar to solution for the hydrogen atom in spherical polar coordinates with the single nucleus at the center of the sphere. The result for the energy of the ground state is $E = 0.1026$ hartrees $= 2.791$ eV $= 269.3$ kJ mol^{-1} (Hanna, 1981). The bond length is $R = 2.00$ bohr where 1.000 bohr $= 0.5292$ Å. The *total energy* for this simple system, 0.6026 h, is the bond energy plus the energy of the hydrogen atom, 0.5000 h.

Experimental. The *vibrational spectrum* of an ideal harmonic oscillator would consist of one line at frequency ν corresponding to $\Delta E = h\nu$, where ΔE is the distance between levels on the vertical energy axis in Fig. 10-1a. In the harmonic oscillator, ΔE is the same for a transition from one energy level to an adjacent level. A *selection rule* $\Delta n = \pm 1$, where n is the *vibrational quantum number, requires* that the transition be to an adjacent level.

The H_2^+ ion is not, however, a perfect harmonic oscillator. Its spectrum consists of many lines because its vibrational levels get closer together as the vibrational energy increases, as in Fig. 10-1b. The H_2^+ ion displays *anharmonicity*. As oscillations become more energetic, an internuclear distance R is reached, seen on the right of Fig. 10-1b, at which further separation of the nuclei results in only an infinitesimal increase in energy E. Beyond this limit, *dissociation* has occurred

$$H_2^+ \rightarrow H + H^+ \tag{10-1}$$

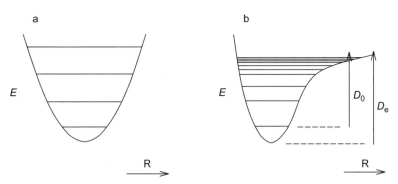

Figure 10-1 The Potential Well and Energy Levels of (a) a Perfect Harmonic Oscillator and (b) an Anharmonic Oscillator Resembling $H_2{}^+$. R is the internuclear distance, D_0 is the dissociation energy and $-D_e$ is the bond energy.

The *dissociation energy* D_0 is the energy that must be put into the system to break the bond, bringing about reaction (10-1). The *extrapolated dissociation energy* or *bond energy* D_e is defined slightly differently. If, as is usually done, we set the zero of energy at the level of the free, unbound atom H and ion H^+, D_e is the amount of energy released when the system $H + H^+$ goes to H_2^+ *at the bottom of the potential well* in Fig. 10-1b. Because energy is coming out of the system, the bond energy is a negative number.

A simple mathematical manipulation of the dissociation energy of H_2^+, as determined from its absorption spectrum, yields the bond energy. Planck's law, $\Delta E = h\nu$, permits us to calculate the energy difference between the lowest level and the next higher level from its spectroscopic line at 2191 cm^{-1}. This is the highest frequency line because the lowest is the largest of all energy *spacings* in Fig. 10-1b. We can also measure the second energy increment, which corresponds to the spectral peak of next lower frequency, the third, and so on, corresponding to the gradual diminution of energy spacing in Fig. 10-1b. The series approaches zero. The sum of all energies of transition is the dissociation energy.

Exercise 10-4

What is the energy difference in electron volts and kilojoules per mole between the ground state of H_2^+ and its lowest vibrationally excited state? The vibrational spectrum of H_2^+ has lines at 2191, 2064, 1941, 1821, 1705, 1591, 1479, 1368, 1257, 1145, 1033, 918, 800, 677, 548, 411, 265, 117 cm^{-1}.

Solution 10-4

The gap between the states with vibrational quantum numbers $n = 0$ and $n = 1$ (ground state and lowest vibrationally excited state) corresponds to the highest energy line, 2191 cm^{-1}, hence

$$2191(1.240 \times 10^{-4}) = 0.2717\,\text{eV} = 26.20\ \text{kJ mol}^{-1}$$

where 1.240×10^{-4} is the conversion factor from cm^{-1} to eV.

Exercise 10-5

What is the sum of all the energy gaps as determined from the vibrational spectrum of H_2^+?

Solution 10-5

The sum is 21331 cm^{-1} $= 2.645$ eV $= 255.1$ kJ mol^{-1}. This is a first approximation to the dissociation energy.

If we plot the energy of each transition as a function of the quantum number of the vibrational state to which the system is excited, we have what is called a *Birge–Spooner plot* (Fig. 10-2). Two corrections are needed to convert the result of Exercise 10-5, which is essentially an integration under the Birge–Spooner function, to the bond energy of H_2^+. One correction is the small amount of energy not accounted for between the highest vibrational quantum number $n = 18$ and $E = 0$ in Fig. 10-2. This energy is represented by the small triangle between the rightmost end of the curve and the extrapolation to the $E = 0$ axis. It is 46.2 cm^{-1}, and it is added to the result of Exercise 10-5.

The second correction is much larger. The residual energy that the molecule ion has in the ground state above the D_e at the equilibrium bond length is the *zero point energy*, ZPE.

$$E(ZPE) = D_e - D_0 \qquad (10\text{-}2)$$

We must add the amount of energy at the bottom of the "bowl" in Fig. 10-1b to the sum from Exercise 10-5. This energy is one-half a quantum at the wavenumber extrapolated one-half quantum number below $n = 0$ (see Problems).

When corrected for both the energy unaccounted for at the low end of the Birge–Spooner plot, and $E(ZPE)$,

$$\Delta E_{vib} = 22505 \text{ cm}^{-1} = 2.791 \text{ eV} = 269.3 \text{ kJ mol}^{-1}$$

Recall that the result of the exact theoretical calculation is 2.791 eV $= 269.3$ kJ mol^{-1}.

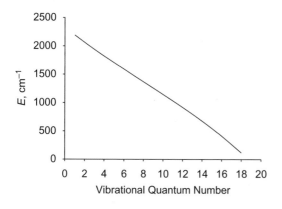

Figure 10-2 BirgeSpooner Plot of the Energy Increment Between Vibrational Energy Levels vs. the Vibrational Quantum Number.

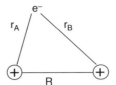

Figure 10-3 The Hydrogen
Molecule Ion, H_2^+.

A zero point energy is found in every chemical bond; therefore, it will be a crucial part of all of our future calculations. The existence of an irreducible ZPE satisfies the Heisenberg uncertainty principle because the molecule does not exist precisely at the potential energy minimum in the ground state. We do not know the exact positions of the two nuclei on the lowest horizontal line in Fig. 10-1b; we only know that they are separated by a distance that is somewhere on the line. (Actually, a complete quantum mechanical treatment allows internuclear distances that go slightly beyond the ends of the horizontal line at D_0.)

Approximate Theoretical. The simplest molecular orbital problem is that of the hydrogen molecule ion (Fig 10-3). H_2^+ is a preliminary example of all molecular orbital problems to come, which, although they may be very complicated, are elaborations on this simple example.

Assuming the Born–Oppenheimer approximation, we are looking at a problem of one electron in the field of two singly-positive nuclei at some fixed distance R

$$\left[-\tfrac{1}{2}\nabla^2 - \frac{1}{r_A} - \frac{1}{r_B}\right]\Psi = E_{el}\Psi \tag{10-3}$$

There are many possible values of R, each of which leads to a unique value of the electronic energy E_{el}. Thus, although R is constant for any single calculation under the Born–Oppenheimer approximation, the total energy of the system, including internuclear repulsion energy (always positive) is a function of whatever value of R has been selected for the calculation

$$E_{bond} = E_{el} + \frac{1}{R} \tag{10-4}$$

$E_{bond}(R)$ has a minimum at the equilibrium bond length.

Under the LCAO approximation,

$$\psi = a_1 e^{-r_A} \pm a_2 e^{-r_B} \tag{10-5}$$

where $a_1 e^{-r_A}$ and $a_2 e^{-r_B}$ are normalized hydrogen 1s wave functions, call them $1s_A$ and $1s_B$. By the variational theorem,

$$E = \int \psi \hat{H} \psi \, d\tau$$

$$= \frac{1}{2(1+S)} \int \left(1s_A + 1s_B\left[-\tfrac{1}{2}\nabla^2 - \frac{1}{r_A} - \frac{1}{r_B}\right]1s_A + 1s_B\right) d\tau \tag{10-6}$$

where S is the overlap integral and $1/[2(1+S)]$ is the normalization constant. The H_2^+ molecule ion is symmetrical so

$$\int 1s_A \left[-\tfrac{1}{2}\nabla^2\right] 1s_A \, d\tau = \int 1s_B \left[-\tfrac{1}{2}\nabla^2\right] 1s_B \, d\tau \tag{10-7}$$

and

$$\int 1s_A \left(-\frac{1}{r_A}\right) 1s_A \, d\tau = \int 1s_B \left(-\frac{1}{r_B}\right) 1s_B \, d\tau \tag{10-8}$$

If we expand Eq. (10-7) and simplify according to the symmetry of the problem, (Richards and Cooper, 1983) the integral breaks up in the way it did for the helium atom excited state

$$E = \frac{1}{(1+S)} \int 1s_A \left[-\tfrac{1}{2}\nabla^2\right] 1s_A \, d\tau + \int 1s_A \left[-\tfrac{1}{2}\nabla^2\right] 1s_B \, d\tau - \int 1s_A \left(-\frac{1}{r_A}\right) 1s_A \, d\tau$$
$$- \int 1s_A \left(-\frac{1}{r_B}\right) 1s_A \, d\tau - 2 \int 1s_A \left(-\frac{1}{r_B}\right) 1s_B \, d\tau \tag{10-9}$$

Evaluation of the integrals is simplified by the observation that

$$1s_A = \frac{1}{\sqrt{\pi}} e^{-r_A} \qquad \text{and} \qquad 1s_B = \frac{1}{\sqrt{\pi}} e^{-r_B}$$

When we perform these integrations, we get

$$E = \frac{J+K}{1+S} \tag{10-10}$$

where

$$J = \left(1 + \frac{1}{R}\right) e^{-2R}$$

$$K = \frac{S}{R} - e^{-R}(1+R)$$

and

$$S = \left(1 + R + \frac{R^2}{3}\right) e^{-R}$$

Exercise 10-6

The minimum for the equilibrium internuclear distance in H_2^+ is 2.49 bohrs in this first approximation. Calculate the dissociation energy of H_2^+ at this distance.

Solution 10-6

$$S = 0.461 \qquad J = 0.00963 \qquad K = -0.1044$$

$$E = -0.065 \text{ hartrees}$$

The energy is negative, which indicates bonding. The dissociation energy neces-sary to separate the ion into H and H^+ is positive by convention. One hartree $= 627.51 \text{ kcal mol}^{-1} = 27.212 \text{ eV} = 2625 \text{ kJ mol}^{-1}$, leading to $E = 170 \text{ kJ mol}^{-1}$. This is 63% of the experimental value. The quantitative result is not good, but the qualitative result is absolutely spectacular: *The chemical bond is a natural result of quantum mechanics!* Bear in mind that Schroedinger was not attempting to explain the chemical bond with his celebrated equation (Schroedinger, 1926); he was explaining the spectrum of the hydrogen atom. The explanation of the chemical bond came later (Heitler and London, 1927), as a consequence of the Schroedinger equation.

Why Is the Calculated Energy Wrong?

In the case of H_2^+, the energy is wrong because the molecular orbital is not a linear combination of atomic orbitals, it is *approximated* by a linear combination of atomic orbitals. Use of scaled atomic orbitals

$$\chi_A = \frac{\eta^{\frac{3}{2}}}{\sqrt{\pi}} e^{-\eta r_A} \qquad \text{and} \qquad \chi_B = \frac{\eta^{\frac{3}{2}}}{\sqrt{\pi}} e^{-\eta r_B}$$

with an *optimized* scale factor $\eta = 1.24$ brings the discrepancy down to 15.8%.

Can the Basis Set Be Further Improved?

Almost from the dawn of quantum mechanics a great deal of effort has been put into devising better basis sets for molecular orbital calculations (for example, Rosen, 1931; James and Cooledge, 1933). Prominent among more recent work is the long series of GAUSSIAN programs written by John Pople's group (see, for example, Pople et al., 1989) leading to the award of the Nobel prize in chemistry to Pople in 1999. GAUSSIAN consists of a suite of programs from which one can select members by means of keywords. We have already seen the keyword STO-3G used in File 10-1 to select the approximation of Slater-type orbitals by three Gaussian functions. Along with individual calculations, GAUSSIAN also permits you to run sequential calculations by executing a *script* that specifies the programs to be run, the order in which they are to be run, and how the various outputs are to be combined to give a final result. This technique has been developed to a high degree of accuracy in the GAUSSIAN family of programs, of which we shall use the principal members, G2 and G3.

Exercise 10-7

Calculate the bond energy of H_2^+ by the G2 method in the GAUSSIAN implementation.

Solution 10-7

Run the H_2^+ input file with the keyword g2.

```
# g2

H2 ion

1 2
h
h 1 1.32
```

The keyword is in the route section, line 1 of the input file. Lines 2, 3, and 4 are blank, comment, blank, respectively. Line 5 designates a charge of 1 and a spin multiplicity of 2 (a doublet). Line 6 specifies one atom as hydrogen, and line 7 specifies the second atom as hydrogen, attached to atom 1 at a distance of 1.32 Å(2.49 bohr). Among several G2 energies printed out in about the last 25 lines of output are

$$E(ZPE) = 0.004373$$

and

$$G2(0 \text{ K}) = -0.597624$$

A distinction must be made between single-point GAUSSIAN calculations or optimizations and the energy output G2(0 K). Individual GAUSSIAN calculations produce the energy D_e coming out of the system when all the nuclei and electrons come together to form a molecule, radical, or ion, in this case,

$$H^+ + H^+ + e^- \rightarrow H^+ + H \rightarrow H_2^+ \tag{10-11}$$

at the bottom of the potential well. G2 and G3 are scripted to produce the energy D_0 (Fig. 10-3) of H_2^+ (or other molecular systems) *in the ground state*, one-half a quantum above the bottom. Thus the G2 output is the total energy of the system (negative) plus $E(ZPE)$, which is positive relative to the bottom of the well,

$$\text{total energy} + E(ZPE) = G2(0 \text{ K})$$

in this case

$$-0.601997 + 0.004373 = -0.597624 \, h$$

One may wonder why it is important to distinguish between and keep track of these two energies D_e and D_0, when it seems that one would do. Actually, both are important. The bond energy D_e dominates theoretical comparisons and the dissociation energy D_0, which is the ground state of the real molecule, is used in practical applications like calculating thermodynamic properties and reaction kinetics.

The total energy is very much larger in magnitude (more negative) than the bond energy because most of the energy in a molecule resides in its atoms. The chemical bond is only a small perturbation on the total energy of the molecular system. The bond energy is the energy of the second step in Eq. (10-1), combination of H^+ with H. Thus we must subtract the energy of the first step, formation of H, from the total to obtain

$$-0.601997 - (-0.500000) = 0.101997 = 0.1020 \text{ hartrees}$$

about 0.6% in error relative to the exact value for the bond energy of 0.1026 h. The equilibrium internuclear distance is 1.04 Å or 1.97 bohr (experimental value 2.00 bohrs). Before we look at the details of basis set improvement and other issues in the GAUSSIAN procedures, let us carry out a calculation on a more difficult problem, that of the hydrogen molecule.

Hydrogen

Even though the problem of the hydrogen molecule H_2 is mathematically more difficult than H_2^+, it was the first molecular orbital calculation to appear in the literature (Heitler and London, 1927). In contrast to H_2^+, we no longer have an exact result to refer to, nor shall we have an exact energy for any problem to be encountered from this point on. We do, however, have many reliable results from experimental thermochemistry and spectroscopy.

Like H_2^+, H_2 has a simple Hamiltonian

$$\hat{H} = -\tfrac{1}{2}\nabla_1^2 - \tfrac{1}{2}\nabla_2^2 - \frac{1}{r_{1A}} - \frac{1}{r_{1B}} - \frac{1}{r_{2A}} - \frac{1}{r_{2B}} + \frac{1}{r_{12}} + \frac{1}{R} \qquad (10\text{-}12)$$

where the subscripts 1 and 2 refer to electrons, A and B refer to the nuclei, and R is the internuclear distance, which we shall take to be constant in any single calculation. This problem appears to be nothing more than an extension of H_2^+, but, because of the $1/r_{12}$ term in the Hamiltonian, it is insoluble. We have only the experimental equilibrium bond length of 1.400 bohrs (.7408 Å) and the bond energy of 1.74 h (Hertzberg, 1970), measured by the spectroscopic method in the section on determining molecular energies above in this chapter, for comparison.

Reading the output for H_2 is similar to H_2^+ as well. The optimized bond distance is

```
- - Stationary point found.
        - - - - - - - - - - - - - - - - - -
        !   Optimized Parameters   !
        !   (Angstroms and Degrees) !
- - - - - - - -                    - - - - - - - - - -
! Name    Definition      Value      Derivative Info.        !
- - - - - - - - - - - - - - - - - - - - - - - - - - - - - -
! R1        R(2,1)        0.7301     -DE/DX =    -0.0001     !
```

which is shorter than it is for H_2^+ as expected in a molecule with two bonding electrons rather than just one. The discrepancy between the experimental length and the calculated bond length is a little over 1.4%. The bond energy is

$$G2(0 \text{ K}) - E(\text{ZPE}) = -1.16635 - 0.00946 = -1.17581 \text{ h}$$
$$D_e = -1.17581 - (-1.00000) = -0.17581 \text{ h}$$

where the electronic energy of two isolated hydrogen atoms is taken into account in the second equation. The discrepancy between the calculated result and experiment is slightly over 1%.

Despite these similarities between the G2 calculations of D_e for H_2 and H_2^+, there is a profound difference that is only hinted at in the single-point energies in Table 10-1. In the first group of results we see that all the calculations, with the exception of the last one, give the same answer. In the lower block of results, for H_2, this is not the case. Some results are duplicated and some are not. The results are more mixed.

Table 10-1 Energies of of H_2 and H_2^+ Calculated by Various High-Level Basis Sets.

H_2^+

MP2/6-311G(d, p) = −0.6011593	QCISD/6-311G(d, p) = −0.6011593
MP4/6-311G(d, p) = −0.6011593	MP2/6-311 + G(d, p) = −0.6011593
MP4/6-311 + G(d, p) = −0.6011593	MP2/6-311G(2df, p) = −0.6011593
MP4/6-311G(2df, p) = −0.6011593	MP2/6-311 + G(3df, 2p) = −0.6018074

H_2

MP2/6-311G(d, p) = −1.1602718	QCISD/6-311G(d, p) = −1.1683162
MP4/6-311G(d, p) = −1.1677248	MP2/6-311 + G(d, p) = −1.1602718
MP4/6-311 + G(d, p) = −1.1677248	MP2/6-311G(2df, p) = −1.1602718
MP4/6-311G(2df, p) = −1.1677248	MP2/6-311 + G(3df, 2p) = −1.1627639

In addition to the mixed results in Table 10-1, the G2 calculation for H_2 produces an energy that is lower than the experimental value, in contradiction to the rule that variational procedures reach a least *upper bound* on the energy. Some new factors are at work, and we must look into the structure of the G2 procedure in terms of high-level Gaussian basis sets and electron correlation.

Gaussian Basis Sets

Gaussian-Type Orbitals, GTOs. In G2 and G3, an effort is made to extend the basis set to its practical limit of accuracy. We have seen, in the case of STO-nG basis sets, that more contributing Gaussian functions make for better agreement between calculated and known energies, but there is a point of diminishing return, beyond which further elaboration produces little gain. The same is true of the Gaussian n-xxG basis sets, for example, 6-31G, except that there are more of them and the notation can be daunting at first. In using many Gaussians to express the space part of a molecular orbital, one finds that demands on computer resources increase roughly as the fourth power of the number of basis functions. This rule brings about a practical limit on the complexity of the basis set. Thus basis set efficiency becomes an important factor in molecular methods because an efficient computer protocol enables us to treat more difficult and more interesting problems. One also finds that the basis functions fall into natural groups that are replicated from one calculation to the next; hence, a way of using computer resources efficiently is by treating each group as though it were a single function.

Contracted Gaussian-Type Orbitals, CGTOs. Each natural group of basis functions can be treated as a unit called a *contracted Gaussian*. Constituent Gaussians making up a contracted Gaussian are called *primitive Gaussians* or simply *primitives*. Optimizing all the parameters in all of the primitives on each run is a big job (remember the 4th power rule). A better strategy is to optimize each group once and for all

$$\chi_j = \sum_i c_{ji} g_i(\alpha, r) \tag{10-13}$$

where $g_i(\alpha, r)$ is the Gaussian, for example

$$g_i(\alpha, r) = \left(\frac{2\alpha}{\pi}\right)^{3/2} e^{-\alpha r^2} \tag{10-14}$$

in the simple case of the 1s orbital. The groups are then combined to find the desired molecular orbital

$$\psi_i = \sum_j a_{ji} \chi_j = \sum_j a_{ji} \sum_i c_{ji} g_i(\alpha, r) \tag{10-15}$$

If we start out with 36 primitives (by no means a large number in this context) and segment them into groups of 6, we have reduced the problem sixfold.

Split-Valence Basis Sets. In split-valence basis sets, inner or core atomic orbitals are represented by one basis function and valence atomic orbitals are represented by two. The carbon atom in methane is represented by one 1s inner orbital and 2(2s, $2p_x$, $2p_y$, $2p_z$) = 8 valence orbitals. Each hydrogen atom is represented by 2 valence orbitals; hence, the number of orbitals is

$$1 + 8 + 8 = 17$$

In the 6-31G basis, the inner shell of carbon is represented by 6 primitives and the 4 valence shell orbitals are represented by 2 contracted orbitals each consisting of 4 primitives, 3 contracted and 1 uncontracted (hence the designation 6-31). That gives us $4(4) = 16$ primitives in the valence shell. The single hydrogen valence shells are represented by 2 orbitals of 2 primitives each. That gives us a total of

$$6 + 4(4) + 4(2 \times 2) = 38 \text{ primitives}$$

which we verify by running the program in the 6-31G basis and finding

```
17 basis functions     38 primitive gaussians
```

The reason the inner shell of carbon is represented by 6 primitives in this basis is that the cusp in the 1s orbital is difficult to approximate with Gaussians that have no cusp.

Basis sets can be further improved by adding new functions, provided that the new functions represent some element of the physics of the actual wave function. Chemical bonds are not centered exactly on nuclei, so *polarized* functions are added to the basis set leading to an improved basis denoted p, d, or f in such sets as 6-31G(d), etc. Electrons do not have a very high probability density far from the nuclei in a molecule, but the little probability that they do have is important in chemical bonding, hence *diffuse functions*, denoted + as in 6-311 + G(d), are added in some very high-level basis sets.

COMPUTER PROJECT 10-1 | *Gaussian Basis Sets: The HF Limit*
The input file for an STO-3G calculation of the bond distances, energies, and other molecular properties of the isolated water molecule in the gaseous state at 0 kelvins is

```
# sto-3g

water

0 1
O
H 1 .74
H 1 .74 2 105.
```

where the bond distances and the bond angle are fairly close to experimental values. This is a single-point calculation because it is not optimized. The objective of this computer project is to examine the influence of increasingly high-level basis sets on the calculated total energy of the water molecule.

Procedure. Calculate and record the STO-3G energy of water by running the file above. Carry out the optimization with `sto-3g opt` in the route section. Repeat with `3-21G`, `3-21G opt`, `6-21G`, `6-21G opt`, `4-31G`, `4-31G opt`, `6-31G`, `6-31G opt`, `6-311G`, `6-311 opt`, `6-311 + G`, `6-311 + G opt`, `6-311 + G(d) opt`, `6-311 + G(d,p) opt`, `6-311 + + G(d,p) opt`, and `6-311 + + G(3df,2p) opt` in the route section. Tabulate and correlate your output results. Is the calculation approaching a limiting value? Strictly speaking, the Hartree–Fock limit (HF limit) is the limit at an infinitely high-level Gaussian basis set, which is, of course, unobtainable. The `6-311++G(3df,2p)` calculation is, for practical reasons, as high as one goes in G2 calculations. Estimate about where you think the HF limit would be found if it could be calculated. Repeat the calculation one last time at the G2 level with `g2` in the route section. Calculate E_e from the G2 result. The *correlation energy* is $E_e - E(\text{HF limit})$. Estimate the correlation energy for the water molecule based on this series of calculations.

Electron Correlation

A very important difference between H_2 and H_2^+ molecular orbital calculations is *electron correlation*. Electron correlation is the term used to describe interactions between electrons in the same molecule. In the hydrogen molecule ion, there is only one electron, so there can be no electron correlation. The designators given to the calculations in Table 10-1 indicate first an electron correlation method and second a basis set, for example, MP2/6-31G(d,p) designates a Moeller–Plesset electron correlation extension beyond the Hartree–Fock limit carried out with a 6-31G(d,p) basis set.

Exercise 10-8

Compare the energy found using 6-31G in the route section of the H_2^+ file in the GAUSSIAN implementation with the result found using MP2/6-31G in the route section. Repeat this comparison for H_2.

Solution 10-8

For the H_2^+ ion, the results are the same $\text{HF} = -0.5768653\,\text{MP2} = -0.5768653$. For the H_2 molecule, which has electron correlation, the results are not the same $\text{HF} = -1.1267553\,\text{MP2} = -1.1441366$. (Your F and P may come out $H\Phi = -1.1267553\,\text{M}\prod 2 = -1.1441366$.)

Each of the basis sets described in Computer Project 10-1 assumes one anti-symmetrized Slater determinant and only approaches the Hartree-Fock limit, which, as we have seen, is a long way from the desired result for chemical applications. Releasing restrictions on an electron (or any confined particle) lowers its energy. Suppose to our ground-state determinant we add an antisymmetrized determinant representing one valence electron in the first excited state. The constraints on that electron are reduced because it can be in either the ground or excited state; hence, the energy of the system is lowered. If we can add one antisymmetrized determinant, we can add more than one. The wave function is now a sum

$$\psi = b_0\psi_0 + \sum_n b_n\psi_n \tag{10-16}$$

where ψ_0 is the Hartree-Fock antisymmetrized determinant and ψ_n are antisymmetrized determinants of various excited states. If the sum is infinite, all possible determinants are included and this procedure is called *full configuration interaction* or *full CI*. Full CI is not possible in a practical sense, but we would like to get as close to it as possible by using a truncated sum in Eq. (10-16). *Truncated CI* methods include CIS for single substitution [$n = 1$ in Eq. (10-16)], CID for double substitution, CIT for triple and QCISD(T) for quadratic singles, doubles, and (possibly triples) excitation. In Table 10-1 different MP CI levels produce the same results for H_2^+, which has no electron correlation, but MP2 and MP4 results are different for H_2, which has electron correlation.

A *Moeller–Plesset CI* correction to ψ is based on perturbation theory, by which the Hamiltonian is expressed as a Hartree–Fock Hamiltonian perturbed by a small perturbation operator $\hat{P}$ through a minimization constant λ

$$\hat{H} = \hat{H}_{HF} + \lambda\hat{P}$$

After expansion of $\hat{P}$ as a power series (Foresman and Frisch, 1996) Moeller–Plesset theory results in a correction for the wave function and the energy. The energy correction is always negative, which "improves" our calculation, but the Moeller–Plesset (MP) procedure is not a variational procedure, does not produce a least upper limit, and can overcorrect the energy. This is part of the explanation of why the G2 bond energy for H_2 in the section on the hydrogen molecule above in this chapter is lower than the experimental value. Gaussian basis sets containing component MP CI calculations are denoted, for example, MP2/6-31G or, if polarization functions are also included, MP2/6-31G(d,p).

G2 and G3

By systematically applying a series of corrections to approximate solutions of the Schroedinger equation the Pople group has arrived at a family of computational protocols that include an early method G1, more recent methods, G2 and G3, and their variants by which one can arrive at thermochemical energies and enthalpies of formation, $\Delta_f E^0$ and $\Delta_f H^{298}$, that rival experimental accuracy. The important thing is that the corrections are not unique to the molecule but are unique to the computational system, G1, G2, G3(MP2), etc., and can be applied to any molecule. We shall treat the G2 method in some detail and then treat the G3 method as an extension of it.

G2. The objectives are to obtain:

1. An equilibrium geometry.
   The geometry is obtained at the MP2/6-31G(d) level.
2. A total electronic energy E_e.
   The electronic energy is obtained by correcting the MP4/6-311G(d,p) energy.
3. A set of harmonic frequencies leading to E(ZPE).
4. The ground state energy E_0.
   The ground state energy is obtained as $E_0 = E_e + E$(ZPE).

Corrections to the MP4/6-311G(d,b) Energy. Higher-level basis functions, if they are prudently chosen, should be better than lower-level functions. Thus the energy of, for example, a diffuse function, $E[\text{MP2/6-311} + \text{G(d,p)}]$ should be lower (more negative) than the same function in which diffuse electron density is not taken into account $E[\text{MP2/6-311G(d,p)}]$, provided that the levels of electron correlation are the same. All other things being equal, the diffuse function $(+)$ improves the calculated energy by some small amount. A plausible correction to any chosen base energy would be the difference $E[MP2/6\text{-}311+G(d,p)]-E[MP2/6\text{-}311G(d,p)]$.

G2 theory, along with other members of the Gn family, rests on the assumption that both basis set corrections and electron correlation corrections are *additive*. The total energy of a molecular system E_0 by the G2 method, is found by arbitrarily selecting a starting point, $E[MP4/6\text{-}311G(d,p)]$ in this case, and adding to it corrections found by applying various higher-level basis functions and higher levels of electron correlation to the system. Once we know what the various corrections are, we shall be able to construct G2 energies of startling accuracy for a wide variety of molecules, radicals, and ions. It is this achievement that won the Nobel prize for John Pople.

To obtain the G2 value of E_0 we add five corrections to the starting energy, $E[MP4/6\text{-}311G(d,p)]$ and then add the zero point energy to obtain the ground-state energy from the energy at the bottom of the potential well. In Pople's notation these additive terms are

$$E_0 = E[MP4/6\text{-}311G(d,p)] + \Delta E(+) + \Delta E(2d,f) + \Delta + \Delta E(QCI)$$
$$+ HLC + E(ZPE) \tag{10-17}$$

where

$$\Delta E(+) = E[MP4/6\text{-}311 + G(d,p)])] - E[MP4/6\text{-}311G(d,p)]$$
$$\Delta E(2d,f) = E[MP4/6\text{-}311G(2df,p)] - E[MP4/6\text{-}311G(d,p)]$$
$$\Delta = E[MP2/6\text{-}311 + G(3df,2p)] - E[MP2/6\text{-}311G(2df,p)] - E[MP2/$$
$$6\text{-}311 + G(d,p)] + E[MP2/6\text{-}311G(d,p)]$$
$$\Delta E(QCI) = E[QCISD/6\text{-}311G(d,p)] - E[MP4/6\text{-}311G(d,p)]$$
$$HLC = 4(-0.00500)$$
$$E(ZPE) = 0.04269$$

The second, third, and fourth corrections to $E[MP4/6\text{-}311G(d,p)]$ are analogous to $\Delta E(+)$. The zero point energy has been discussed in detail (scale factor 0.8929; see Scott and Radom, 1996), leaving only HLC, called the "higher level correction," a purely empirical correction added to make up for the practical necessity of basis set and CI truncation. In effect, thermodynamic variables are calculated by methods described immediately below and HLC is adjusted to give the best fit to a selected group of experimental results presumed to be reliable.

Substituting calculated values for each of these energies in the case of methane,

$$\Delta E(+) = -40.4053269 - (-40.4050234) = -0.00030$$
$$\Delta E(2d,f) = -40.4246603 - (-40.4050234) = -0.01964$$
$$\Delta = -40.4056666 - (-40.3976534) - (-40.3795243)$$
$$+ (-40.379232) = -0.00772$$
$$\Delta E(QCI) = -40.4058874 - (-40.4050234) = -0.00086$$
$$HLC = -0.02000$$
$$E(ZPE) = +0.04269$$

The result of these corrections for methane is

$$E_0 = -40.40502 - 0.00030 - 0.01964 - 0.00772 - 0.00086 - 0.02000 + 0.04269$$
$$= -40.41085 \text{ hartrees}$$

where all of the calculated energies are selected from the block of data at the end of the G2 output file

```
MP2/6-311G(d,p) = -40.379232
QCISD/6-311G(d,p) = -40.4058874
MP4/6-311G(d,p) = -40.4050234
MP2/6-311+G(d,p) = -40.3795243
MP4/6-311+G(d,p) = -40.4053269
MP2/6-311G(2df,p) = -40.3976534
MP4/6-311G(2df,p) = -40.4246603
MP2/6-311+G(3df,2p) = -40.4056666
```

The calculated result as we have found it by this dissection of the G2 method agrees with the final result of the G2 script

$$G2 = -40.4108546$$

G3. G3 theory (Curtiss et al., 1998) is very similar to G2 except that certain refinements have been added to improve accuracy and computational efficiency. The Pople group has devised a new basis function for the largest calculation called, appropriately enough, G3large.

G2(MP2) and G3(MP2). Depending on your system, G2 and G3 calculations may take a prohibitive amount of time or memory for a fairly modest increase in molecular complexity. Don't forget that *ab initio* calculations are *much* more demanding than semiempirical calculations, so a "large molecule" in this field may have only a few heavy (nonhydrogen) atoms. If you do research in this field, you will soon run into the problem of long run times and you may get error messages indicating that you have exceeded memory. For this reason, several members of the G-*n* family have been written so as to use reduced basis set extensions, resulting in shorter run times and smaller demands on memory space (Curtiss, 1999). Most noteworthy are the G2(MP2) and G3(MP2) modifications, which do not use basis set extensions above the second order or MP2 level. Somewhat surprisingly, these methods give thermodynamic results that are almost as accurate as the parent methods G2 and G3 and may even be slightly better for some subsets of molecules, for example, hydrocarbons.

Energies of Atomization and Ionization

The energy of atomization of a ground state molecule at 0 K, for example, methane, is the energy of the reaction

$$CH_4(g) \rightarrow C(g) + 4 H(g)$$

which we can calculate because we can calculate the ground state energies E_0 for all of the components of the reaction

$$\Delta E_0 = E_0(C) + 4E_0(H) - E_0(\text{methane})$$
$$- 37.78432 - 4(0.50000) - (-40.41085) = 0.62653 \text{ hartrees}$$
$$= 393.15 \text{ kcal mol}^{-1} \tag{10-18}$$

This number is experimentally accessible and is 392.5 kcal mol^{-1} (Chase et al., 1985).

The ionization energy

$$CH_4(g) \rightarrow CH_4(g)^+$$

is calculated for the methyl cation in a similar way

$$-39.94759 - (-40.41085) = 0.46326 \text{ h} = 12.61 \text{ eV}$$

The experimental value is 12.62 eV and is positive for ionization of a stable molecule.

We are now in a position to calculate the energy change of any reaction in the gaseous state at 0 K

$$\Delta E_0 = \sum E_0(\text{products}) - \sum E_0(\text{reactants}) \tag{10-19}$$

provided we have the computational resources to calculate E_0 for all of its components.

COMPUTER PROJECT 10-2 | *Larger Molecules: G2. G2(MP2), G3, and G3(MP2)*

Calculate the energy change E_0 for isomerization of cyclopropane to propene in the gaseous state at 0 K

cyclopropane   propene

using the G2, G2(MP2), G3, and G3(MP2) procedures.

Procedure. Depending on your system, the run times, especially G2 run times, for these two molecules may be too long. If you have exclusive use of a system or if you can make a congenial sharing arrangement, run them overnight. (A good deal of research in this field happens while people are asleep.) It is good practice to work up to the G-*n* calculations starting with simpler single-point calculations, for example, STO-3G and 6-31G, to get an idea of the time requirements you will be facing and to debug any small failings of your input file before committing it to a long run.

The experimental enthalpy of isomerization of cyclopropane is $\Delta_{isom}H^{298} = -8.0 \pm 0.2\,\text{kcal mol}^{-1}$. Why is this enthalpy of isomerization negative? Evidently, there is not much difference between computed values of $\Delta_{isom}E_0$ and $\Delta_{isom}H^{298}$ as obtained from the thermodynamic equations

$\Delta_{isom}E_0 = $ [G2 (0 K)(propene)] $-$ [G2 (0 K)(cyclopropane)]

$\Delta_{isom}H^{298} = $ [G2 Enthalpy(propene)] $-$ [G2 Enthalpy (cyclopropane)]

We shall discuss this difference in the section on thermodynamic functions below.

The GAMESS Implementation

High-level molecular orbital calculations can be carried out with the freeware program GAMESS [*General Atomic and Molecular Electronic Structure System* (Schmidt et al. 1993, 1998)]. Input files can be written from the **save** command of PCMODEL just as GAUSSIAN input files are. Input files are copied to the working filename INPUT before a run, and output files are designated filename.out. We ran our programs from a separate GAMESS directory. GAMESS is written for professionals, so it is not quite as user friendly as the commercial program GAUSSIAN, but it is by no means beyond the student level.

Exercise 10-9

Run a single point STO-3G calculation of the total energy of H_2O at the MM3 geometry in the GAMESS implementation. Compare your result with the identical calculation in the GAUSSIAN implementation. Repeat the calculation using the double zeta valence (DZV) and triple zeta valence (TZV) basis sets in the GAMESS implementations. Comment on the relative energies calculated by single, double, and triple zeta basis sets.

Solution 10-9

Using PCMODEL, draw H_2O. Minimize using the MM3 force field. Save to the filename water.inp (or some such) in the GAMESS format. Copy to your GAMESS directory. Copy to filename INPUT and be sure that PUNCH has been renamed to PUNCH.OLD or has been erased entirely. Run using GAMESS.EXE > FILENAME.OUT. The INPUT file

```
$CONTRL SCFTYP = RHF COORD = CART $END
     $BASIS GBASIS = STO NGAUSS = 3 $END
$DATA
water
     Cn 1
H    1.0      -1.012237    0.210253    0.097259
O    8.0      -0.260862    0.786229    0.119544
H    1.0       0.489699    0.209212    0.142294
     $END
```

leads to

FINAL ENERGY IS −74.9610273642 AFTER 13 ITERATIONS

Repeat in the GAUSSIAN implementation. The .gif file

```
# sto-3g

wa

0 1
H    1.0    −1.012237    0.210253    0.097259
O    8.0    −0.260862    0.786229    0.119544
H    1.0     0.489699    0.209212    0.142294
```

leads to

7 basis functions   21 primitive gaussians

and

SCF Done: E(RHF) = −74.9610207086   A.U. after
                                          4 cycles

Double zeta valence or triple zeta valence calculations can be carried out by putting DZV or TZV in place of STO NGAUSS = 3 in the second line of the INPUT file in the GAMESS implementation. The calculated energies become progressively lower (better) for double and triple zeta basis sets

FINAL ENERGY IS −76.00923
FINAL ENERGY IS −76.02007

but they approach a limit.

COMPUTER PROJECT 10-3 | The Bonding Energy Curve of H_2: GAMESS

Plot the curve of the bond energy of H_2 vs. internuclear distance for the H_2 molecule using the STO-3G, double zeta valence (DZV), and triple zeta valence (TZV) basis sets in the GAMESS implementation.

Procedure. Carry out the STO-3G single point calculations on H_2 in a way similar to that of Exercise 10-9 at interatomic distances of 0.4 to 1.2 Å at intervals of 0.1 Å. This is best done in the z-matrix format, for example,

```
$CONTRL SCFTYP = RHF COORD = ZMT $END
$BASIS GBASIS = STO NGAUSS = 3 $END
$DATA
```

```
hydrogen
Cn 2

H     1.0
H     1.0        1.0
$END
```

for a bond distance of 1.0 Å, where z-matrix format is signified in the input as COORD = ZMT. This input file results in

```
TOTAL ENERGY = -1.0661086701
```

as part of the output data block. Double and triple zeta basis sets are obtained by replacing STO NGAUSS = 3 with DZV or TZV.

The Thermodynamic Functions

The output of a G3(MP2) calculation for propene, for example,

```
G3MP2(0 K) = 117.672791
```

which we have called E^0[G3(MP2)], is the energy of the molecule in the ground state and in the gas phase at 0 K relative to isolated nuclei and electrons. From a theoretical point of view, the problem is solved, but for practical purposes, we would like to have the enthalpy of formation in the standard state $\Delta_f H$ and the Gibbs free energy of formation in the standard state $\Delta_f G$ at some other temperature, most importantly, 298 K.

First, we would like to change the reference state from the isolated nuclei and electrons to the elements in their standard states, C(graphite) and $H_2(g)$ at 298 K. This leads to the energy of formation at 0 K $\Delta_f E_0$, which is identical to the enthalpy of formation $\Delta_f H_0$ at 0 K. The energy and enthalpy are identical only at 0 K. Next we would like to know the enthalpy change on heating propene from 0 to 298 K so as to obtain the enthalpy of formation from the isolated nuclei and electrons elements H^{298}. This we will convert to $\Delta_f H^{298}$ from the elements in their standard states at 298 K. From that, with the absolute entropy S of propene at 298, we arrive at the standard Gibbs free energy of formation $\Delta_f G^{298}$, which leads to the equilibrium constant at 298 K for reactions involving propene. Classical thermodynamic equations permit these conversions to be carried out in a straightforward way, but because the heat capacities C_P and absolute entropies S are usually not known over a wide range of temperatures, the power of statistical thermodynamics is also brought to bear.

The entire procedure can be carried out in steps. We find the ground-state energy of formation of propene at 0 K from C and H atoms in the gaseous state

$$C(g) + 3H(g) \rightarrow C_3H_6(g)$$

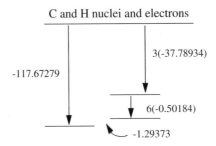

Figure 10-4 The Energy of Formation of $C_3H_6(g)$ from 3 C(g) and 6 H(g).

by subtracting the energy of formation of C(g) and H(g) in the ground state at 0 K relative to their constituent isolated nuclei and electrons, from the energy of formation of propene(g) in the ground state relative to its constituent isolated nuclei and electrons (Fig. 10-4).

The result of this calculation is -1.29373 hartrees $= -811.82851 \, \text{kcal mol}^{-1}$. To this we add the energy of formation of C(g) and H(g) from the elements in the standard state, C(graphite) and $H_2(g)$ (Fig. 10-5).

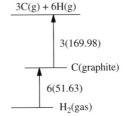

Figure 10-5 Formation of Gaseous Atoms from Elements in the Standard State.

We now know the energy of the propene thermodynamic state {propene(g)} relative to the state {3 C(g) and 6 H(g)} and the energy of the thermodynamic standard state of the elements relative to the same state {3 C(g) and 6 H(g)}, which is opposite in sign to the summed energies of formation of 3 C(g) and 6 H(g). The energy difference between these thermodynamic states is

$$-811.82851 + 509.94 + 309.78 = 7.89 \text{ kcal mol}^{-1}$$

which is the energy of formation at 0 K $\Delta_f E_0$ of propene(g) (Fig. 10-6).

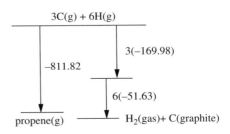

Figure 10-6 Energies Leading to the Energy of Formation of Propene (g).

The same series of calculations starting with

```
G3MP2 Enthalpy = -117.667683
```

lead to $\Delta_f H° = \Delta_f H^{298}$, except that the enthalpies of formation of gaseous atoms are 169.73 kcal mol^{-1} from C(graphite) and 101.14 kcal mol^{-1} for H$_2$(gas) at 298.15 K. These give

$$\Delta_f H^{298}(\text{propene}(g)) = 4.29 = \text{kcal mol}^{-1}$$

as compared to the experimental value of 4.78 ± 0.19 kcal mol^{-1}.

The remaining question is how we got from `G3MP2(0 K) = -117.672791` to `G3MP2 Enthalpy = -117.667683`. This is not a textbook of classical thermodynamics (see Klotz and Rosenberg, 2000) or statistical thermodynamics (see McQuarrie, 1997 or Maczek, 1998), so we shall use a few equations from these fields opportunistically, without explanation. The definition of heat capacity of an ideal gas

$$C_V = \left(\frac{\partial E}{\partial T}\right)_V$$

leads to

$$\int dE = \int C_V \, dT$$

and

$$E^{298} = E_0 + \int_0^{298} C_V \, dT$$

To evaluate this integral, we must know C_V as a function of temperature, and usually this is not known.

Statistical thermodynamics tells us that C_V is made up of four parts, translational, rotational, vibrational, and electronic. Generally, the last part is zero over the range 0 to 298 K and the first two parts sum to 5/2 R, where R is the gas constant. This leaves us only the vibrational part to worry about. The vibrational contribution to the heat capacity is

$$E_{\text{vib}} = RT \frac{x}{e^x - 1}$$

where

$$x = \frac{\Delta E_{\text{vib}}}{k_B T}$$

We don't know ΔE_{vib} but we can approximate it from the vibrational spacing of the bond vibrations in the harmonic oscillator approximation.

When these four (or three) contributions are summed for a molecule such as propene, we have the thermal correction to the energy `G3MP2(0 K)`. The result is `G3MP2 Energy` in the G3(MP2) output block. To this is added *PV*, which is equal to *RT* for an ideal gas, in accordance with the classical definition of the enthalpy

$$H = E + PV = E + RT$$

The sum of the energy correction for heating the molecule from 0 to 298 K plus *RT* is called the *thermal enthalpy correction* (TCH) and yields

```
G3MP2 Enthalpy = −117.667683
```

in the output block of the G3(MP2) calculation

```
G3MP2(0 K) = −117.672791
G3MP2 Energy = −117.668627
G3MP2 Enthalpy = −117.667683
G3MP2 Free Energy = −117.697857
```

One can obtain THC for the G2 or G3 family of calculations by taking `G3MP2 Enthalpy − G3MP2(0 K)`.

From the third law of thermodynamics, the entropy $S = 0$ at 0 K makes it possible to calculate S at any temperature from statistical thermodynamics within the harmonic oscillator approximation (Maczek, 1998). From this, ΔS of formation can be found, leading to $\Delta_f G^{298}$ and the equilibrium constant of any reaction at 298 K for which the algebraic sum of $\Delta_f G^{298}$ for all of the constituents is known. A detailed knowledge of ΔS, which we already have, leads to K_{eq} at any temperature. Variation in pressure on a reacting system can also be handled by classical thermodynamic methods.

One can now see why there is not much difference between computed values of $\Delta_{isom} E_0$ and $\Delta_{isom} H^{298}$ as obtained from the thermodynamic equations in Computer Project 10-2

$$\Delta_{isom} = [\text{G2 (0 K)(propene)}] − [\text{G2 (0 K)(cyclopropane)}]$$
$$\Delta_{isom} H^{298} = [\text{G2 Enthalpy(propene)}] −$$
$$[\text{G2 Enthalpy(cyclopropane)}]$$

The difference between the energy of a molecule at 0 K and its enthalpy at 298 depends on the thermal contribution due to vibration at the two temperatures. If the molecule in question is rigid, with few vibrational degrees of freedom, this contribution will be small, as it is for propene and cyclopropane. For larger molecules with a good deal of vibrational freedom, the difference will be correspondingly larger.

Koopmans's Theorem and Photoelectron Spectra

In studying molecular orbital theory, it is difficult to avoid the question of how "real" orbitals are. Are they "mere" mathematical abstractions? The question of reality in quantum mechanics has a long and contentious history that we shall not pretend to settle here but Koopmans's theorem and photoelectron spectra must certainly be taken into account by anyone who does.

Koopmans proposed that the orbital structure of a cation M^+ ought to be nearly the same as that of the molecule that engenders it, so that the amount of energy necessary to remove electrons from a stable molecule by hitting it with high-energy photons

$$h\nu + M \rightarrow M^+ + e^-$$

ought to be equal and opposite to the energy of the orbitals they come from. (Note that we have been using Koopmans's theorem implicitly in our thermochemical calculations.) If more than enough energy is supplied to a molecule to drive electrons from one or more molecular orbitals, different excess energies E_{excess} should be imparted to them according to the binding energy they had in their orbitals

$$E_{\text{in}} - E_{\text{orb}} = E_{\text{excess}}$$

The excess energies can be measured for a known E_{in} by essentially a stopping potential method, giving a spectrum. This spectrum is then matched with calculated orbital energies (eigenvalues) derived from molecular orbital calculations.

Exercise 10-10

The measured energy spectrum of ethylene is shown in Fig. 10-7.

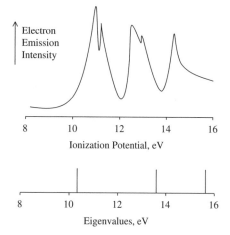

Figure 10-7 Photoelectron Spectrum of Ethylene. Energies of the highest three eigenvalues, converted to eV, are shown below the spectrum.

The occupied eigenvalues of ethylene according to a 6-31G calculation are

```
Alpha  occ.  eigenvalues --   -11.23781 -11.23620 -1.03132
                              -0.77980 -0.63720
Alpha occ. eigenvalues --   -0.57707 -0.49601 -0.37194
```

Match the energies of the three highest orbitals with the peaks in Fig. 10-7.

Solution 10-10

The highest eigenvalues have the smallest (negative) energies in the third line of occupied eigenvalues. Converted to electron volts (conversion factor 27.21, with a change in sign), they are 15.7, 13.5, 10.1 eV, respectively. Quantitatively, the match isn't as good as we might wish. Nevertheless, we have sound evidence of three molecular orbitals with energies in the vicinity of the three highest 6-31G eigenfunctions. Remember that the orbital structure of the cation is not really the same as that of the neutral molecule; that is an approximation.

Larger Molecules I: Isodesmic Reactions

Granting that absolute energy calculations may be very accurate by G2 and G3 methods, they are also very demanding of computer resources. Long ago Warren Hehre (1970) suggested a method for determining relative energies by using lower-level molecular orbital calculations in such a way that the error cancels across an *isodesmic reaction*. An isodesmic reaction is a reaction in which the number of bonds and bond types are the same on either side of the reaction but their arrangement is different. This permits determination of the enthalpy of reaction and thus the enthalpy of formation of one component of the reaction provided that all the others are known.

Because the $\Delta_f H^{298}$ values of many small molecules are known (Pedley, 1986) to within 0.1 or 0.2 kcal mol^{-1}, one need not start with atoms in the hypothetical formation reaction as in the sections on energies of atomization and ionization and the thermodynamic function above. One can build up the target molecule from smaller *molecules* rather than from atoms. Suppose again, for illustrative purposes, that we make believe we don't know $\Delta_f H^{298}$ of propene but we do know $\Delta_f H^{298}$ of the simpler C2 hydrocarbons ethene and ethane along with $\Delta_f H^{298}$ of methane. An isodesmic reaction containing these enthalpies is

$$CH_2=CH_2(g) + CH_3-CH_3(g) \rightarrow CH_3CH=CH_2(g) + CH_4(g) \qquad (10\text{-}20)$$

Note that, for thermochemical purposes, there is no requirement that we can actually carry out the reaction. Systematic computational errors will, in some measure, cancel between the right and left sides of isodesmic reactions (10-20), giving an estimate of the $\Delta_f H^{298}$. GAMESS calculations at the STO-3G level lead to total energies of

$$-77.073955 + (-78.306180) \rightarrow -115.660299 + (-39.726864)$$

obtained from the ENERGY COMPONENTS section of the output file. The calculated enthalpy of reaction is

$$\Delta H^{298}(\text{reaction}) = \sum v_p \Delta_f H^{298}(\text{products}) - \sum v_r \Delta_f H^{298}(\text{reactants})$$
$$= 7.028 \text{ millihartrees(mh)} = 4.41 \text{ kcal mol}^{-1}. \qquad (10\text{-}21)$$

where v_p and v_r are the appropriate stoichiometric coefficients. The calculated $\Delta_r H^{298}$ above and the *experimental* $\Delta_f H^{298}$ for $CH_4(g)$, $CH_2{=}CH_2(g)$, and $CH_3{-}CH_3(g)$, which are -17.90, 12.54, and -20.08 kcal mol$^{-1,}$ respectively (Pedley, 1986), leave only the unknown $\Delta_f H^{298}$ (propene) in the equation

$$4.41 = \Delta_f H^{298}(\text{propene}) - 17.90 - (12.54 - 20.08)$$
$$\Delta_f H^{298}(\text{propene}) = 5.95 \text{ kcal mol}^{-1}$$

This compares with the G3(MP2) value of $\Delta_f H^{298}$ (propene) $= 4.29$ kcal mol^{-1} and the corresponding experimental value of $\Delta_f H^{298}$ (propene) $= 4.78 \pm 0.19$ kcal mol^{-1}.

Exercise 10-11

Carry out a calculation of $\Delta_f H^{298}$ (propene) at the 6-31G MP2 level of theory in the GAMESS implementation.

Solution 10-11

The required TOTAL ENERGY entries are

$$-117.29857 - 40.27913 - (-78.18420 + (-79.38560))$$
$$= -0.00790 \text{ h} = -4.96 \text{ kcal mol}^{-1}$$
$$\Delta_f H^{298}(\text{propene}) = 4.96 - 10.36 = -5.40 \text{ kcal mol}^{-1}$$

as contrasted to the G3(MP2) value of $\Delta_f H^{298}$ (propene) $= 4.29$ kcal mol^{-1} and the experimental value of $\Delta_f H^{298}$ (propene) $= 4.78 \pm 0.19$ kcal mol^{-1}.

A second issue that arises in relation to isodesmic reaction enthalpies is why they should exist at all. If all we are doing is rearranging bonds, shouldn't the summed bond energies be the same on either side of the reaction? Not really. A negative 6-31G MP2 enthalpy of 5 kcal mol^{-1} for the reaction

$$CH_2{=}CH_2(g) + CH_3{-}CH_3(g) \rightarrow CH_3CH{=}CH_2(g) + CH_4(g)$$

tells us that the thermodynamic state on the right containing propene is more stable than the thermodynamic state consisting of the simpler molecules on the left. We recognize this enthalpy difference as the "hyperconjugation" stabilization that a methyl group exerts on an α double bond, and we find that it is about 5 kcal mol^{-1} in the 6-31G MP2 model chemistry. At this level of accuracy, distinctions among the terms energy, enthalpy, and free energy are usually not made and they are used as though they were synonymous. Other stabilization or destabilization enthalpies are reflected in isodesmic reactions, for example, the enthalpy $\Delta_r H^{298}$ of the isodesmic reaction

$$(10\text{-}22)$$

$$3(-234.18629) \rightarrow -231.82430 + 2(-235.39571)$$
$$\Delta_r H^{298} = -0.05685\,h = -35.7 \text{ kcal mol}^{-1}$$

is a measure of the "resonance energy" of benzene.

COMPUTER PROJECT 10-4 | *Dewar Benzene*

In the mid-nineteenth century, the empirical formula of benzene, C_6H_6, was known but its structural formula was not. Two proposed structures

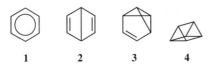

are called Kekule benzene (cyclohexatriene) and Dewar benzene after the chemists who proposed them. Neither formula is in accordance with the relative stability of actual benzene, which is given in formula **1** below

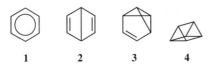

Within the last decade or so, these three remarkable isomers of benzene (**2–4**) have been synthesized (with considerable difficulty). The purpose of this computer project is to obtain the energies, enthalpies, or Gibbs free energies of compounds (**1–4**) and rank them according to energy on a vertical scale with the highest at the top.

Procedure

A. Obtain the energies of benzene (**1**), Dewar benzene (**2**), benzvalene (**3**), and prismane (**4**), all of which have the empirical formula C_6H_6, in either the GAUSSIAN or GAMESS implementation and at a level of theory [6-31G(d), etc.] of your choosing. Your choice of implementation and level will likely be dictated by the power of the computer system you have. Construct a graph showing the energies of the four isomers on a vertical scale. Comment on the graph you obtain (see Li et al., 1999).

B. Dewar benzene (**2**) exists as *cis* and *trans* isomers. Draw structures of the two forms, construct the appropriate input files and determine the *cis-trans* isomerization energy of (**2**).

3,3′-Bicyclopropenyl

is also an isomer of benzene. Obtain the energy of 3,3′-bicyclopropenyl, locate the corresponding point on the energy diagram from Part A, and use this result to speculate on the origin of the strain energy evident in prismane.

Larger Molecules II: Density Functional Theory

A functional is a function of a function. Electron probability density ρ is a function $\rho(\mathbf{r})$ of a point in space located by radius vector $\mathbf{r}$ measured from an origin (possibly an atomic nucleus), and the energy E of an electron distribution is a function of its probability density, $E = f(\rho)$. Therefore E is a functional of $\mathbf{r}$ denoted $E = [\rho(\mathbf{r})]$.

The first Hohenberg–Kohn theorem states that, for a nondegenerate ground state, there is a one-to-one mapping among ρ, V, and ψ_0

$$\rho(\mathbf{r}) \leftrightarrow V(\mathbf{r}) \leftrightarrow \psi_0 \tag{10-23}$$

where V is the potential energy, and ψ_0 is the wave function at a given potential, that is, ψ_0 is a functional of V and of ρ

$$\psi_0 = \psi_0[V] = \psi_0[\rho] \tag{10-24}$$

All properties, in particular the energy, are functionals of $\mathbf{r}$ because

$$E[\mathbf{r}] = \int_{-\infty}^{\infty} \psi_0[\mathbf{r}] E \psi_0[\mathbf{r}] d\tau \tag{10-25}$$

Density Functional Methods. The Kohn–Sham equations are

$$\mathbf{K}\psi_i = E_i\psi_i \tag{10-26}$$

where $\mathbf{K}$ is an operator

$$\mathbf{K} = \{-\nabla_i^2 - \sum Z_I/r_{Ii} + \int \rho(r_2)/r_{12}dr_2 + V^{XC}(\mathbf{r}_1)\} \tag{10-27}$$

analogous to the Fock operator in Hartree–Fock theory [Eqs. (9-1)–(9-10)] for electron 1 in the vicinity of electron 2 where Z_I is the nuclear charge. In the **K** operator as written above, the exchange part of the Hartree–Fock operator is conspicuous by its absence and a new term $V^{XC}(\mathbf{r}_1)$ appears in its place. The Kohn–Sham equations are one-electron equations and ψ_i is a one-electron space orbital such that

$$\rho(\mathbf{r}) = \sum |\psi_i|^2 \tag{10-28}$$

The first three terms in Eq. (10-26), the electron kinetic energy, the nucleus-electron Coulombic attraction, and the repulsion term between charge distributions at points $\mathbf{r}_1$ and $\mathbf{r}_2$, are classical terms. All of the quantum effects are included in the exchange-correlation potential V^{XC}

$$V^{XC} = \frac{\delta E^{XC}}{\delta \rho} \tag{10-29}$$

a *functional* derivative (Atkins and Friedman, 1997). The sum of the three classical energies in Eq. (10-26) plus the exchange-correlation energy E^{XC} is the total energy.

E^{XC} can be treated as the sum of two parts, the exchange energy and the correlation energy, $E^{XC} = E^X + E^C$. Each of the parts can be treated under the *local density approximation* or with *gradient* functionals.

(1) One approach, using a local density approximation for each part, has $E^{XC} = E_S + E_{VWN}$, where E_S is a Slater functional and E_{VWN} is a correlation functional from Vosko, Wilk, and Nusair (1980). Both functionals in this treatment assume a *homogeneous* electron density. The result is unsatisfactory, leading to errors of more than 50 kcal mol^{-1} for simple hydrocarbons.

(2) Gradient functionals do not assume constant charge (electron) density, but treat variation of charge density in space. Combining two gradient functionals as in the BLYP approximation, $E^{XC} = E_B + E_{LYP}$ where E_B is from Becke (1988) and E_{LYP} is the Lee, Yang, and Parr (1988) functional, brings about a dramatic improvement in agreement with experiment, reducing the average difference between calculated and experimental values to less than 3 kcal mol^{-1} for the test compounds acetylene, ethylene, and ethane.

The notation B3LYP denotes a *3-parameter empirical* functional that expresses two parts of the exchange-correlation energy $E^{XC} = E^X + E^C$, the first part being local and the second part a gradient approximation (Foresman and Frisch, 1996; Baerends and Gritsenko, 1997). The first part is further broken down into a local density approximation to the exchange energy

$$E_{LDA}^X = -\tfrac{3}{2}(3/4\pi) \int \rho^{4/3} d\tau \tag{10-30}$$

plus a term that corrects the difference between the Hartree–Fock exchange energy and the local density approximation using an adjustable parameter c_0 multiplied into the difference between the Hartree–Fock exchange energy and E_{LDA}^X. This product enters into the exchange-correlation energy as

$$E^X = E_{LDA}^X + c_0(E_{HF}^X - E_{LDA}^X) + c_X \Delta E_{B88}^X \qquad (10\text{-}31)$$

where

$$\Delta E_{B88}^X = E_{LDA}^X - \gamma \int \frac{\rho^{4/3}x}{(1 + 6\gamma \sinh^{-1} x)} d\tau \qquad x = \rho^{-\frac{4}{3}}|\nabla \rho| \qquad (10\text{-}32)$$

and ΔE_{B88}^X is a gradient correction from Becke (1988).

A similar thing is done with the second part of the B3LYP hybrid, which is also comprised of two terms

$$E^C = E_{VWN3}^C + c_c(E_{LYP}^C - E_{VWN3}^C) \qquad (10\text{-}33)$$

the local density approximation to E_{VWN3}^C due to Vosko, Wilk, and Nussair (1980) corrected by the Lee, Yang, and Parr term E_{LYP}^C, which enters as the correction $E_{LYP}^C - E_{VWN3}^C$ premultiplied by an adjustable parameter c_c. B3LYP is arguably the best estimate of E^{XC} in current use; it produces agreement with experiment that is within 1.3 kcal mol^{-1} for the three simple test hydrocarbons methane, acetylene, ethylene and ethane.

In hybrid DFT-Gaussian methods, a Gaussian basis set is used to obtain the best approximation to the three classical or one-electron parts of the Schroedinger equation for molecules and DFT is used to calculate the electron correlation. The Gaussian parts of the calculation are carried out at the restricted Hartree–Fock level, for example 6-31G or 6-311G(3d,2p), and the DFT part of the calculation is by the B3LYP approximation. Numerous other hybrid methods are currently in use.

The most obvious practical difference between density functional theory (DFT) calculations and the G-*n* family calculations is that DFT calculations are *single-point, single-electron* calculations whereas each of the G-*n* family of calculations consists of a suite of calculations, each utilizing a Gaussian basis set and a post Hartree–Fock extension to arrive at the total energy (E_0) of a molecule. Because they are single-point calculations, we might expect that *other things being equal*, DFT calculations will be less demanding of computer resources than the G-*n* family of calculations. Indeed they are, but the saving is not as great as one might expect because integrals (10-30) and (10-32) and others like them (Foresman and Frisch, 1996) cannot be solved to give a simple form. They are solved numerically over a closely spaced grid in 3-space, a method that can be time-consuming. Moreover, there are ancillary calculations to be carried out as described in the procedure section of Computer Project 10-5.

If the B3LYP run time for calculating E_0 of H_2O is arbitrarily taken as 1, O_2 (triplet) and CO_2, run times scale as 1.5 and 2.5 for the B3LYP calculational procedure and 2, 7.5, and 15 for the same three molecules calculated by G2 (Pan et al., 1999). Taking the B3LYP run time as 1 for methane, the run times for methane, ethane, propane, and cyclobutane scale roughly as 1, 14, 54, and 137 for B3LYP calculations and 13, 163, 929, and 2351 for G2. These ratios are somewhat less favorable to DFT if the geometry minimization time is counted in, but they become more favorable to DFT calculations on larger molecules.

COMPUTER PROJECT 10-5 | Cubane

Cubane, a hypothetical molecular curiosity for many years, has been synthesized and is receiving attention because it is a highly energetic molecule, storing angular strain energy in its distorted sp^3 bonds. In principle, at least, the strain energy can be released in a stepwise fashion by adding hydrogen across edges of the cube, one edge at a time until the strain-free molecule 3,4-dimethylhexane is reached. If you have access to a power system, determine the enthalpy change of the reaction sequence in Fig. 10-8 by the G3(MP2) method, thereby estimating the strain energy of cubane. Most of the structures represent molecules that have not been isolated, but two experimental checkpoints do exist, a value of $\Delta_f H^{298} = 148.7 \pm 2.0$ kcal mol^{-1} for cubane and -50.7 ± 0.2 kcal mol^{-1} for 3,4-dimethylhexane, the end product of this sequential hydrogenation.

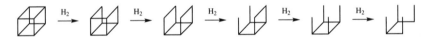

Figure 10-8   Sequential (Hypothetical) Hydrogenation of Cubane to 3,4-Dimethylhexane.

Procedure.   Start with an optimized geometry using, for example, the MM3 minimization of PCMODEL. The default keyword b3lyp in the GAUSSIAN implementation will result in a rapid but inaccurate STO-3G calculation. Despite the inadequacy of STO-3G calculations on an absolute basis, they show trends and are useful for determining enthalpies of isomerization or hydrogenation, both of which are isodesmic.

For $\Delta_f H^{298}$ calculations from B3LYP theory, one must correct for zero point energies and make a thermal correction for the enthalpy change from 0 to 298 K. These ancillary corrections can be found from the **-Thermochemistry-** section by using the freq keyword in the appropriate model chemistry. Basis sets given in Computer Project 10.6.1 can be combined to form a compound keyword. The compound keyword 6-31G b3lyp is recommended for this project.

PROBLEMS

1. Write a program in BASIC to calculate the dissociation energy of H_2^+. This can be done by filling in an appropriate data block using one or more DATA statements.
2. As an interesting variation on this experiment, one can try reading in the experimental data from an external file. The student should do some outside

reading on advanced BASIC and should include a discussion of external file handling with this laboratory report.

3. The molecule HgH has vibrational lines at 1204, 966, 632, and 172 cm^{-1}. Construct the Birge–Spooner plot for this molecule and find its dissociation energy D_0 and bond energy D_e.

4. The first five vibrational energy levels of HCL are at 1482, 4367, 7149, 9827, and 12 400 cm^{-1}. Find the dissociation energy and bond energy of HCl.

5. Diatomic molecules, which are anharmonic oscillators, produce vibrational spectra that not only decrease in energy for the higher transitions but decrease in intensity as well, so that the principal line is for the transition from the ground state to the first excited state. Using the G2 calculated bond strength for H_2, predict the wavelength of the predominant line in the vibrational spectrum of H_2.

6. Sketch the hydrogen molecule system (2 protons and 2 electrons) and verify the Hamiltonian 10.3.1.

7. Carry out a series of calculations comparable to those in Computer Project 10-1 on the hydrogen molecule. Estimate the correlation energy from the GAUSSIAN calculations.

8. Write a program in BASIC to calculate $\Delta_f E^{298}$ from the output of G3MP2.

9. Write a program in BASIC to calculate $\Delta_f H^{298}$ from the output of G3MP2.

10. Combine the answers to Problems 8 and 9 to calculate both $\Delta_f E^0$ and $\Delta_f H^{298}$.

11. Repeat the calculation in Exercise 10-7 using the G3 method in the GAUSSIAN implementation. What is the % difference between G2 (0 K) and G3 (0 K)?

12. Increase the dimension of a one-dimensional box containing an electron from a $= 1.0$ Å to a $= 1.1$ Å (from 1.9 bohr to 2.1 bohr). What happens to the energy of the system? What is the % change?

13. Repeat the analysis of the G2 calculation in the section on G2 and G3 in this chapter for the acetylene molecule.

14. Calculate E_0[G2] for the methyl cation CH_4^+. Check your result against the value used in the section on energies of atomization and ionization in this chapter.

15. What is the energy of atomization of H_2^+ in the STO-3G approximation? Carry out the calculation in the GAUSSIAN implementation.

16. What is the energy of atomization of H_2^+ in the STO-3G approximation? Carry out the calculation in the GAMESS implementation.

17. What is the energy of atomization of methane in the STO-3G approximation? Carry out the calculation in both the GAUSSIAN and GAMESS implementations.

18. Calculate the G2 value of E_0 for H(g) and C(g) for use in the section on thermodynamic functions in this chapter.

19. Run the GAMESS input file for Exercise 10-9 using the commands GAMESS. EXE > FILENAME.OUT. Erase PUNCH and run the same input file using gamess > fi or gamess > fi. Does it run? Try several other combinations of upper and lower case letters in the run command. Try leaving out the space before >.

20. Water has a photoelectron spectrum with peaks at 539.7 32.2, 18.5, 14.7, and 12.6 eV. Using the method of Exercise 10-10, match the Hartree–Fock energies of H_2O calculated at the 6-31G level in the GAMESS implementation. Is the fit better than it is in Exercise 10-10? Why is one peak so far from the others?

21. Use the experimental values of the enthalpies given in the section on isodesmic reactions along with the isodesmic reaction

$$2CH_3-CH_3(g) \rightarrow CH_3CH_2CH_3(g) + CH_4(g)$$

to determine the $\Delta_f H^{298}$ of propane(g). The experimental value is -25.02 ± 0.12 kcal mol^{-1}

22. Dopamine (DOPA)

is one of a group of psychoactive substances that includes adrenaline. The electronic structure of this molecule promises to be complicated because it has two electronegative oxygens and an electronegative nitrogen interacting across a benzene ring with mobile electrons. Draw an electron map of DOPA showing regions of relative negative charge (use color if you like). Comment on which part of the molecule is likely to interact with the brain, causing psychoactivity. Write a short essay on the psychoactive properties of DOPA and mechanisms proposed for it from your outside reading.

23. What is the average energy release per bond on breaking bonds in cubane? Compare this with the energy released on hydrogenation of ethylene.

24. Butyric acid,

is found in rancid butter, stale sweat, and organic chemistry laboratories. Plot the energy of acetic, propanoic, and butyric acids calculated at the 6-31G MP2 level in the GAMESS implementation and find the equation of the curve you obtain.

25. Based on the equation found in Problem 23, estimate the total energy of n-pentanoic acid by extrapolation to 5 carbon atoms. Carry out the calculation at the 6-31G MP2 level in the GAMESS implementation and determine the % difference between the GAMESS calculation and the extrapolated estimate.

Bibliography

Akimoto, H.; Sprung, S. L.; Pitts, J. N., 1972. *J. Am. Chem. Soc.* **94**, 4850.

Alberty, R. A.; Silbey, R. J., 1996. *Physical Chemistry*, 2nd. ed. Wiley, New York.

Allinger, N. L., 1976. *Adv. Phys. Org. Chem.* **13**, 1.

Allinger, N. L.; Chen, K.; Lii, J-H., 1996. *J. Comp. Chem.* **17**, 642.

Allinger, N. L.; Dodziuk, H.; Rogers, D. W.; Naik, S. N., 1982. *Tetrahedron* **38**, 1593.

Allinger, N. L.; Lii, J.-H., 1989. *J. Am. Chem. Soc.* **111**, 8566.

Allinger, N. L.; Yuh, Y. H.; Lii, J.-H., 1989. *J. Am. Chem. Soc.* **111**, 8551.

Anderson, J. M., 1966. *Mathematics for Quantum Chemistry.* Benjamin, New York.

Atkins, P. W., 1998. *Physical Chemistry*, 6th ed. W. H. Freeman, New York.

Atkins, P. W.; Friedman, R. S., 1997. *Molecular Quantum Mechanics*, 3rd ed. Oxford Univ. Press, Oxford.

Balam, L. N., 1972. *Fundamentals of Biometry.* Halsted Press (Wiley), New York.

Barrante, J. R., 1998. *Applied Mathematics for Physical Chemistry*, 2nd ed. Prentice-Hall, Englewood Cliffs, NJ.

Barrow, G. M., 1996. *Physical Chemistry*, 6th ed. WCB/McGraw-Hill, New York.

Becke, A. D., 1988. *Phys. Rev. A* **38**, 3098.

Baerends, E. J.; Gritsenko, O. V., 1997. *J. Phys. Chem.* **101**, 5383.

Benson, S. W., 1976. *Thermochemical Kinetics*, 2nd ed. Wiley, New York.

Computational Chemistry Using the PC, Third Edition, by Donald W. Rogers
ISBN 0-471-42800-0   Copyright © 2003 John Wiley & Sons, Inc.

Benson, S. W.; Cohen, N., 1998. In *Computational Thermochemistry*, lrikura, K. K.; Frurip, D. J. eds., ACS Symposium Series 677; American Chemical Society, Washington, DC.

Brady, J. E.; Holum, J. R., 1993. *Chemistry.* Wiley, New York.

Brown, J. B., 1998. *Molecular Spectroscopy.* Oxford Univ. Press, Oxford.

Burkert, U.; Allinger, N. L., 1982. *Molecular Mechanics* ACS Publ. No. 177, American Chemical Society, Washington, DC.

Carley, A. F.; Morgan, P. H., 1989. *Computational Methods in the Chemical Sciences.* Halsted Press (Wiley) New York.

Carnahan, B., Luther, H. A. and Wilkes, J. O., 1969. *Applied Numerical Methods.* Wiley, New York.

ccl.net Victor Lobanov, 1996, University of Florida http://ccl.net/cca/software/MS-DOS/mopac_for_dos/index.shtml

Chase, M. W. Jr., 1998. *NIST-JANAF Themochemical Tables*, 4th ed., *J. Phys. Chem. Ref. Data, Monograph* **9**, 1.

Chase, M. W. Jr.; Davies, C. A.; Downey, J. R. Jr.; Frurip, D. J.; McDonald, R. A.; Syverud, A. N., 1985. *J. Phys Chem. Ref. Data*, 14 Suppl. 1.

Chatterjee, S. and Price, B., 1977. *Regression Analysis by Example.* Wiley, New York.

Chesnut, D. B.; Davis, K. M., 1996. *J. Comput. Chem.* **18**, 584.

Chirlian, P. M., 1981. Microsoft FORTRAN. Dilithium Press, Beaverton, OR.

Christian, S. D.; Lane, E. H.; Garland, F., 1974. *J. Chem. Ed.* **51**, 475.

Christian, S. D.; Tucker, E. E., 1982. *American Laboratory* **14**(9) 31.

Conant, J. B.; Kistiakowsky, G. B., 1937. *Chem Rev.* **29**, 181.

Cox, J. D. and Pilcher, G., 1970. *Thermochemistry of Organic and Organometallic Compounds.* Academic Press, London.

Curtiss, L. A.; Redfern, P. C.; Raghavachari, K. Rassolov, V.; Pople, J. A., 1999. *J. Chem. Phys.* **110**, 4703.

Dence, J. B., 1975. *Mathematical Techniques in Chemistry*, Wiley, New York.

Dewar, M. J. S., 1969. *The Molecular Orbital Theory of Organic Chemistry*, McGraw-Hill, New York.

Dewar, M. J. S., 1975. *Science* **187**, 1037.

Dewar, M. J. S.; Thiel, W., 1977. *J. Am. Chem. Soc.* **99**, 4899.

Dickson, T. R., 1968. *The Computer and Chemistry.* Freeman, San Francisco.

Dirac, P. A. M. *Proc. R. Soc. Lond.* 1929. **A123**, 714.

Ebbing, D. D.; Gammon, S. D., 1999. *General Chemistry.* Houghton Mifflin, Boston.

Ebert, K.; Ederer, H.; Isenhour, T. L., 1989. *Computer Applications in Chemistry.* VCH Publishers, New York.

Eğe, S., 1994. *Organic Chemistry.* 3rd ed. Heath, Lexington, MA.

Eisenberg D. S.; Crothers, D. M., 1979. *Physical Chemistry with Applications to the Life Sciences.* Benjamin/Cummings, Menlo Park, CA.

Emsley, J., *J. Chem. Soc.*, 1971. (London) Section A, 2702.

Ewing, G. W., 1985. *Instrumental Methods of Chemical Analysis*. McGraw-Hill, New York.

Fock, V. A., 1930. *Z. f. Physik.* **61**, 126.

Foresman, J. B.; Frisch A., 1996. *Exploring Chemistry with Electronic Structure Methods* 2nd ed. Gaussian Inc. Pittsburgh, PA.

Gerhold, G., et al., 1972. *Am. J. Phys.* **40**, 918.

Greenwood, H. H., 1972. *Computing Methods in Quantum Organic Chemistry.* Wiley Interscience, New York.

Grivet, J-P., 2002. *J. Chem. Ed.* **79**, 127.

Halgren, T. A.; Nachbar, R. B., 1996. *Merc Molecular Force Field I–V, Computational Chemistry* **17**, 490.

Hanna, M. W., 1981. *Quantum Mechanics in Chemistry.* 3rd ed. Benjamin, Menlo Park, CA.

Hehre, W. J.; Radom, L.; v. R. Schleyer, P.; Pople, J. A., 1986. *Ab Initio Molecular Orbital Theory.* Wiley, New York.

Heitler, W.; London, F., 1927. *Z. f. Phys.* **70**, 455.

van Hemelrijk, D. V.; van den Enden, L. Geise, H. J.; Sellers, H. L., Schafer, L., 1980. *J. Am. Chem. Soc.* **102**, 2189.

Hertzberg, G., 1970. *J. Mol. Spectroscopy* **33**, 147.

Huckel, E., 1931. *Z. f. Physik* **70**, 204.

Huckel, E., 1932. *Z. f. Physik* **76**, 628.

Irikura, K. K. *Essential Statistical Thermodynamics* in *Computational Thermochemistry,* In Irikura, K. K.; Frurip, D. J. Eds., 1998. *Computational Thermochemistry.* American Chemical Society, Washington, DC.

James, H. M.; Cooledge, A. S., 1933. *J. Chem. Phys.* **1**, 825.

Jensen, F., 1999. *Introduction to Computational Chemistry.* Wiley, NY.

Jurs, P. C., 1996. *Computer Software Applications in Chemistry,* 2nd ed. Wiley, New York.

Kapeijn, F.; van der Steen, A. J.; Mol, J. C., 1983. *J. Chem. Thermo.* **15**, 137.

Kar, M.; Lenz, T. G.; Vaughan, J. D., 1994. *J. Comp. Chem.* **15**, 1254.

Kauzmann, W., 1966. *Kinetic Theory of Gases.* Benjamin, New York.

Kistiakowsky, G. B. and Nickle, A. G., 1951. *Disc. Far. Soc.* **10**, 175.

Klotz, I. M.; Rosenberg, R. M. 2000. *Chemical Thermodynamics,* 6th ed. Wiley, New York.

Kreyszig, E., 1988. *Advanced Engineering Mathematics,* 6th ed. Wiley, New York.

Laidler, K. J.; Meiser, J. H. 1999. *Physical Chemistry.* Houghton Mifflin, Boston.

Lee, C; Yang, W.; Parr, R. G., 1988. *Phys Rev B,* **37**, 785.

Levine, I. N. 1991. *Quantum Chemistry,* 4th ed. Prentice Hall, New York.

Lewis, G. N.; Randall, M.; Pitzer, K. S.; Brewer, L., 1961. *Thermodynamics,* 2nd ed. McGraw-Hill, New York.

Li, Z.; Rogers, D. W.; McLafferty, F. J.; Mandziuk, M.; Podosenin, A. V., 1999. *J. Phys. Chem.* **103**, 426.

Lii, J.-H.; Allinger, N. L., 1989. *J. Am. Chem. Soc.* **111**, 8576.

Lowe, J. P. 1993. *Quantum Chemistry*, 2nd ed. Academic Press, San Diego, CA.

MathSoft Engineering & Education, Inc., 101 Main St. Cambridge, MA 02142-1521.

McQuarrie, D. A., 1983. *Quantum Chemistry*. University Science Books, Sausalito, CA.

McQuarrie, D. A.; Simon, J. D., 1999. *Molecular Thermodynamics*. University Science Books, Sausalito, CA.

McWeeny. R., 1979. *Coulson's Valence*. Oxford Univ. Press, Oxford.

Meloan, M. G. and Kiser, R. W., 1963. *Problems and Experiments in Instrumental Analysis*. Merill, Columbus, OH.

Millikan, R. S. et al., 1949. *J. Chem. Phys.* **17**, 1248.

Montgomery, J. A.; Ochterski; J. W.; Petersson, G. A., 1994. *J. Chem. Phys*, **101**, 5900.

Mortimer, R. G., 1999. *Mathematics for Physical Chemistry*, 2nd ed. Academic Press, San Diego, CA. [This book contains an introduction to computer use with brief comments, references and sources to BASIC, Excel, graphics, curve fitting, and Mathematica.]

Murrell, J. M., Kettle, S. F. A. and Tedder, J. M., 1985. *The Chemical Bond*, 2nd ed. Wiley, New York.

Nevins, N.; Chen, K.; Allinger, N. L., 1996. *J. Comp. Chem.* **17**, 669.

Nevins, N.; Lii, J-H.; Allinger, N. L., 1996. *J. Comp. Chem.* **17**, 695.

Norris, A. C., 1981. *Computational Chemistry*. Wiley, New York.

Ochtersky, 2000. *Thermochemistry in Gaussian*; help@gaussian.com

Pan, J.-W.; Rogers, D. W.; Mc Lafferty, F. J., 1999. *J. Molecular Structure (Theochem.)* **468**, 59.

Pariser, R.; Parr, R. G., 1953. *J. Chem. Phys.* **21**, 466, 767.

Parr, R. G.; Yang, W. 1989. *Density Functional Theory of Atoms and Molecules*. Oxford Univ. Press, New York.

Pauling. L., 1960. *The Nature of the Chemical Bond*. Cornell Univ. Press, Ithaca, NY.

Pauling, L. and Wilson, E. B., 1935. *Introduction to Quantum Mechanics*. McGraw-Hill, New York. Reprinted (1963). Dover, New York.

PCMODEL v 8.0, 1993–2002. Serena Software, Box 3076, Bloomington, IN 47402-3076.

Pedley, J. B.; Naylor, R. D.; Kirby, S. P., 1986. *Thermochemical Data of Organic Compounds*, 2nd ed. Chapman and Hall, London.

Pullman, A.; Pullman, B., 1955. *Cancerisation par les Substances Chimiques et Structure Moleculaire*. Masson, Paris.

Pullman, B. *The Modern Theory of Molecular Structure* (Engl. Ed.) 1962. Dover, New York.

Nyden, M. R.; Petersson, G. A., 1981. *J. Chem. Phys.* **75**, 1843.

Petersson, G. A., 2000. *Perspectives on "The activated complex in chemical reactions"* [Eyring, H. (1935) *J. Chem. Phys.* **3**, 107.] *Theor Chem. Acc.* **103**, 190.

Petersson, G. A.; Al-Laham M. A., 1991. *J. Chem. Phys.* **94**, 6081.

Petersson, G. A.; Tensfeldt, T.; Montgomery, J. A., 1991. *J. Chem. Phys.* **94**, 6091.

Pople, J. A., 1953. *Trans. Faraday Soc.* **49**, 1375.

Pople, J. A., et al., 1989. *J. Chem. Phys.* **90**, 5622 and numerous papers in this series.

Rice, J. R., 1983. *Numerical Methods, Software and Analysis.* McGraw-Hill, New York.

Rioux, F., 1987. *Eur. J. Phys.* **8**, 297, See also *Advanced Chemistry Collection* Special Issue 2nd ed. Division of Chemical Education, American Chemical Society, 2001.

Rogers, D. W., 1983. *BASIC Microcomputing and Biostatistics.* Humana, New Jersey.

Rogers, D. W., 1988. *American Laboratory* **20** (10), 122.

Rogers, D. W.; Angelis, B. P.; Mc Lafferty F. J., 1983. *American Laboratory* **15**(11), 54.

Rogers, D. W.; McLafferty, F. J., 2001. *J. Org. Chem.* **66**, 1157

Rosen, N. 1931. *Phys. Rev.* **38**, 2099.

Salem, L., 1966. *The Molecular Orbital Theory of Conjugated Systems.* Benjamin, New York.

Saunders, M., 1987. *J. Amer. Chem. Soc.* **109**, 3150.

Scheid, F., 1968. *Numerical Analysis.* Schaum's, McGraw-Hill, New York.

Schlecht, M. F., 1998. *Molecular Modeling on the PC.* Wiley-VCH, New York.

Schmidt, J. C.; Gordon, M. S. 1998; *Ann. Rev. Phys. Chem.* **49**, 233.

Schmidt, M. W.; Baldridge, K. K.; Boatz, J. A.; Elbert, S. T.; Gordon, M. S.; Jensen, J. H.; Koseki, S.; Matsunaga, N.; Nugyen, K. A.; Su, S.; Windus, T. L.; Dupuis, M.; Montgomery, J. A., 1993. *J. Comp. Chem.* **14**, 1347.

Schroedinger, E., 1926. *Ann. der Phys.* **79**, 361, 489; **80**, 437; **81**, 109.

Schroedinger, E., 1928. *Abhandlungen zur Wellenmechanik.* Barth, Leipzig.

Schroedinger, E., 1928. *Collected Papers on Wave Mechanics.* Blackie & Son, London and Glasgow.

Schwartz, J. T., 1961. *Introduction to Matrices and Vectors.* Dover, New York.

Scott A. P.; Radom, L., 1996. *J. Phys Chem.* **100**, 16502.

SigmaPlot for Windows 5.0 © SPSS Inc., 1986–1999.

Slater, J. C., 1931. *Phys. Rev.* **38**, 1109.

Smith, W. B., 1996. *Introduction to Theoretical Organic Chemistry and Molecular Modeling.* VCH, New York.

Starzak, M. E., 1989. *Mathematical Methods in Chemistry and Physics.* Plenum, New York.

Stewart, J. J. P., 1990. *Computer-Aided Molecular Design* **4**, 1.

Stewart, J. J. P., 1993. MOPAC Manual (7th ed.).

Streitwieser, A., 1961. *Molecular Orbital Theory for Chemists.* Wiley, New York.

Streitwieser, A.; Heathcock, C. H., 1981. *Introduction to Organic Chemistry*, 2nd ed. Macmillan, New York.

Stull, D. R.; Westrum, E. F.; Sinke, G. C., 1969. *The Chemical Thermodynamics of Organic Compounds.* Wiley, New York.

TableCurve, © Jandel Scientific, SYSTAT Software Inc., 510 Canal Boulevard, Suite F, Richmond, CA.

TableCurve Windows v 1.0 User's Manual, 1992. AISN Software, Jandel Sci, San Rafiel, CA.

Thiel, W., 1998. *Thermochemistry from Semiempirical MO Theory,* In Irikura, K. K.; Frurip, D. J. Eds., 1998. *Computational Thermochemistry.* American Chemical Society, Washington, DC.

Turner, R. B.; Garner, R. H., 1957. *J. Amer. Chem. Soc.* **80**, 1424.

Vosko, S. H.; Wilk L.; Nusair, M., 1980. *Can J. Phys.* **58**, 1200.

www.gaussian.com

www.mathcad.com Mathcad 2001.

www.msg.amesiab.gov;GAMESS free ware

www.planaria-software.com Planaria 2002.

www. rahul.net/rhn/basic Chipmunk BASIC homepage; freeware.

www.truebasic.com TrueBasic Inc., PO Box 5428, West Lebanon, NH 03784-5428; free demos

www.Spss.com; go to sigmaplot

www.webbook.nist.gov National Institute of Standards and Technology database.

Wentworth, W. E., 2000. *Physical Chemistry: A Short Course.* Blackwell Science, Malden, MA.

Wheland, G. W., 1955. *Resonance in Organic Chemistry.* Wiley, New York.

Young, D. C., 2001. *Computational Chemistry: A Practical Guide for Applying Techniques to Real World Problems.* Wiley, New York.

Young, H. D., 1962. *Statistical Treatment of Experimental Data.* McGraw-Hill, New York.

Young, H. D.; Friedman, R. A., 2000. *Sears and Zemansky's University Physics*, 10th ed. Addison Wesley Longmans, San Francisco.

Zerner, M. C., 2000. *Theor. Chem. Acc.* **103**, 217.

APPENDIX

A

Software Sources

This appendix is a brief and incomplete introduction to software useful computational chemistry for the PC. The order below is approximately the order in which the programs are used in the text.

QBASIC is available as part of the DOS 6.0 operating system. Though DOS 6.0 has been supplanted by more complicated operating systems, it is still available at a modest price (<$50). Two companies offering DOS 6.0 at this writing are *www.bigclearance.com* and *www.buycheapsoftware.com*. Computer software companies have a tendency to come and go, so a good strategy for locating sources is to consult a recent copy of *PC Magazine* or an equivalent publication. Help in using QBASIC can be found by executing the online DOS Help command. Many sources on the BASIC language exist, for example, Coan, J. S., *Advanced Basic,* Hayden Book Co. Inc., 1977.

True BASIC. Several modestly priced versions (starting <$50) of *True BASIC* are available from its authors at True BASIC Inc., 1523 Maple St., Hartford, VT 05047-0501 (*www.truebasic.com*). True BASIC is transportable to many operating systems, including Unix and Linux. Detailed tutorial manuals are available, including one on numerical methods.

Computational Chemistry Using the PC, Third Edition, by Donald W. Rogers
ISBN 0-471-42800-0.   Copyright © 2003 John Wiley & Sons. Inc.

BASIC programs referred to in the text can be found at the Wiley website (*www.wiley.com*). They were written by the author and may be copied and modified in any way you please.

Mathcad is a product of MathSoft Engineering & Education, Inc., 101 Main Street, Cambridge, MA 02142-1521. Tel: +1-617-444-8000, Fax: +1-617-444-8001 (*www.mathcad.com*).

   This software is available to academic institutions at reduced price from Academic Superstore, Suite A110, 223 W. Anderson Ln., Austin, TX 78752.

TableCurve is available from Systat Software Inc., 501 Canal Boulevard, Suite F, Richmond, CA 94804-2028 (*http://www.systat.com/products/tablecurve2d/*). A free trial version of *TableCurve* is available at this web site. Systat offers other curve-fitting and statistical software.

SigmaPlot is available from SPSS Science, 233 S. Wacker Dr. 11th Floor, Chicago, IL 60606-6307 (*www.sigmaplot.com*). These companies have been "acquisitioned and merged" in the way that big-time business moguls so love to do. You may have to follow a trail to find the current name of the program and company you want.

Excel is part of Microsoft Works, **Microsoft**

PCMODEL v 8.0 1993–2002 is available from Serena Software, Box 3076, Bloomington, IN 47402-3076 (*www.serenasoft.com*).

Molecular Mechanics
Academic and other nonprofit institutions can get MM3 from *qcpe.chem.indiana. edu*. The commercial source is Tripos Inc., 1699 South Hanley Road, St. Louis, MO 63144. For resources on MM3 and MM4 see references 1–3 in Langley, C. H. and Allinger, N. L., *J. Phys. Chem.* **2003**, *107*, 5208–5216.

Tinker (J. W. Ponder, Washington University School of Medicine, St. Louis, MO) is available at *dasher.wustl.edu/tinker.*

Huckel MO ccl.net
go to ccl.net → MS-DOS → Huckel-MO-Calculator → hmo10.zip and unzip
Copyright © 1996 by Ajit J. Thakkar.

Semiempirical
MOPAC (freeware) Victor Lobanov, 1996, University of Florida **ccl.net** (*http://ccl.net/cca/software/MS-DOS/mopac_for_dos/index.shtml*)
go to ccl.net → MS-DOS → mopac_for_dos → mopac_for_dos.zip and unzip
See Stewart, J. J. P., *Computer-Aided Molecular Design* **1990**, *4*, 1.
Arguslab (*www.planaria-software.com*).

Ab Initio

GAUSSIAN Copyright © 1988, 1990, 1992, 1993, 1995, 1998 Gaussian, Inc. All Rights Reserved. (copyright © 1983 Carnegie Mellon University). Gaussian is a federally registered trademark of Gaussian, Inc.

Gaussian 98. Revision A.4, Frisch, M. J.; Trucks, G. W.; Schlegel, H. B.; Scuseria, G. E.; Robb, M. A.; Cheeseman, J. R.; Zakrzewski, V. G.; Montgomery, J. A. Jr.; Stratmann, R. E.; Burant, J. C.; Dapprich, S.; Millam, J. M.; Daniels, A. D.; Kudin, K. N.; Strain, M. C.; Farkas, O.; Tomasi, J.; Barone, V; Cossi, M; Cammi, R.; Mennucci, B.; Pomelli, C.; Adamo, C.; Clifford, S.; Ochterski, J.; Petersson, G. A.; Ayala, P. Y.; Cui, Q.; Morokuma, K.; Malick, D. K.; Rabuck, A. D.; Raghavachari, K.; Foresman, J. B.; Cioslowski, J.; Ortiz, J. V.; Stefanov, B. B.; Liu, G.; Liashenko, A.; Piskorz, P.; Komaromi, I.; Gomperts, R.; Martin, R. L.; Fox, D. J.; Keith, T.; Al-Laham, M. A.; Peng, C. Y.; Nanayakkara, A.; Gonzalez, C.; Challacombe, M.; Gill, P. M. W.; Johnson, B.; Chen, W.; Wong, M. W.; Andres, J. L.; Gonzalez, C.; Head-Gordon, M.; Replogle, E. S.; and Pople, J. A. Gaussian, Inc., Pittsburgh, PA, 1998.

GAMESS http://www.msg.ameslab.gov/gamess/GAMESS
go to http://www.msg.ameslab.gov/gamess/GAMESS
Follow the path How to get GAMESS → PC → PC GAMESS etc.

Note that this program is not exactly "freeware," but it is "a site license at no cost," which means (I think) that you can't package it and sell it to someone who's not hip enough to get it for himself.

See (a) Schmidt, M. W.; Gordon, M. S. *Ann. Rev. Phys. Chem.* **1998**, *49*, 233–266 (b) Schmidt, M. W.; Baldridge, K. K.; Boatz, J. A.; Elbert, S. T.; Gordon, M. S.; Jensen, J. H.; Koseki, S.; Matsunaga, N.; Nugyen, K. A.; Su, S.; Windus, T. L.; Dupuis, M.; Montgomery, J. A. *J. Comp. Chem.* **1993**, *14*, 1347–1363.

Index

Ab initio, 241, 277, 299
Absolute entropy, 24
Absorbance, 53, 83, 88
Absorptivity, 83
Adamantane, 168
Algorithm, 2, 47
Allinger, N. L., 102, 112
Allyl, 189, 192, 216, 253
AM1, 279
AMBER, 112, 114
Amplitude, 95
    constant, 135
Angular frequency, 94,133
Anharmonicity, 116, 301
Antibonding
    orbital, 175
Antisymmetrized
    orbital, 270, 273, 275
    wave function, 256
Antisynchronous mode, 137
Approximate theoretical energies, 304
Arguslab, 282
Arithmetic mean, 61, 62, 70
Aromaticity, 156, 219
Asynchronous motion, 137

Atomic
    coordinates, 102, 104, 107
    orbitals, 22
    units, 173
Atomization enthalpy, 57

b vector, 46
B3LYP, 328
BASIC, 6
Basis, 75
    function, 175
    set, 90, 175, 202, 309, 310, 311
    set improvement, 306
Beer's law, 83
Bending mode, 116
Benson, S. W., 57
Benzene, 157, 225, 291, 326
Benzopyrene, 292
Beyond Huckel theory, 231
Bicyclo[3.3.0]octane, 165
Bicyclohexatriene, 225
Bicyclopropenyl, 327
Binary solution, 77
Biphenyl, 225
Birge Sponer plot, 303

Computational Chemistry Using the PC, Third Edition, by Donald W. Rogers
ISBN 0-471-42800-0.   Copyright © 2003 John Wiley & Sons. Inc.

Blackbody radiation, 2
Block matrices, 143
Bohr theory, 178
Boltzmann
    constant, 74
    distribution, 151
Bond
    additivity, 57
    angle, 98
    energies of hydrocarbons, 89
    energy, 145, 302, 307, 309
    enthalpies, 56
    enthalpies of hydrocarbons, 56
    length, 300
    order, 214, 253
Bonding
    energy curve of H_2, 318
    orbital, 175, 211
Born Oppenheimer approximation, 172, 264
Bosons, 266
But-1-ene, 128, 168, 218
But-2-ene, 148
Butadiene, 190, 215, 218
Butadienyl, 190
Butane, 123, 125
    conformational mix, 125
Butyric acid, 332

Calculated energy, 306
Calculation of K_{eq} at 298 K, 164
Calibration surfaces, 80 *ff*
    not passing through the origin, 88
Carcinogenesis, 291
Cartesian space, 98, 142, 173
CBS-4, 241
CGTOs, 310
Charge
    correlation, 274
    densities, 211, 253
Cholesterol, 17
CI, 256, 312
CID, 312
CIS, 312
cis-But-2-ene, 148
CIT, 312
CLS, 13
Coefficient
    matrix, 45
    of determination, 70
Column matrix, 40
Complete set, 242
Complex
    conjugate, 42, 266

matrices, 42
    plane, 294
Confidence level, 17
Configuration, 156, 178
    interaction, 255, 256
Conformable, 32
Conformation, 120, 125
    anti, 121, 125
    gauche, 121, 125
Conformational mix, 126, 151, 152
    search, 127
    space, 166
Conrotatory, 227
Conservation of orbital symmetry, 227
Conservative system, 95
Contracted Gaussian type orbitals, 310
Contractions, 242
Convergence, 2, 6
Cookie, 293
Core, 222, 243
    potential, 176
Correlation energy, 312
Coulomb integral, 183
Coulombic energy, 124
Coupled mases, 141
Coupling, 131
    forces, 143
Cramer's rule, 50, 64
Cross terms, 128
Cubane, 330
Curve fitting, 59, 73
Cycloalkanes, 55
Cyclopentadienone, 227
Cyclopentene, 164
Cyclopropane, 316
Cyclopropenone, 226
Cyclopropyl, 211

Debye, 189
Decalin, 290
    cis and *trans*, 290
Define function, 12
Degeneracy, 126
Degenerate, 160
Degree, 37, 68
    of freedom, 71
Delocalization energy, 215, 216
Density functional theory, 299, 327, 329
Determinant, 50, 58, 134, 185
Dewar benzene, 326
DFT, 327, 329
Diagonal matrix, 40, 140
Diagonalization, 187

Diatomic molecule, 286
Differential equation, 94
Diffuse functions, 311
Dihedral driver, 160
Diketocyclohexnes, 289
Dimethylcyclobutene, 228
Dimethylcyclohexadiene, 228
Dipole moment, 124, 213, 226, 258, 289
Disrotatory, 227
Dissociation, 301
    energy, 302, 307
Distribution, 19, 60
    function, 19
Division of matrices, 34
Do loop, 6
Dopamine (DOPA), 332
Doublet, 281
DZV, 317, 319

Efficiency and machine considerations, 13
EHT, 221
Eigenfunction, 39, 170, 253
Eigenvalue, 38, 42, 169, 170, 187, 193, 195, 209, 324
    equation, 39
    by diagonalization, 187
Eigenvector, 201, 203, 206, 209, 254
Electrochemical cell, 67
Electron, 267
    correlation, 312
    spin, 267
Elements of the secular matrix, 232
Ellipse, 43
Empirical model, 97
Energy, 195
    corrections (G2), 313
    equation 114
    levels, 195
    of atomization and ionization, 315
    of atoms and ions, 240
    of formation, 319, 320
    of larger molecules, 289
Enthalpy, 144
    of atomization, 89
    of formation, 144, 321
    of isomerization of *cis*- and *trans*-2-butene, 148
    of reaction, 147
    of reaction at temperatures $\neq$ 298 K, 150
Entropy, 24
    and heat capacity, 162
Error analysis, 86
    vector, 90

Ethylene, 100, 154, 177, 187, 252
ethylene.xyz, 108
Even function, 120
Exact Theoretical Energies, 301
Excel, 25
Exchange, 183
    correlation, 328
    energy, 328
    gradiaent functional, 328
    integral, 183, 273
    operation, 266
    symmetry, 266
Expectation value of the energy, 178
Experimental energies, 301
Extended Huckel theory, 219 *ff*
    Hoffman's EHT method, 221
    Wheland's method, 219

F operator, 249
False minima, 158
Fermions, 266
Fock
    equation, 276
    matrix, 279
    operator, 277
Force
    constant, 94, 114, 132
    field, 93, 109
FORTRAN, 101, 103
Fourier series, 119
Free energy and equilibrium, 163
Free valency index, 217
Full CI, 312
Full statistical method, 161
Fulvene, 226
Functional, 328

G2, 307, 313
    corrections, 314
G2(MP2), 313, 315
G3, 307, 313
G3(MP2), 313, 315
GAMESS, 317, 318, 324, 325
GAUSSIAN, 240, 243, 244, 299
GAUSSIAN94-W, 292
Gaussian ©, 299
Gaussian
    approximation, 182, 245
    basis sets, 309, 311
    distribution, 15
    elimination, 47, 48, 54
    function, 10
    type orbitals, 309

Gauss-Jordan elimination, 49
Gauss-Seidel iteration, 50, 54
Generalized coordinates, 287
Geo File, 102
Geometry of small molecules, 110
Global minimization, 158
Global MM, 127
GTOs, 309
GUI Interface, 112, 127

h2o.xyz, 110
Hamiltonian, 263, 275, 308
    operator 169, 174, 176, 233, 235, 238
    radial, 179 *ff*, 198
Harmonic oscillator, 93, 97 132, 142, 285
Hartree (unit), 293
    equation, 263, 265
Hartree Fock equation, 273, 276
    limit, 299, 311
Heat capacity, 25, 29, 150
Heat
    of formation, 144
    of hydrogenation, 154
    of hydrogenation of ethene, 154
Helium, 174, 235, 236, 239, 264, 273
Hertz (unit), 94
Hessian
    eigenvector, 144
    matrix, 140, 142
HF, 284
    limit 311
HI, 284
HLC, 314
HMO, 224, 229
    spectroscopic transitions, 197
    matrix, 194
Hohenberg Kohn theorem, 327
HOMO, 197, 199
Homogeneous simultaneous equations, 185
HOMO-LUMO transitions, 251, 257
Hooke's law, 94
Huckel, 169 *ff*
    coefficient matrix, 207
    matrix, 210
    method, 172, 176, 183
    molecular orbital theory, 169, 201
    theory and the LCAO Approximation,
        183
Hybrid DFT, 329
Hydrocarbons, 56
Hydrogen, 281, 282, 308
    atom, 171, 243
    molecule ion, 171, 304

Hydrogenation, 147
    of ethylene, 154

Importance of the least equation, 38
Independent particle approximation, 175
Indistinguishability, 266
Information loss, 60
Intensity, 3
Interactive, 12
Intercept, 65
Internal coordinates, 96
Inverse matrix, 87
Ionization energy, 76, 316
    of hydrogen, 76
Ionization potential, 236
    first row atoms, 241
Isodesmic Reactions, 324
Isomerization, 147
Iteration, 99
Iterative methods, 1

Jacobi Method, 191

k matrix, 287
Kekule structure, 218
KF, 73, 79
    solvation, 73
Kinetic energy operator, 173
Koopman's theorem, 323

LCAO, 177, 183, 278
Least equation, 37
Least squares, 19, 60 *ff*
    minimization, 61
Linear
    combination, 136
        of atomic orbitals, 177
    curve fitting, 73
    functions, 62, 63
        not passing through the origin, 63
        passing through the origin, 62
    independence, 45
    nonhomogeneous simultaneous equations, 45
    operations, 52
    transformation, 41
Linearly dependent equations, 185
LUMO, 197
Lyman series, 76

Machine efficiency, 13
Mass spectra, 54, 55
Mass weighting, 141
Mathcad, 28, 49, 55, 84, 182, 197, 208, 239

Matrix, 31
   addition, 31
   algebra, 31
   as operator, 207
   complex, 42
   diagonalization, 51
   element, 31
   formalism for two masses, 138
   inversion, 51
   inversion and diagonalization, 51
   mechanics, 39
   multiplication, 33
   powers and roots, 35
   rank, 38
   transformation, 41
Maxwell Boltzmann distribution, 20
Median speed, 21
Medical statistics, 17
Methane, 89
Method of least squares, 60
Methylenecyclopropene, 226
Methylenepentadiene, 225
Minimal, 100, 108
   basis, 242
Minimization, 63, 99, 105
   parameter, 63
minimize, 150, 154
MM3, 100 *ff*, 107, 117, 148, 154, 157, 162
   parameters, 117 *ff*
MM4, 147
MMFF94, 112, 127
MNDO, 279
Molality, 78, 79
Molecular energies, 301
   mechanics, 93, 98, 131
   orbital, 175, 299
   speeds, 19
Moment of inertia, 106, 108
MOPAC, 281, 283
MP2, 313
MP4, 313
Multivariate, 45
   least squares analysis, 80

Naphthalene, 226, 290, 291
NDDO, 279
Newton-Raphson
   generalization, 144
   method, 7
Nitrogen, 284
Nodes, 171
Nonhomogeneous vector, 46, 185
Nonsingular matrix, 51

Normal
   coordinates, 136, 285, 288
   curve, 15
   equations, 64, 82
   modes, 137
   modes of motion, 136
Normalized, 16
Number density, 3
Numerical integration, 9, 24

Odd function, 120
Operator, 173
Optimization, 99, 143, 292, 300, 308
Orbital, 179, 323
   approximations, 176, 237, 265
Orthogonal matrix, 40
Orthonormal transform, 206
Orthonormality, 184
Output file, 103
Overdetermined set, 81
Overlap integral, 183, 220

PALA, 91
Parameterization, 113, 280
   semiempirical, 281
Parameters, 97, 117, 251
Partial molal volume, 77
   of $ZnCl_2$, 77
Partial molar volume, 78, 79
Particle in a box, 170
Partition function, 146
   contribution, 146
Pauli principle, 255, 267
PCMODEL, 112, 127, 149, 155, 283
Permutations, 271
Peroxide complexes, 52
Phenantherene, 291
Photoeolectron spectruma, 323
Photon, 2, 3
Pi electron calculations, 155
Planck radiation law, 4
PM3, 279
Polarized functions, 311
Polarographic reduction, 225
Polyatomic molecules, 97, 288
Polynomial, 8, 68
   equations, 36
   higher degree, 68
POP, 151
Pople, J. A., 306
Population, 14, 215
   energy increments, 151
Postmultiplication, 33

Potassium fluoride, 74
Potential energy, 97, 120,122, 126, 174
    function, 123
    well, anharmonic, 302
    well, harmonic, 302
Powers and roots of matrices, 35
PPP, 248
    -SCF, 248, 249, 256
Premultiplication, 33
Primitives, 242, 310
Principal axes, 43
Prismane, 326
Probability density, 23
Propene, 111, 316, 325
Pyridine, 228

QCISD(T), 312
QMOBAS, 194
QSIM, 18
Quadratic functions, 65
Quadrature, 9
Quantum mechanics, 23
    number, 171

Radiation density, 3
Rank, 37, 38
Rayleigh frequency, 3
Reduced mass, 95
Regression, 70
Reliability
    of fitted parameters, 70
    of fitted polynomial parameters, 76
Residual, 69, 86
Resonance, 155
    energy, 157, 217
        of benzene, 157
    integral, 185
Root, 6, 7, 139, 187, 234, 254, 274
Roothaan Hall equations, 278
Rotamers, 128
Rotation matrix, 188, 191
Row operations, 51
Rydberg equation, 76

Sample, 14
Scalar, 33
Scale factor, 306
SCF, 231, 236, 241
    ultraviolet spectral peaks, 256
    dipole moments, 258
    energies of first row atoms and ions, 240
    matrix, 252
Schroedinger equation, 169

approximate solutions, 172
    exact solutions, 170
Secular determinant, 6, 186, 203, 274
    equations, 134, 185, 203
    matrix, 7, 186
        elements of, 232
Semiempirical, 248, 263
    small molecules HF to HI, 284
    methods, 248
    model, 97
        approximations, 279
SHMO, 223, 225
Sigmaplot, 25
Significant figures, 84
Similar matrices, 42, 192
Similarity transformation, 287
Simpson's rule, 9, 10
Simultaneous analysis, 52
    analysis by visible spectroscopy, 83
    equations, 51, 64
    probabilities, 60
    spectrophotometric analysis, 52
Singular matrix, 38, 46
Slater determinant, 255, 269
Slater orbital, 221
Slater-type orbital, 237, 238
Slope, 65
    matrix, 83
Solvation, 73
Space, 3
Span, 44
Special matrices, 39
Spectrophotometry, 52
    transitions, 197, 253
Spectrum, blackbody, 3
Speed, 19
Spin, 268
    correlation, 273
Spinorbital, 255, 268
Spline fit, 27 *ff*
Split valence basis sets, 310
Spreadsheet, 25
Standard deviation, 14, 72
    of the regression, 77
Standard hydrogen electrode, 68
Statement number, 5
Stationary point, 300, 308
Statistical criteria for curve fitting, 69
Statistics, 14
Steric energy, 98 104, 161
STO-1G "Basis Set", 243
STO-2G, 245
STO-3G, 241 247, 300

Stochastic search, 159, 166
STO-xG, 240, 242
Strain energy, 158
Strainless
   bond energies, 145
   parameters, 145
Stretching mode, 115
Styrene, 168, 225
Sums in the energy equation, 115
Superpositions, 135
Symmetric matrix, 40
Symmetry, 305
Synchronous mode, 137
System, 13

TableCurve, 24, 69, 70, 71
Taylor
   series, 141
   expansion, 115
THC, 322
Thermodynamic functions, 319
TINKER, 108 *ff*, 148
TMOBAS, 194
Tobacco smoke, 291
TORS, 151
Torsion, 118
   modes of motion, 153
Total energy (GAMESS), 325
Trace, 40, 221
trans-But-2-ene, 148
Transformation matrix, 41
Transpose, 39
Triangular matrix, 40
Tridiagonal matrix, 40
TrueBASIC, 6, 196
Tryptophan, 88
Two-Mass Problem, 95
Tyrosine, 88
TZV, 317, 319

Uncertainty, 87
Uncoupled equations, 136
Unit matrix, 34
Unitary matrix, 42
Unitary transformation, 42
Upper bound, 309
Upper triangular matrix, 48

Valence bond approximation, 177
van der Waals Energy, 122
Variance, 86
Variational
   calculation, 236
   method, 178, 181
   treatment of the hydrogen atom, 181
VB, 177
Vector, 40
   space, 44, 201
Velocity, 19
VESCF, 156
Vibration
   of nitrogen, 284
Vibrational spectrum, 301, 302
Virtual orbitals, 256

Wave
   function, 22
   mechanics, 39
What's going on here?, 42
Why so much fuss about coupling?, 143
Wien's law, 4, 6

Zero point energy, 162, 303
ZPE, 303
Z-Score, 18

β, 198, 257
γ_{ij}, 257
π-electron, 176